PLUMBERS and PIPE FITTERS LIBRARY

Welding · Heating · Air Conditioning

by Charles McConnell

THEODORE AUDEL & CO.
a division of
HOWARD W. SAMS & CO., INC.
4300 West 62nd Street
Indianapolis, Indiana 46268

SECOND EDITION

FIRST PRINTING—1977

International Standard Book Number: 0-672-23257-X
Library of Congress Catalog Card Number: 77-75528

Foreword

Plumbing and pipe fitting play major roles is the construction of all types of residential, commercial, and industrial buildings. Of all the building trades, the plumbing trade is most essential to the health and well-being of the community in general, and to the occupants of the buildings in particular. It is an obligation and responsibility for each and every plumber to uphold the vital trust placed in him for the proper installation of the plumbing materials and equipment.

Each plumbing installation is governed by the rules and regulations set forth in the local plumbing codes that have been adopted from standards established at a local, state, or federal level. In addition, each installation is subject to inspection by a licensed inspector to insure compliance with all the rules and regulations. The requirement that all those persons engaged in installation of plumbing pass an examination for a license is another practice that indicates the tremendous importance of this phase of the building trades.

The books in this series have been written to aid those persons who wish to become plumbers as well as those who are already actively engaged in this occupation. This, the second of three volumes, deals with heating systems in which plumbing plays a part, air-conditioning systems, and brazing and welding. Heating and air conditioning are natural companions of the plumbing industry and, as such, have been adopted by many plumbers as a part of their normal business. Welding and brazing are both being used extensively in many plumbing and pipe-fitting installations and should therefore be of concern to nearly every person in the plumbing trade.

The author and publishers express their thanks to the following companies for inclusion of texts and illustrations:

The Union Carbide Corporation — Linde Division — *The Oxyacetylene Handbook*

Hoffman Specialty ITT — *Hoffman Steam Heating Systems, Design Manual and Engineering Data*

Contents

CHAPTER 1

Brazing and Welding

Brazing and welding are now being extensively used in the plumbing and pipe-fitting trades. Brazing has taken the place of many operations that formerly were performed by soldering. The brazing process is usually simpler and faster, and results in stronger joints. Welding is being employed in many installations, especially where larger-diameter pipes are concerned. Installation time and costs can be materially reduced by using this method of joining pipe and fittings. Every plumber and pipe fitter should be familiar with the basic principles and techniques of brazing and welding.

BRAZING

(Courtesy NCG, Division of Chemetron Corp.)

Brazing operations are specifically different from those applicable to fusion welding. In brazing, the brazing alloy is applied at a temperature *below* the melting temperature of the metal being brazed. In brazing a joint, the procedure is an adhesion of a brazing alloy to the surface and into the porous structure of the metal which is in the process of being brazed. In almost all cases, a brazed joint is at least as strong as the brazed metal, if not stronger.

Nonferrous Metals

Such nonferrous metals as those in the copper and copper-zinc category braze well with a low-temperature brazing alloy or a silver alloy. Copper and copper-zinc metals have a melting temperature of approximately 1670°F. The melting point of the brazing alloy must be below the melting temperature of the metal.

Aluminum and aluminum-alloy brazing requires some experience by the operator. Aluminum brazing alloy melts at approximately 200°F, which is under the melting temperature of aluminum in its pure state. The brazing of aluminum alloys is difficult because the differential in melting temperatures is under 200°F.

Ferrous Metals

In brazing most ferrous metals, the bonding temperatures must not exceed 1400°F, which is below the critical temperature of these metals. This critical point should not be exceeded for the following reasons:

Cast-Iron—Cast-iron contains a high percentage of silicon and carbon. A majority of the carbon content is in graphitic form that will, when heated to temperatures in excess of 1400°F, flow to the outer surface, preventing the brazing alloy from penetrating the porous structure and making it difficult to obtain good surface adhesion. Silicon has the same effect.

Malleable Iron—Malleable iron is a heat-treated form of cast-iron, and when heated above the critical zone (1575°F), the structure returns to that of cast iron. When this occurs, the metal is no longer ductile and will not be resistant to shock.

Steel—When heated to a temperature above 1500°F, steel will alloy with the copper in the brazing material and become very brittle. Copper, when properly alloyed with steel, produces a high tensile strength, but when alloyed by brazing, the temperature is too uneven to obtain uniform distribution of the copper in the steel, thereby creating not strength, but weakness.

Stainless Steel

Stainless steels are readily brazed with silver alloys, but are difficult to braze with copper-zinc alloys. While the brazing of stainless steels is successful, there is an element of danger that should be considered before brazing is performed. That is, when stainless steel is heated slowly and cooled slowly, there is a great possibility of forming carbides at the grain boundary, and carbide precipitates are very brittle. Therefore, stainless steels (unless stabilized) should not be brazed if the joints are to be subjected to shock or vibrating stress.

Brazing Flux

Flux is used in all brazing operations except with copper-to-copper joints. Do not rely on the chemical action of the flux alone to properly clean scale, dirt, and foreign matter from the metal. Follow the instructions on the flux container for good-quality brazing.

Joint Preparation

Before attempting brazing, be sure that all joints are clean and properly prepared. Joints should have close fits and be uniform in shape. The use of excessive amounts of brazing alloys is expensive; it will result in excessive warpage and stress in some operations, and often results in low physical properties.

Brazing With Silver Alloys

Silver brazing alloys are available in a variety of silver content, and almost all metals (except magnesium and aluminum) can be brazed with one of the alloys. Certain procedures must be followed in order to obtain quality results.

A high capillary attraction to metals is provided by silver alloys. As a result, a close clearance between metal parts is necessary for a good joint. Specifically, .002″ to .003″ clearance makes a stronger joint than do larger clearances, which require a greater amount of alloy. Not only are proper fits necessary, but clean joints are a must (fingerprints from handling will often cause poor adhesion and an inferior joint). Carbon tetrachloride or trichlorethylene is a good solution for final cleaning after all parts have been cleaned of foreign matter.

Note: Proper precautions must be taken to provide adequate ventilation when certain types of cleaners are used. For example, carbon tetrachloride is toxic, and is dangerous when used in confined spaces. If carbon tetrachloride is subjected to heat, deadly phosgene gas is created. Thoroughly dry all parts cleaned with tetrachloride in order to prevent the creation of this deadly gas. Before using any cleaner, read the label carefully.

After final cleaning, the parts are then fluxed with a brush and assembled. Certain types of joints may be made by inserting shims or ring stock in or on the joint after being fluxed. A tip of ample size to produce a soft flame should be used for the heating, applying the heat evenly to the entire joint to avoid distortion by overheating some parts and underheating others. When the flux changes to a liquid state, the alloy will flow freely to all parts of the joint.

SOCKET BRAZING

(Courtesy of Kennedy Valve Mfg. Co., Inc.)

Some piping and valves are provided with brazing sockets designed to permit a pipe-to-pipe or pipe-to-valve connection by means of a silver brazed joint. These joints have high tensile strength and will hold firm at any temperature to which the valve, pipe, or tubing can be safely subjected.

Silver-brazed joints resist corrosion since the brazing alloy fuses with the bronze of the valve and the material of the pipe or tubing. This results in a bond of greater strength and corrosion resistance than either the valve or the tubing.

Brazing socket valves (Fig. 1) have threadless, smooth socket ends, precision machined to close tolerances, into which standard pipe-sized outside-diameter brass or copper pipe or tubing is fitted. At the inner end of the socket, a square shoulder acts as a stop to limit pipe insertion.

The accurately controlled clearance between the pipe and valve socket assures the capillary action of the molten brazing alloy. When the pipe and end of the valve are properly heated, the brazing alloy is drawn into the socket by this capillary action, flowing to the full depth and between the shoulder end of the pipe. This bonds the entire contacting surfaces of the pipe and valve socket.

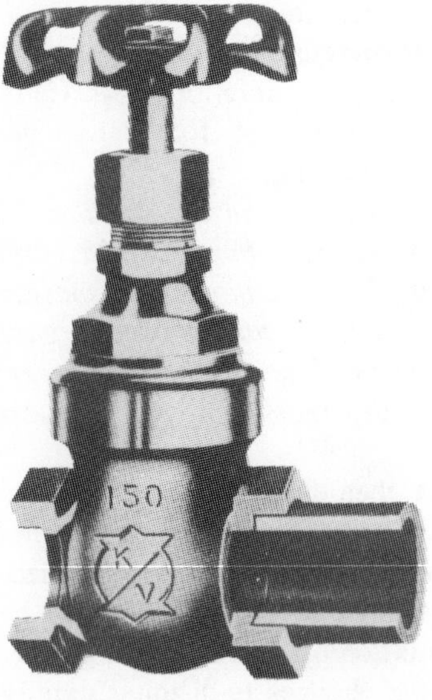
150

Fig. 1. A typical standard bronze gate valve with brazing sockets.

Courtesy *The Kennedy Valve Mfg. Co.*

Valves of this type are generally suitable for use in any nonferrous piping installation in industrial plants, power plants, oil refineries, chemical plants, hospitals, ships, etc. They are also suitable for use in systems handling oil, hot or cold water, compressed air, acid, Freon, carbon dioxide, and many other liquids and gases.

Cleanliness is the first requirement in the installation of socket brazing valves. The connecting pipe or tubing must be cut squarely, reamed, burred, and polished with emery cloth. The valve sockets and the faces of the hex ends must be wiped clean and polished with emery cloth. The polished surfaces should be coated with flux and the pipe inserted in the socket until it butts against the shoulder. See Fig. 2.

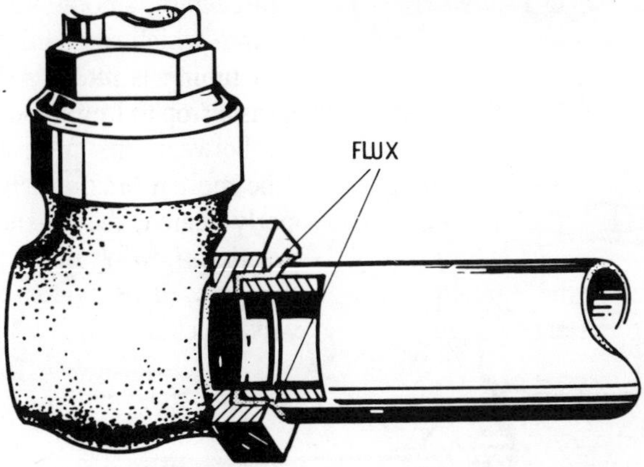
FLUX

Courtesy *The Kennedy Valve Mfg. Co.*

Fig. 2. Flux is applied to pipe end, hex and face, and socket.

Open the valve and apply a flame to the pipe about 1 inch from the valve and directed away from the valve. Move the flame around the pipe until the flux changes to beads (Fig. 3). Then move the flame to the valve hex, but still directed toward the pipe. When the condition of the flux indicates that the valve end is heated sufficiently, apply fluxed brazing alloy to the joint (Fig. 4). Sweep the flame across the valve hex and pipe while the molten brazing alloy is drawn into the socket.

Remove the flame and allow the joint to cool. If the joint has been properly made, there will be an unbroken fillet of brazing alloy completely around the edge of the socket.

A low-silver brazing alloy is recommended for this type of brazing. This type of alloy contains only about 15% silver, will start to melt at 1185°F, and will be completely liquid and free flowing at 1300°F.

Courtesy *The Kennedy Valve Mfg. Co.*

Fig. 3. Heat is applied until the flux forms beads.

Courtesy *The Kennedy Valve Mfg. Co.*

Fig. 4. Brazing alloy is applied to the heated area, where it melts and is drawn into the joint by capillary attraction.

OXYACETYLENE WELDING

Oxyacetylene welding is a process of joining metals by heating to the melting point the surface to be joined, and then fusing together the molten metal. The heat used for this process is a flame produced by the burning of a proper mixture of acetylene and oxygen gases. During the heating process, a filler metal (welding rod) is melted into a prepared groove of the material to be welded.

Oxygen and acetylene are used to produce the welding flame. The combination of these two gases gives a higher temperature (approximately 6000°F) than any other gas flame that could safely be used. For complete combustion a proportion of 2-1/2 parts oxygen to 1 part acetylene is required. However, only 1 part of oxygen is required to be supplied through the torch because 1-1/2 parts are supplied by the natural air surrounding the flame.

In lighting the torch prior to welding, the acetylene is always turned on first and ignited. The torch will burn with an orange flame and give off black smoke. The flame is now receiving only the 1-1/2 parts of oxygen from the air around it, so combustion is incomplete. When the oxygen is turned on at the torch handle, the additional 1 part of oxygen is provided, and the flame will turn blue with a white inner core, indicating maximum heat and good combustion.

Oxygen

Oxygen is an odorless, tasteless, and colorless gas which will combine with almost every element known. In combination with certain other substances and under certain conditions, it will produce flame and heat. Oxygen in itself will not burn, but is required for burning to occur. Its property of supporting combustion makes it very valuable in the welding process. Combined with acetylene, it produces one of the hottest gas flames available.

> WARNING: *Oxygen under pressure should* NEVER *be permitted to come in contact with grease or oil. The mixture is very explosive and could result in serious injury. That is why the warning* "USE NO OIL" *is displayed on the regulators for oxygen cylinders.*

Oxygen also results in a rapid rate of oxidation when in contact with hot steel. Considering this factor, no attempt should ever be made to heat a pipe or container holding compressed oxygen.

Acetylene

Acetylene, which has a characteristic odor and is colorless, is composed of carbon and hydrogen atoms in equal proportions. It is a combustible gas and can form explosive mixtures with oxygen or air.

When calcium carbide is brought into contact with water, a reaction takes place, resulting in the liberation of acetylene gas. Manufacturers state that, because of one of its inherent properties, acetylene gas should never be used at a pressure in excess of 15 psi. This maximum pressure is sufficient for all cutting and welding operations.

Welding Rods

An important part in the quality of the finished weld is the welding rod which is melted into the welded joint. Welding rods of good quality are designed to permit free-flowing metal which will fuse easily with the metal being welded. In this way, good sound welds of correct composition are produced. There are a variety of welding rods available. They are made for cast-iron, carbon steel, bronze, aluminum, and stainless steel, as well as other metals.

Welding Flux

Welding and brazing fluxes are used for the following general purposes:

1. To prevent the molten metal from absorbing and reacting with atmospheric gases.
2. To protect and clean the surfaces of the base metal while the welding is taking place.
3. To purify the molten metal during the welding by removing oxides which are formed.

Flux compositions vary according to the purpose for which they are to be used. Compositions usually are in paste or powder form, and they should be used as recommended by the manufacturer's instructions on the container.

SETUP AND OPERATION OF EQUIPMENT

Always fasten the cylinders to be used in an upright position so they cannot be knocked or pulled over. If cylinders are not on a suitable cylinder cart, they should be securely fastened, with chain or equivalent, to a workbench, wall or post. Acetylene cylinders should never be stored or used in other than a vertical position.

"Crack" the cylinder valves. Stand so that the gas leaving the cylinder outlet will not be directed onto your face or clothing. Open the valve quickly about one quarter of a turn, then close it immediately. This will clear the valve outlet opening of accumulated dust or dirt which might, if not blown out, mar the seat of the regulator nipple or be carried into the regulator.

PRECAUTION: *Never "crack" a fuel gas cylinder valve near other welding or cutting work in progress, or near sparks, flame, or other possible source of ignition. The correct way to crack a cylinder valve is shown in Fig. 5.*

OXYGEN

Courtesy *Union Carbide Co.*

Fig. 5. "Cracking" a cylinder valve.

Connect the oxygen regulator to the oxygen cylinder and the acetylene regulator to the acetylene cylinder. If the acetylene regulator and the acetylene cylinder have different threads, it will be necessary to use an adapter between the regulator and the cylinder. Two quite different acetylene cylinder connections are used in the U. S., the CGA 510 connection has left hand threads, internal on the cylinder outlet; the CGA 300 connection has right hand threads, external on the cylinder outlet.

Tighten both regulator connections firmly with a wrench, as shown in Fig. 6. Wrenches with two or more fixed openings, designed specifically for use with gas welding and cutting apparatus, are available from all apparatus suppliers.

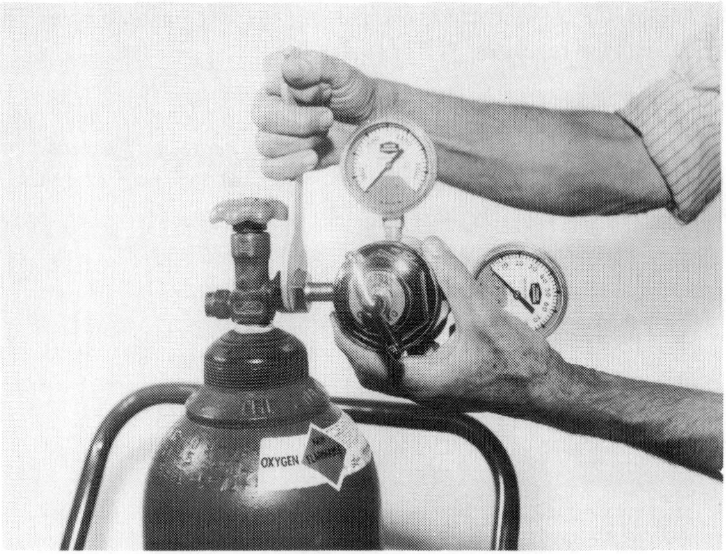
OXYGEN

Courtesy *Union Carbide Co.*

Fig. 6. Tightening regulator connection with a wrench.

PRECAUTION: *Should it ever be necessary to retighten the regulator union nut after the outfit has been set up and the cylinder valves opened, be sure to close the cylinder valve before tightening the nut.*

Rotate the pressure adjusting screw of each regulator to the left (counter-clockwise) until it turns freely. The valve in the regulator must be closed before cylinder pressure is applied to the regulator. This is shown in Fig. 7.

Open each cylinder valve SLOWLY. Stand where you can see the cylinder pressure gauge, hand on the regulator, but do *not* stand directly in front of the regulator gauge faces. It is especially important that the oxygen cylinder valve be opened only very slightly at first, and that you wait until the cylinder pressure gauge has stopped moving before opening the valve fully. The acetylene cylinder valve should furnish sufficient operating pressure at approximately 3/4 of a turn open, the acetylene cylinder valve should never be opened more than one and one-half turns. When opening the cylinder valves stand to one side as shown in Fig. 5.

If the acetylene cylinder valve requires a wrench to open it, always leave the wrench in place on the valve while the valve is open. This is to allow the valve to be closed quickly if an emergency should arise. This wrench is shown, on the valve, in Fig. 8.

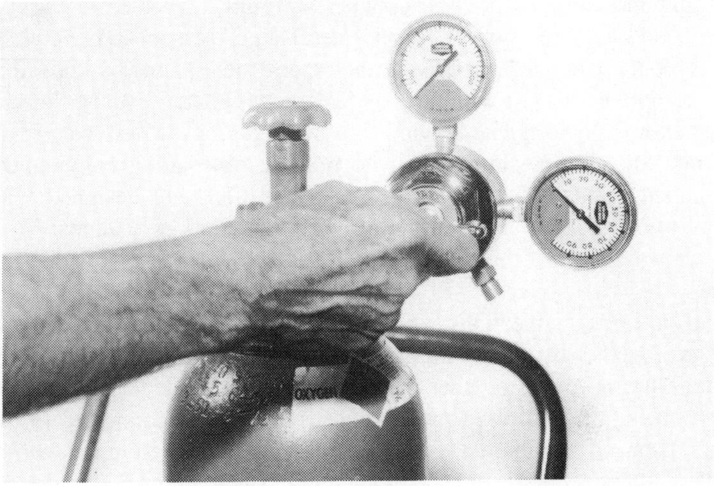

Courtesy *Union Carbide Co.*

Fig. 7. The valve in the regulator must be closed before cylinder pressure is applied to the regulator.

Courtesy *Union Carbide Co.*

Fig. 8. If the acetylene cylinder valve requires a wrench to open it, always leave the wrench in place on the valve while the valve is open.

Connecting Gas Supplies To The Torch

Always use hose and hose connections made specifically for gas welding and cutting purposes. Oxygen hose has a green cover; acetylene hose has a red cover. Never interchange oxygen and acetylene hose. Do not use acetylene hose with propane fuel unless you know that it is acceptable for use with propane. (Hose with natural rubber liner is satisfactory for acetylene service, but not for propane service.)

Make up all connections dry; do not use pipe fitting compounds, thread lubricants, oil, or grease. All connections are designed with metal-to-metal seals. They do not require lubricants or sealants. They must always be made up wrench tight, as shown in Fig. 9, not hand tight.

Never force connections which do not fit. If you cannot run the threads together by hand with ease, either the threads are damaged, or you are trying to put together parts that were not made to go together.

When the oxygen and acetylene hoses have been connected to the torch, connect the other ends of the hoses to their respective regulators.

Test all connections for leaks. One quick way to do this is to turn the oxygen and acetylene regulator screws in (clock-wise) to obtain an oxygen pressure of approximately 25 psi and an acetylene pressure of 7 to 10 psi with the torch valves closed. Close the cylinder valves and

Courtesy *Union Carbide Co.*

Fig. 9. Tightening a hose connection with a wrench.

observe the gauges. Any leakage between the cylinder valves and the torch will be signalled by a drop in the gauge pressures. A soap suds solution can be applied to all connections between the cylinder valves and the torch; a leakage at any of these points will be shown by bubbles arising from the soap suds solution. If no leakage is shown using the soap suds solution, the leakage may be internal in the regulator or through the throttling valves on the torch. A leak through the valves in the torch may be found by dipping the torch tip in water. If bubbling occurs, open and close the torch valves rapidly to free and blow out any foreign matter which may have lodged in the needle valve openings. If the leakage is still apparent after these measures have been taken, the leaking regulator or torch should be sent to the supplier for repair.

Lighting the Flame

When lighting the flame, always follow the manufacturer's instruction for the specific torch being used. In general, the procedure for lighting a torch is this: (as shown in Fig. 10)

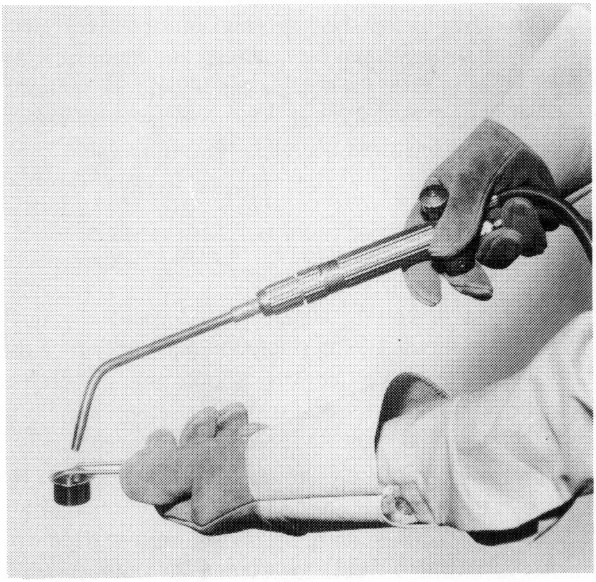

Fig. 10. Lighting a torch with a friction lighter.

1. Open the acetylene valve on the torch about one-half turn.
2. *Immediately* light the flame with a friction lighter. NEVER USE A MATCH.
3. Reduce the acetylene flow, by throttling the torch acetylene valve, until the flame just starts to produce black smoke around its edges; then increase acetylene flow just enough to get rid of the black smoke.
4. Open the torch oxygen valve slowly until the desired flame is obtained.

Flame Adjustment

If oxygen and acetylene working pressures have been set in accordance with the manufacturer's recommendations for the size of the welding tip or cutting nozzle in use, the lighting procedure outlined above will usually produce a neutral flame which has about the right characteristics for most welding and cutting purposes. However, if the flame, when adjusted to neutral, burns away from the tip (if there is a gap between the tip and the flame), reduce the flow of oxygen slightly, using the torch oxygen valve, and then reduce the acetylene flow until the flame is again neutral. If this procedure must be repeated two or three times to eliminate the gap between tip and flame, the acetylene working pressure is probably too high and should be reduced. If the neutral flame produced by following the recommended lighting procedure is too "soft," open the acetylene throttling valve slightly and readjust the flame to neutral by opening the oxygen throttling valve slightly.

Backfire and Flashback

If the welding tip or cutting nozzle should accidently contact the work, or if the flame should be interrupted momentarily by a droplet of slag, the torch flame may go out with a loud pop. A torch that has backfired may be relighted at once; it may even relight itself if the tip is directed into molten or hot metal. If a torch backfires repeatedly without coming into contact with the work the cause may be improper operating pressures, a loose tip or nozzle in the torch, dirt on the cutting nozzle seats, or use of too large a welding tip. When an oversized tip is used and the flame is throttled back to attempt to compensate for the oversized tip, backfiring often occurs.

If the flame goes out and burns back within the torch, this usually produces a hissing or squealing noise. If this should happen, shut the torch off immediately. This is called a flashback and indicates either something is wrong with the torch or with the operation of the torch. After a flashback, *always* allow a torch to cool before attempting to relight it. *Always* check the operating pressures before attempting to relight it. Allow the oxygen, *not* acetylene, to flow through the torch before relighting; the oxygen flow will help clear out any soot accumulation. When these procedures have been followed, relight the torch, and if the flame appears normal, proceed with the work. If there is a second flashback, remove the torch from service and return it to the supplier or repair station to be checked and repaired.

Stopping Work

When stopping work, close the torch fuel valve first, then the oxygen valve. Closing the acetylene valve first reduces the chance of allowing unburned fuel gas to escape and be ignited accidentally. Release all pressure from the torch, hose, and regulators. First close both cylinder valves, then open both torch valves. Finally, release the pressure adjusting screws on both regulators (turning them counter-clockwise until they are free) and close both torch valves. Before disconnecting a regulator from a cylinder, always release all pressure from the regulator.

THE OXYACETYLENE FLAME

The tool used in oxyacetylene welding is not the torch, it is the *flame*. The only purpose of the torch is to provide the gas mixture which will produce the right type of flame for the work being done. There are three types of oxyacetylene flames:

1. A neutral flame.
2. A flame containing excess oxygen (oxidizing).
3. A flame containing excess acetylene (carburizing)

The neutral flame, shown in Fig. 11, is produced when the ratio of oxygen to acetylene, in the mixture leaving the torch, is almost exactly one-to-one. A neutral flame usually has no effect on the metal being welded. It will not oxidize the weld metal, neither will it cause an increase in the carbon content of the weld metal.

The oxidizing flame, shown in Fig. 12, results from burning a mixture which contains more oxygen than required for a neutral flame. An oxidizing flame oxidizes or "burns" some of the metal being welded.

The carburizing flame, shown in Fig. 13, is created when the proportion of acetylene in the mixture is higher than that required to produce a neutral flame. When a carburizing flame is used to weld steel it will cause an increase in the carbon content of the metal being welded.

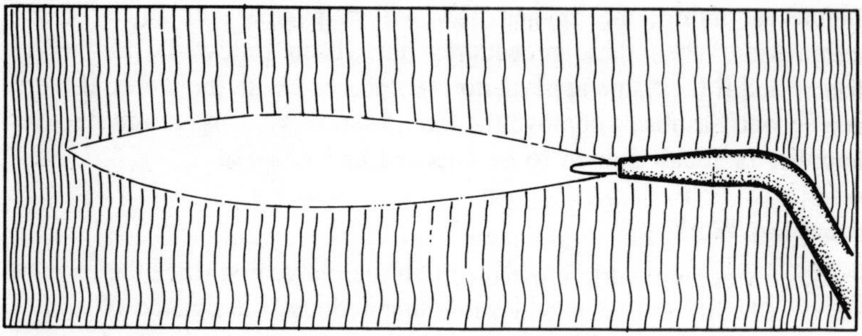

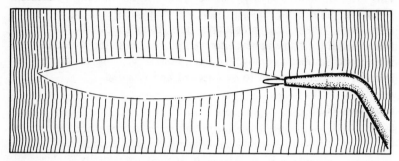

Fig. 11. A neutral welding flame.

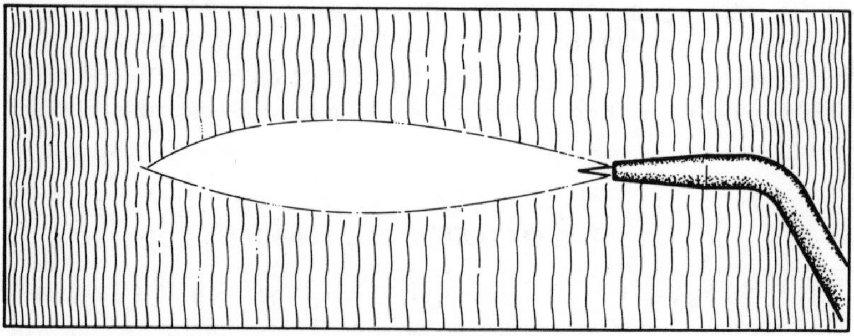

Fig. 12. An oxidizing welding flame.

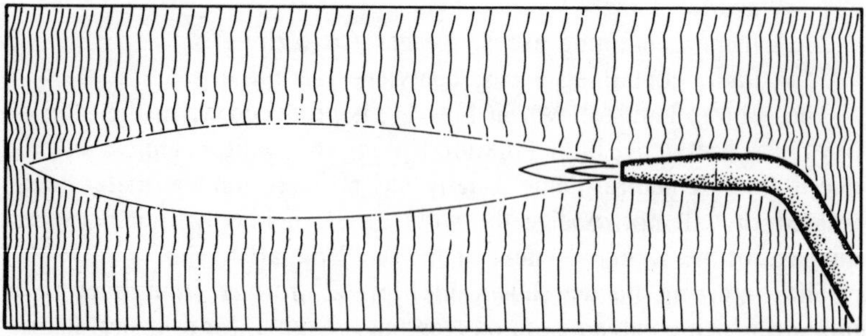

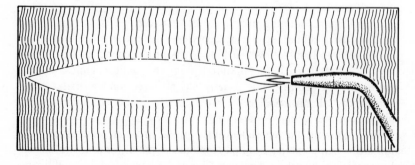

Fig. 13. A carburizing welding flame.

A neutral flame is the proper type flame for most types of welding. It is very hard to distinguish the difference between a true neutral flame and a *slightly* oxidizing flame. The difference between a neutral flame and a flame with a slight excess of acetylene is very apparent, the flame with an excess of acetylene will have a definite "feather." When lighting the torch the flame should be started with an excess of acetylene, then increase the oxygen flow until the acetylene feather just disappears. If the flame is too large for the work being done, reduce the oxygen flow first, to produce a feather, then cut back on the acetylene flow until the feather just disappears. Instructions for some types of welds may call for an oxidizing flame; some types of welds may call for a carburizing flame. The implication here is that an excess of one type of gas will not cause any trouble with the weld but that an excess of another gas may cause trouble.

WELDING TORCHES

A welding torch consists of a handle containing throttling valves for the oxygen and acetylene and a *tip.* Fig. 14, shows one type of torch which is widely used by plumbers and pipe fitters in their work. Various sizes of tips are available for welding a variety of pipe sizes.

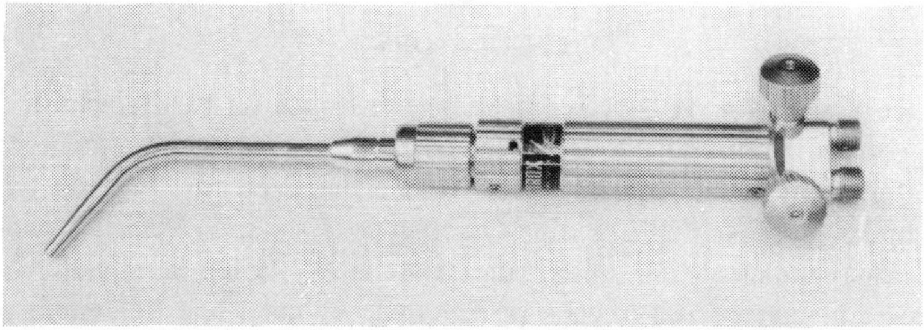

Courtesy *Union Carbide Corp.—Linde Division*

Fig. 14. One type of torch used by plumbers and pipe fitters.

THE CUTTING TORCH

The cutting torch shown in Fig. 15, is made to fit the handle of the welding torch shown in Fig. 14. The throttling valves on the handle control the oxygen and acetylene used in the cutting torch, in addition there is a throttling valve to control only the oxygen used when cutting. The preheat flame of the torch is adjusted to a neutral flame using the same procedure as outlined for the welding torch, the throttling valve

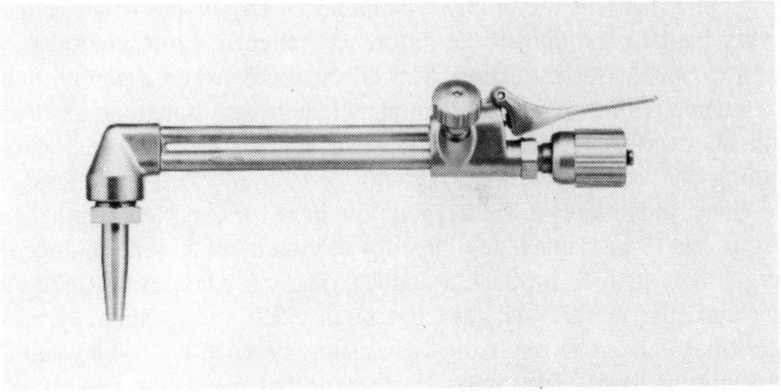

Courtesy *Union Carbide Corp.—Linde Division*

Fig. 15. A cutting torch used with the welding torch shown in Fig. 14.

for the lever-operated cutting valve is closed while adjusting the preheat flame. When the preheat flame is adjusted, the *throttling valve* on the cutting torch is opened and the lever-operated *cutting valve* is depressed or opened until the proper oxygen flow for cutting is obtained. Setting the cutting torch is not as complicated as it might seem, the skilled welder can adjust the cutting torch in a matter of seconds.

REGULATORS

Regulators for oxygen and acetylene are made in two types: single stage regulators and two-stage regulators. The two-stage regulator is preferred by most welders; it delivers a constant flow regardless of cylinder pressure, will stand up better in constant service, and requires less maintenance than single stage regulators. When a regulator has been connected to an oxygen or acetylene cylinder and tightened, the cylinder valves should be opened as shown in Fig. 7. They should always be opened *slowly* to prevent damage to the regulator and the regulator gauges. The valve on the oxygen cylinder should be fully opened when in use; the valve seats when fully opened to prevent leakage around the stem. The acetylene valve should only be opened slightly, 3/4 to 1-1/2 turns, so that it may be closed quickly in the event of an accident. The adjusting valves on the regulators, both oxygen and acetylene, should be closed (turned counter-clockwise until they are free) before turning the cylinder valves on. Fig. 7 shows how the pressure adjusting screw or valve should be released (closed) before turning the cylinder valve on. A two stage regulator is shown in Fig. 16.

OXWELD
UNION CARBIDE

Courtesy *Union Carbide Corp.—Linde Division*

Fig. 16. A two-stage regulator with gauges.

PIPE WELDING

Welded pipe is used extensively in heating and air conditioning systems. There are certain techniques and methods of welding and aligning pipe for welding, that will help the novice welder to become a skilled pipe welder.

The ends of pipe to be joined by welding should be beveled correctly. Weld type pipe and fittings are made with ends beveled to the correct angle, when pipe is cut on the job the welder must use a cutting torch to bevel the ends of cut pieces. Fig. 17 shows the correct method of making a beveled cut. The beveled cut should be at an angle of approximately 35°; the two cuts placed in position for welding should produce a total angle of approximately 70° to be filled when welding is completed.

Correct alignment of two pieces of pipe to be welded can be insured by placing the two pieces in a section of angle iron while the tack welds are being made, as shown in Fig. 18. When long sections of pipe are to be joined by welding, the use of a welding clamp, shown in Fig. 19, is advised. This type of clamp will secure the two pieces in position and alignment while the tack welds are made. After the first tack weld is

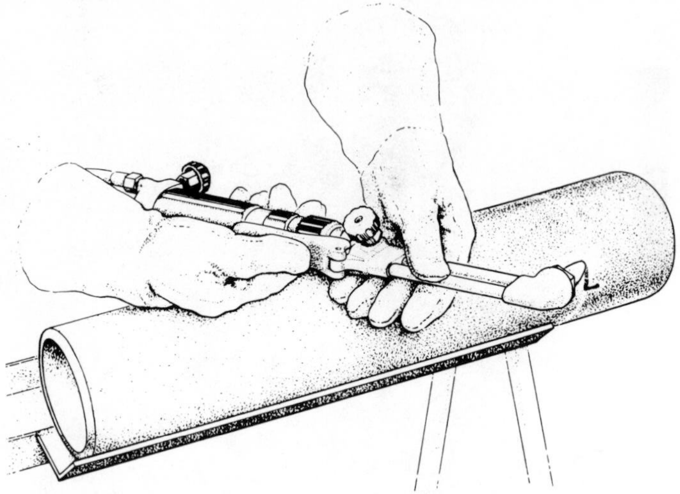

Courtesy *Union Carbide Corp.—Linde Division*

Fig. 17. Correct position for making a beveled cut on pipe.

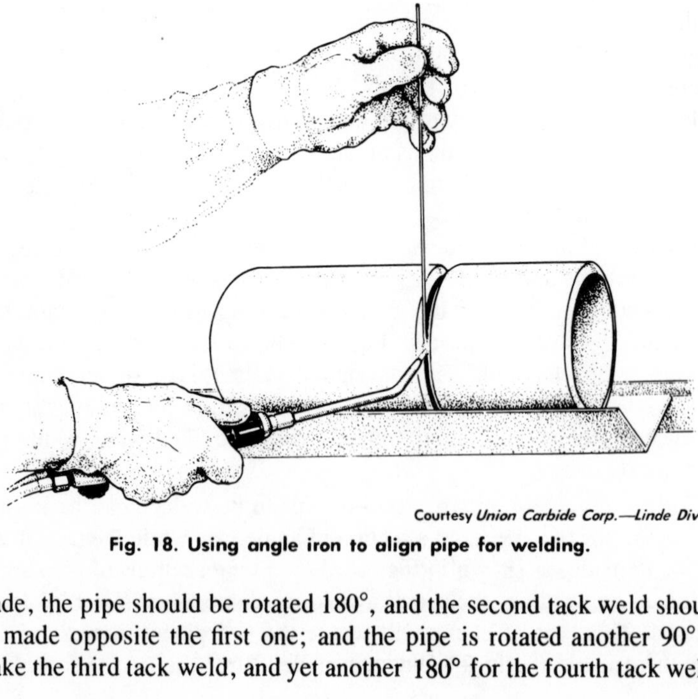

Courtesy *Union Carbide Corp.—Linde Division*

Fig. 18. Using angle iron to align pipe for welding.

made, the pipe should be rotated 180°, and the second tack weld should be made opposite the first one; and the pipe is rotated another 90° to make the third tack weld, and yet another 180° for the fourth tack weld.

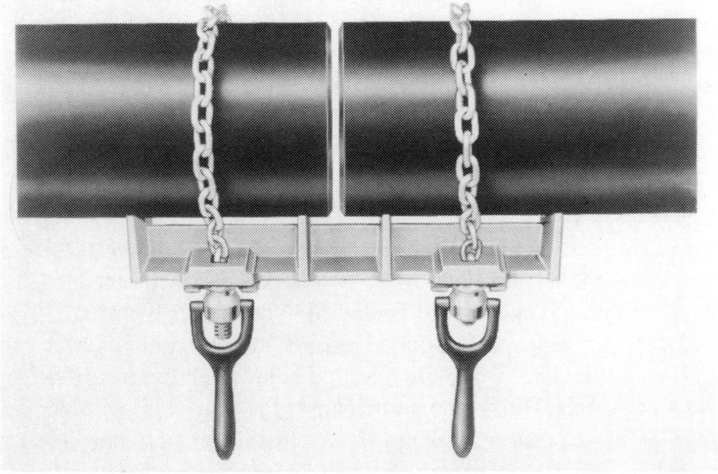

Courtesy *Ridge Tool Co.*

Fig. 19. A pipe alignment clamp.

If the tack welds are made in this sequence the possibility of misalignment due to contraction of the welds will be minimized.

Fig. 20 shows the correct method of starting the finished weld. The weld should be started about an inch away from a tack weld. When

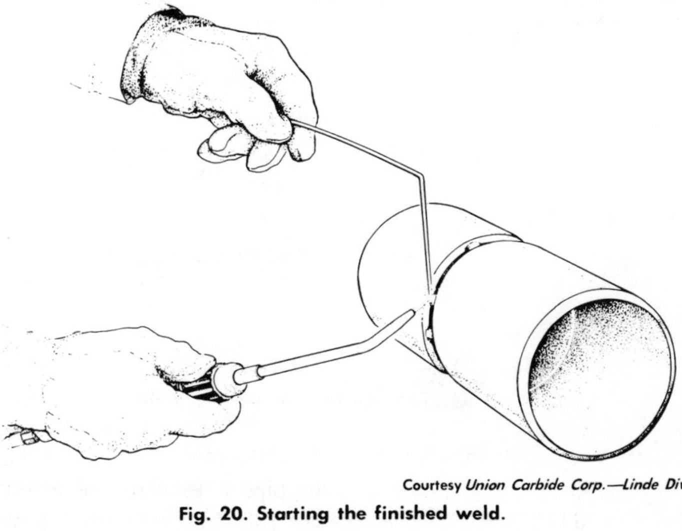

Courtesy *Union Carbide Corp.—Linde Division*

Fig. 20. Starting the finished weld.

working at a bench, a pipe roller is very useful since it permits the weld to be made from one position. A very good pipe roller can be made from a discarded pair of roller skates. As shown in Fig. 20, completing this weld will involve welding in the flat, vertical, and overhead positions. Although the use of a pipe roller will make some welds easier and quicker to make, the skilled pipe welder must be able to work in all positions. The exact starting point of the weld and the points at which the welder should change his position relative to the weld are matters of personal preference.

Fig. 21 shows the root pass of a 4" pipe weld well started. If the pipe wall is more than 1/8th inch thick, two passes are recommended. For thinner-wall pipe only one pass is needed. Note that in both Fig. 20 and Fig. 21 the torch flame is pointed almost, but not quite, toward the centerline of the pipe, while the filler rod is held approximately tangent to the pipe surface. This is the normal angle for pipe welding, regardless of the welding position, except when adjustment of flame angle is required for puddle control in the horizontal and overhead positions. Unless the pipe ends have been perfectly prepared, there will be variations in the width of the joint gap at the root. This may require some careful bridging, by concentrating first on one side of the joint, then on the other. One of the big advantages in oxyacetylene welding is that heat input and filler metal input can be separately controlled. Gaps that can give the are welder trouble are no problem at all to the experienced oxyacetylene pipe welder.

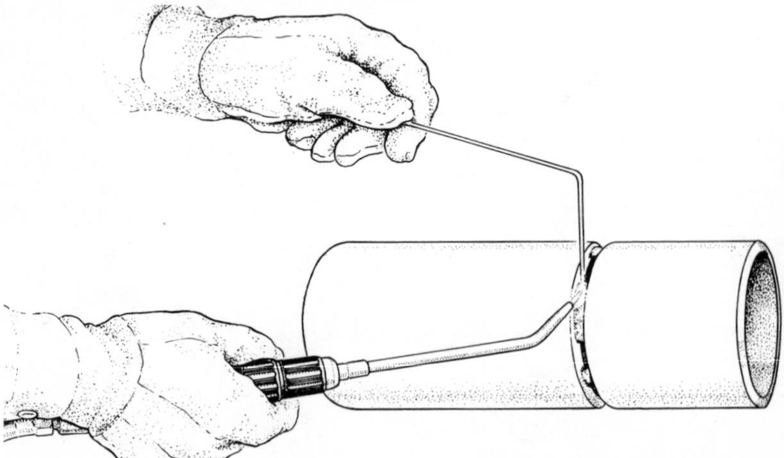

Courtesy *Union Carbide Corp.—Linde Division*

Fig. 21. Continuing the weld, welding toward the top (forehand method).

Proper rod movement is the key to putting the finishing pass on a pipe weld. The rod must rub in the bottom of the puddle and bump against the edges of the joint. The flame must linger a bit at each end of each arc, and be concentrated on the puddle—never on the filler rod. Movement of the flame and rod must be in opposite directions. As shown in Fig. 22, the welder is working in the vertical position. The flame is pointed somewhat away from the centerline of the pipe in order to hold the puddle against the force of gravity.

Fig. 23 shows the welder using the backhand technique. In pipe welding, when the pipe cannot be rotated, there are many occasions when the pipe cannot be rotated, there are many occasions when the ability to switch easily from forehand to backhand technique can save time for the welder. With the backhand technique it is quite easy to make a vertical weld from top to bottom, rather than from bottom to top. The flame is still used to "support" the puddle, but in this case its aim is to keep the puddle from running ahead onto metal which has not yet reached melting temperature. When using the backhand technique the relative movements of flame and rod are quite different from those used in forehand work. The novice welder is advised to practice the backhand technique on flat and vertical pieces of plate before employing the technique on pipe.

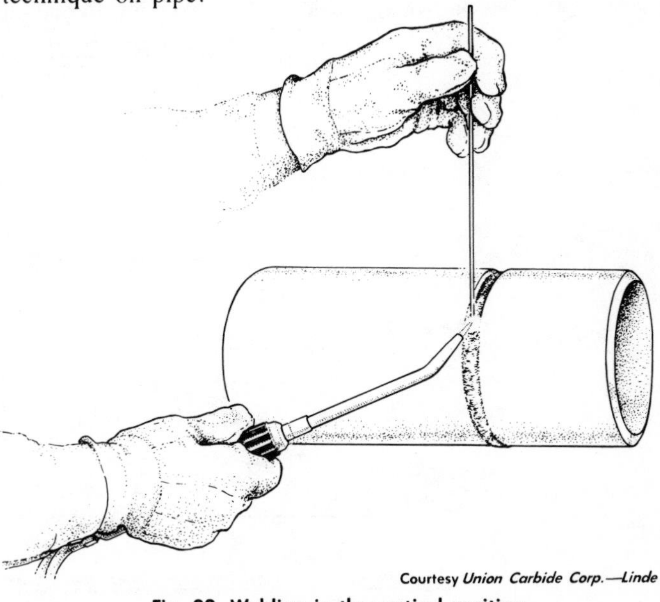

Courtesy *Union Carbide Corp.—Linde Division*

Fig. 22. Welding in the vertical position.

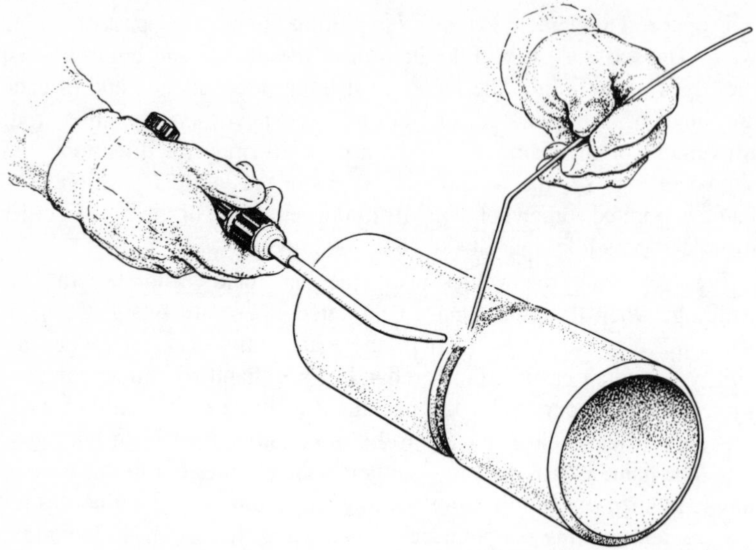

Courtesy *Union Carbide Corp.—Linde Division*

Fig. 23. Finishing the weld, using the backhand technique.

It is often necessary to make a weld in a section of vertical piping. Fig. 24 shows the root pass being made on this type joint. In making the

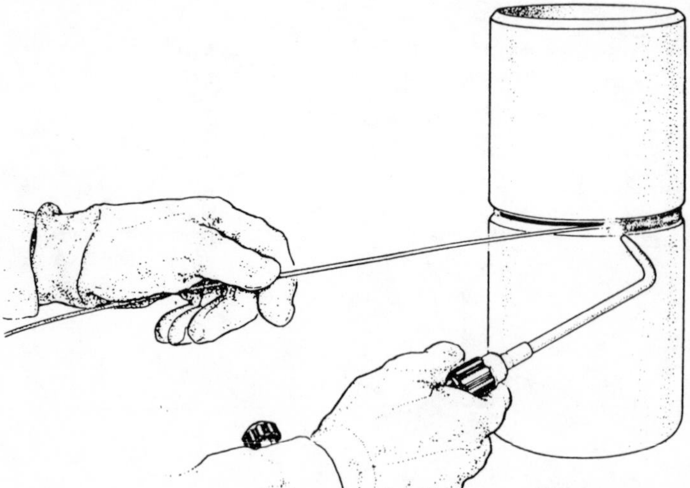

Courtesy *Union Carbide Corp.—Linde Division*

Fig. 24. Making a weld in a section of vertical piping.

root weld, keep the puddle as small as possible, and angle the flame slightly toward the upper side of the joint. The rod should be held in what is considered the normal position for all pipe work, approximately tangent to the circumference of the pipe. When the finish pass is made, the puddle must be slanted slightly with the lower end leading the upper end, and the filler rod must be manipulated to push molten metal up against the top edge of the joint.

Fig. 24 shows the advantages in being able to weld using both the forehand and backhand techniques. If the weld is made using only the forehand method the weld must be started at the bottom of the joint and worked up one side; then the welder must change position, start at the bottom again and work up the other side. If the backhand method has been mastered the welder can start on one side, work down to the bottom using the backhand technique, switch to the forehand method and work up the other side. The ability to switch from forehand to backhand and vice versa can be a big timesaver when welding fixed joints in horizontal pipe runs. As shown in Fig. 25, an adjustable

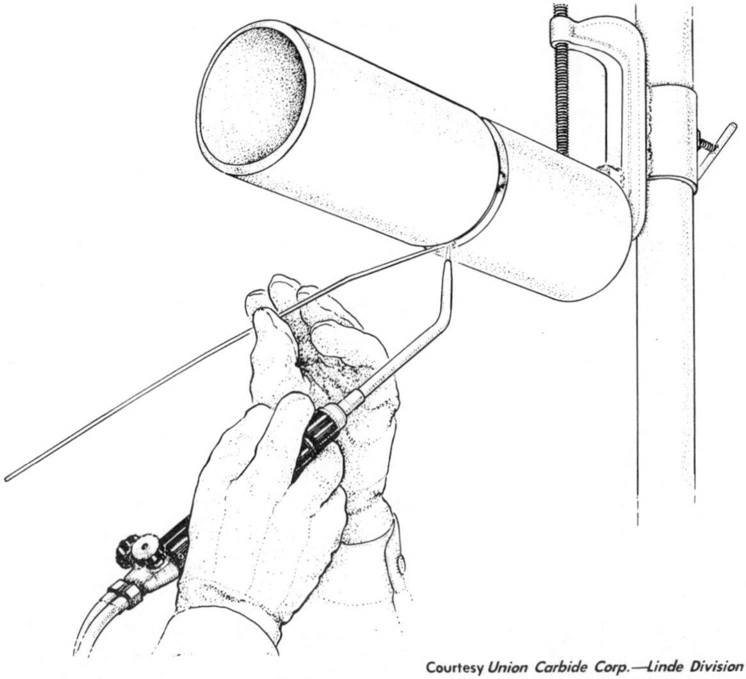

Courtesy *Union Carbide Corp.—Linde Division*

Fig. 25. The ability to make both forehand and backhand welds is very useful when making this type of weld.

support using a sliding clamp to which a C clamp has been welded is excellent for use when practicing overhead welding. Becoming a good pipe welder is largely a matter of having good instruction, good tools and constant practice until the art of welding is mastered.

GENERAL PRECAUTIONS

CAUTION! IMPORTANT! USE NO OIL! Oil, grease, coal dust, and some other organic materials are easily ignited and burn violently in the presence of high oxygen concentrations. These materials should never come in contact with oxygen or oxygen-fuel gas equipment. Lubrication of any type is *never* required with oxygen-fuel gas apparatus.

Always call oxygen by its proper name! Do not call it "air." Never use it as a substitute for compressed air. A serious accident may result if oxygen is used as a substitute for compressed air. Oxygen must never be used in pneumatic tools, in oil preheating burners, to start internal combustion engines, to blow out pipelines, to "dust" clothing or work, for pressure testing or for ventilation.

Never use acetylene at pressures above 15 psi. The use of acetylene at pressures in excess of 15 psi. is extremely hazardous.

Never use torches, regulators, or other equipment that is in need of repair. A "creeping" regulator, leaking torch valves, or leaking hose connections usually require repair by a qualified repair station.

Never connect an oxygen regulator which is not equipped with an inlet filter to a cylinder.

Always use the operating pressures recommended by the manufacturer for the welding head or cutting nozzle in your torch.

Always wear goggles when working with a lighted torch. The choice of shades will be up to the welder and will generally be in the range of shades 4, 5, and 6.

Do not use matches for lighting a torch. Always use a friction lighter or some other source of ignition which will keep your hands away from the torch tip or nozzle.

Wear suitable clothing when using a torch. Fire resistant gauntlet type gloves, pants free from oil or grease, pants without cuffs, and high top shoes should all be worn while operating a torch.

Before starting to weld or cut, check the area to be sure that sparks or slag will not start a fire or damage walls, ceilings, floors, furniture, or equipment.

Never do welding or cutting without adequate ventilation. The four paragraphs which follow are reprinted from *The Oxy-Acetylene Handbook,* published by the Union Carbide Corporation-Linde Division.

1. **When it is necessary to work in a confined space, make certain that the space is adequately ventilated.** *If there are no overhead openings to provide natural ventilation, use an exhaust fan or blower. When screening an area, try to keep the bottom of the screens at least two feet above floor level. Do not take cylinders into a confined space. Test all equipment, including hose, for leaks before taking it into a confined space, and bring it out with you when your work is interrupted for any reason, even for a short time.*

2. **Never flow oxygen into a confined space in order to ventilate it, or to "clear the air".** *Remember that oxygen supports and accelerates combustion, and will cause oil, wood, and many fabrics to burn with great intensity. Clothing saturated with oxygen, or with oxygen enriched air, may burst into flame when touched by a spark.*

3. **Do not weld brass, bronze, or galvanized steel except in a well ventilated location.** *You must protect yourself against breathing the zinc oxide vapors usually generated when these materials are heated to welding temperatures.*

4. **When welding or cutting metals containing or coated with lead, cadmium, beryllium, or mercury, always wear a suitable air-line mask.** *No operator should be considered immune from the effects of these fumes. A straight filter-type mask is inadequate; nothing short of an air supplied respirator is suitable. There must be sure protection against breathing fumes which occur when lead, mercury, beryllium, or their compounds are heated.*

Use particular caution when welding or cutting in dusty or gassy locations. Dusty and gassy atmospheres in some mines and plants call for extra precautions to avoid explosions or fires from sparks, matches, or open flames of any type. Welding or cutting in any such "suspicious" location should be done only when proper precautions are taken, and only after a responsible official has inspected the situation and given approval for the work.

When welding or cutting is done underground, leak testing of equipment set-ups should be performed carefully and frequently. Great care should be taken to protect hoses and cylinders from damage, and to protect timbers and other combustible materials from sparks and hot slag.

Never do any welding or cutting on containers that have held flammable or toxic substances until the containers have been thoroughly cleaned and safeguarded. Complete, detailed procedures are given in American Welding Society booklet A6.0-65 titled "Safe Practices for Welding and Cutting Containers That Have Held Combustibles". Some of the key points in this booklet are:

1. Always start with the assumption that any used drum, barrel, or container may contain flammable (explosive) or toxic (poisonous) residue.
2. No work should be commenced until the container has been cleaned and tested sufficiently to assure that no flammable or toxic solids, liquids, or vapors are present.
3. Bear in mind that some non-poisonous substances can give off toxic vapors when heated.
4. Steam is usually an effective means of removing volatile or readily volatilizable materials from a container. Washing with a strong solution of caustic soda or other alkaline cleaner will remove heavier oils which steam may not volatilize.
5. Whenever possible, the container should be filled with water to within a few inches of the working area before attempting welding, cutting, or intense heating. Care should be taken to provide a vent opening for the release of heated air or steam.
6. When it is impractical to fill the container with water, an inert gas such as nitrogen or carbon dioxide may be used to purge the container of oxygen and flammable vapors. Maintain the inert gas during the entire welding or cutting operation by continuing to pass the gas into the container.

Other Safe Practices To Bear in Mind

Make sure that jacketed or hollow parts are sufficiently vented before heating, welding, or cutting. Air or any other gas or liquid confined inside a hollow part will expand greatly when heated. The pressure created may be enough to cause violent rupture of the part. A metal part which is suspiciously light, is hollow inside, and should be drilled to vent it before heat is applied. After work is complete, the vent hole can be tapped and plugged if necessary.

Bushings in a casting should either be removed or securely fastened in place before heating the casting. Bronze bushings expand more than cast-iron when heated to the same temperature. If a bushing is left in place, the casting may be damaged, or expansion may cause the bushing to fly out, creating a definite hazard. If a bushing cannot be removed, it should be securely fastened in place. Bolting large washers or pieces of plate over the ends of the bushing is the suggested method.

Don't drop stub ends of welding rods on the floor. Put them in a suitable container. Aside from the fire hazard that may be created by carelessly dropping a hot stub end, a serious fall may result from stepping on one. A small container partly filled with water and within easy reach is a good place to dispose of these short ends. Protect cylinders, hose, legs, and feet when cutting. Do not cut material in such a position as will permit sparks, hot metal, or the severed section to fall on or against a gas cylinder, or the gas hoses, or your legs or feet. A suitable metal screen in front of your legs will provide protection against sparks and hot metal.

Preventing Fires

Where welding or cutting has to be done near materials that will burn, take special care to make certain that flame, sparks, hot slag, or hot metal do not reach combustible material and thus start a fire. This is particularly important in the case of cutting operations. Cutting produces more sparks and hot slag than welding, and locations where portable cutting equipment is used must, therefore, be thoroughly safeguarded against fire.

Never use cutting or welding torches where sparks or open flame would be a hazard. Flames are a hazard in rooms containing flammable gas, vapors, liquids, dust, or any material that easily catches fire. It is not safe to use cutting or welding equipment near rooms containing flammable materials unless there is absolutely no chance of sparks passing through cracks or holes in walls or floors, through open or broken windows, or open doorways. There is always the possibility that flammable vapors may escape from such rooms through doors or other openings, making it doubly necessary to keep sparks and flames away.

Before you cut or weld in a new location for the first time, always check with the nearest person in authority, he may know of some serious fire hazard that might otherwise be overlooked.

If the work can be moved, take it to a location where there will be no possibility of setting fires. This must always be done when the metal to be welded or cut is in a place where open flames are barred. This practice may also be sensible in many other locations even if open flames are permitted.

If the work cannot be moved, materials that burn easily should, if possible, be taken a safe distance away. For cutting operations, this distance may be 30 to 40 ft. or more. Floors should be swept clean before the torch is lighted.

If flammable materials cannot be moved, use sheet metal guards, asbestos paper or curtains, or similar protection to keep sparks close in to the work you are doing. Suitable protection to keep back sparks should always be used when it is not possible to move materials that will burn to a safe distance from the cutting or welding work. This also applies if sparks might lodge in wooden parts of the building, or drop through holes or cracks to the floor below. Make sure that the guards are large enough and tight enough so that they do not permit sparks to roll underneath or slide through openings. Curtains should be weighted down against the floor or ground. For weights, use such things as angle iron, pipe, bricks or sand. Use only fire-resisting guards. Do not use tarpaulins for shielding sparks since they may catch fire. Have someone stand by to watch the sparks so that he can give warning if they begin to get beyond the protective guards. It is not reasonable to expect whoever is doing the welding or cutting to watch the sparks since his attention is on the work. In addition, the sparks cannot always be seen easily through goggles.

If welding or cutting over a wooden floor, sweep it clean and wet it down before starting work. Provide a bucket or pan containing water or sand to catch the dripping slag from any cutting that is done.

Before starting to cut off a piece of steel or iron, make sure it will not drop where there is any possibility of starting a fire. This is especially important when working in a high place where sparks or slag might be less likely to cause a fire down below than would a small piece of red hot steel. Pieces can be kept from falling by welding a rod or bar to the piece, and having a helper hold the rod or bar while the cut is made. Or the same thing might be accomplished by tying a chain or other suitable support to the piece.

When welding very close to wooden construction, protect it from direct heat. Wooden beams, partitions, flooring, or scaffolding should always be protected from the direct heat of the flame by sheet metal guards or asbestos. Protective guards should also be used to confine the sparks.

Use the correct oxygen pressure when cutting. An oxygen pressure greater than necessary will only cause extra sparks and increase the slag flow, to say nothing of increasing the oxygen expense.

Store extra cylinders away from important areas. Keep only enough cylinders near the work to insure an adequate supply of gases for the job at hand.

Be ready to put out any fire promptly with fire extinguishers, pails of water, water hose, or sand. When torches are to be used near wooden construction or materials that will burn, take every precaution to prevent fires, but always be prepared to put out fires that may start despite all precautions. In hazardous locations, have a helper, or one or more extra men if necessary, on hand to watch for and be ready to extinguish a fire.

If there is a sprinkler system in the building, maintain this protection without fail while cutting or welding is being done. It is of special importance to make sure that sprinklers are in working order during extensive repairs or building changes. If the sprinkler system must be shut down for a time, have this done when welding or cutting work is not in progress.

If there is a possibility that a smoldering fire may have been started, keep a man at the scene of the work for at least a half hour after the job is through. Have him look carefully for smoke or fire before leaving. This is especially important when cutting torches have been used in locations where sparks may have started smoldering fires in wooden structures or in other slow burning materials. *Never forget* that heavy cutting sparks sometimes fly 25 to 30 ft. or more and hold their heat for several seconds after landing.

When welding or cutting on bridges, structures, or at other outdoor sites, take care to avoid setting fire to grass or brush. Brush should be cleaned out or cautiously burned under or about structures before start of the work. Special care is necessary during a dry spell. A fire extinguisher and water or sand should be available to extinguish any fires started in the course of work. Before leaving the site, examine the premises thoroughly to be sure that sparks have not started smoldering fires.

ARC WELDING

Many different types and sizes of arc-welding machines are available, but all fall into two basic categories—the conventional or *constant-current* welding machine with the drooping volt-ampere curve, and the *constant-voltage* or modified constant-voltage machine with the fairly flat characteristic curve. The conventional machine can be used for manual welding and, under some conditions, for automatic welding.

The constant-voltage machine is used only for semi-automatic and automatic welding.

Another classification of welding machines is based on whether they are rotating types or not. The rotating or generator type is the most universal. This kind of machine gives a wider range of welding-current adjustment and a wider range of electrodes. In shop work, the generator is powered by an electric motor; in the field, by any internal-combustion engine. This type of machine, because it has its own power source, is a good practical machine for emergency work, and when away from available power lines.

Rotating generator types are available to provide either DC or AC. In some cases, particularly when driven by a gasoline or diesel engine, the generator is designed to produce AC, and thus can be used to provide power for lights and tools as well as power for welding. The AC is rectified in some machines to provide DC.

Conventional Single-Operator Welding Machines

This type of machine, called a *constant-current* welder, produces a volt-ampere curve like the one in Fig. 26. This curve is obtained by loading the machine with variable resistance and plotting the terminal voltage for each ampere output. A brief study of the curve will reveal that a machine of this type produces maximum output voltage with no load, and as the load increases, the output voltage decreases. For normal welding conditions, the output voltage is usually from 20 to 40 volts. The open-circuit voltage is between 60 and 100 volts. This high no-load voltage provides for easier arc starting with all types of electrodes. The open-circuit, or no-load, voltage is controlled by the fine adjustment on a dual-control type of welding machine.

The welding current is adjusted by both the range switch for coarse adjustment and a fine adjustment. On this type of machine, the actual arc voltage is controlled by the welding operator and has a direct relationship to the arc length. As the arc length is increased, the arc voltage increases—decreasing the arc length decreases the arc voltage.

A study of the curve will reveal that with a long arc (higher arc voltage), the current output will decrease a limited amount, whereas with a shorter arc (low arc voltage), the current output will increase a limited amount. Thus, the operator can vary the current output or heat of the machine by lengthening or shortening the arc.

On dual-control machines, the slope of the output curve can be varied. In this way, a soft or harsh arc can be obtained. With a flatter

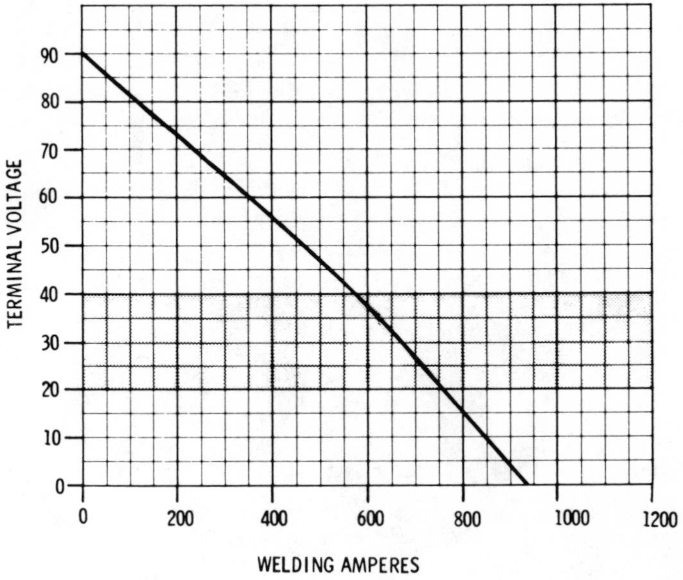
90
80
70
60
50
40
30
20
10
0
TERMINAL VOLTAGE
0
200
400
600
800
1000
1200
WELDING AMPERES

Courtesy *Hobart Brothers Co.*

Fig. 26. Volt-ampere curve of a constant-current welder.

curve, the same change in arc voltage will produce a greater change in current output. This produces a digging arc preferred for pipe welding. With a steeper curve, the same change in arc voltage will produce less of a change in the output current. This is a soft, or quiet, arc useful for sheet-metal welding. In other words, a conventional or constant-current welder with dual controls allows the most flexibility for the welding operator. These machines can be driven by an electric motor or internal-combustion engine.

A typical electric-drive arc welder is shown in Fig. 27. Machines of this type have multirange dual controls which solve the problem of getting just the right heat for each and every job. The large hand-wheel on the front of the control panel enables the operator to choose any one of ten coarse ranges. The volt-amp adjuster in the center of the wheel permits up to 100 fine adjustments for each coarse range, providing 1000 possible welding combinations. This type of welder also has a remote-control device which permits the operator to make adjustments directly at the welding site.

HOBART

Courtesy *Hobart Brothers Co.*

Fig. 27. A portable constant-current arc welder.

Constant-Voltage Welding Machines

A constant-voltage welding machine provides a practically constant voltage to the arc, regardless of the amount of arc current. This type of welder is shown in Fig. 28. The characteristic curve of this type of machine is shown by the volt-ampere curve in Fig. 29.

A constant-voltage welder can only be used for automatic or semiautomatic welding, with one exception. With the proper voltage-dropping ballast resistors, this type of machine can be used to supply power simultaneously to a number of welding operators. However, the constant-voltage power supply was designed specifically for automatic welding. It is a known fact that the burn-off rate of a specific size and type of electrode wire for a given arc voltage is proportional to the welding current. In other words, as the welding current increases, the amount of wire burned off will increase proportionally. This is shown graphically on the curve in Fig. 30. Thus, it can be seen that if a wire is fed into an arc at a specific rate, it will automatically require, or draw, a proportionate amount of current from a constant-voltage power source.

HOBART
Multi-wire

Courtesy *Hobart Brothers Co.*

Fig. 28. A constant-voltage arc welder.

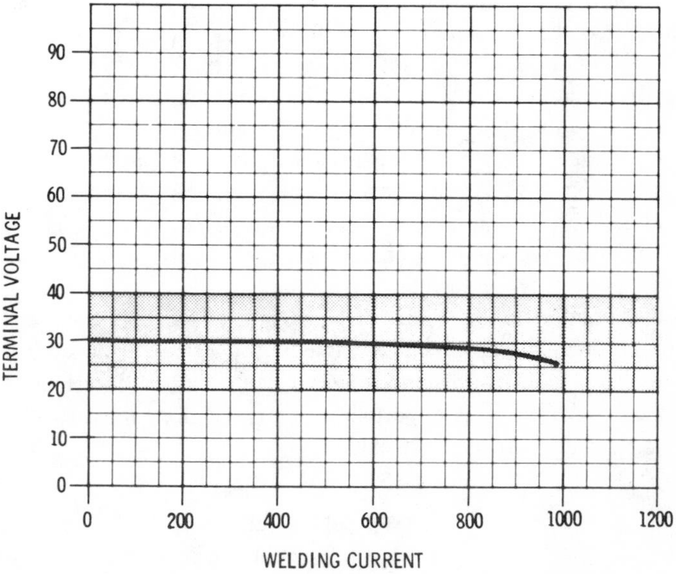
TERMINAL VOLTAGE
90
80
70
60
50
40
30
20
10
0
0
200
400
600
800
1000
1200
WELDING CURRENT

Courtesy *Hobart Brothers Co.*

Fig. 29. Volt-ampere curve of a constant-voltage welder.

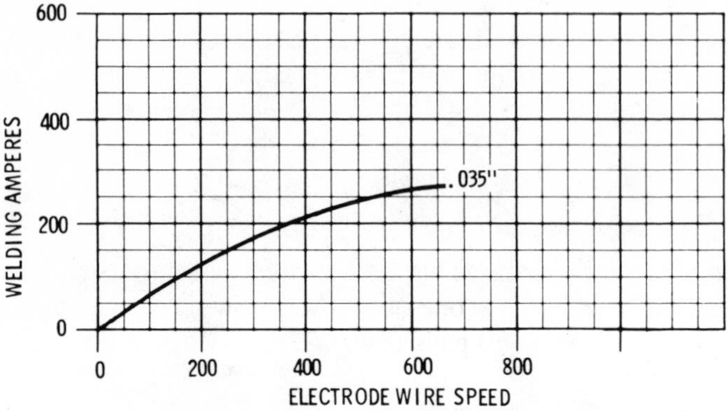
600
400
200
0
WELDING AMPERES
.035"
0
200
400
600
800
ELECTRODE WIRE SPEED

Courtesy *Hobart Brothers Co.*

Fig. 30. Burn off rate curve of an automatic welder.

A constant-voltage welder provides the amount of current required from it by the load imposed on it. In this way, a very basic type of automatic welding control can be employed. The wire is fed into the arc by means of a constant-feed motor. This feed motor can be adjusted to increase or decrease the rate of wire feed. Complicated circuitry is thus eliminated, and the system is inherently self-regulating. If the electrode wire is fed in faster, the current will increase. If it is fed in slower, the current will automatically decrease. The current output of the machine is thus set by the speed of the wire-feed motor.

The voltage of the generator is regulated by an output rheostat on the generator proper. Thus, only two controls maintain the proper welding current and arc voltage in a constant-voltage system. It is almost impossible to have stubbing or burn-back, which is common with the constant-current type of welding machine.

The characteristic curves of constant-voltage machines have a slight inherent droop. This droop can be increased (the slope made steeper) by various methods. Most machines have taps for varying the slope of the characteristic curve. It is important to select the slope most appropriate for the type of work being welded.

PIPE WELDS

Courtesy *Crane Co.*

There are numerous types of welds used for various purposes and different kinds of pipe connections. Those in general use are as follows.

Butt Welds

Two types of butt welds are generally used to join pipe sections. The one in Fig. 31A is for pipe with a wall thickness of 3/4″ or less. For pipe thickness greater than 3/4″, the butt weld shown in Fig. 31B is recommended. Figs. 31C and D show welds utilizing a machined backing ring, and with an inert arc used on the first pass.

Welded Couplings and Pads

Smaller sizes of screwed and socket-weld couplings and connections are generally prepared in one of two ways—welded couplings. (Figs. 32A and B) and welded pads (Fig. 32C).

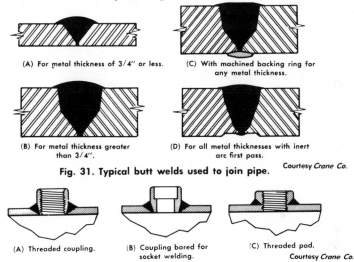

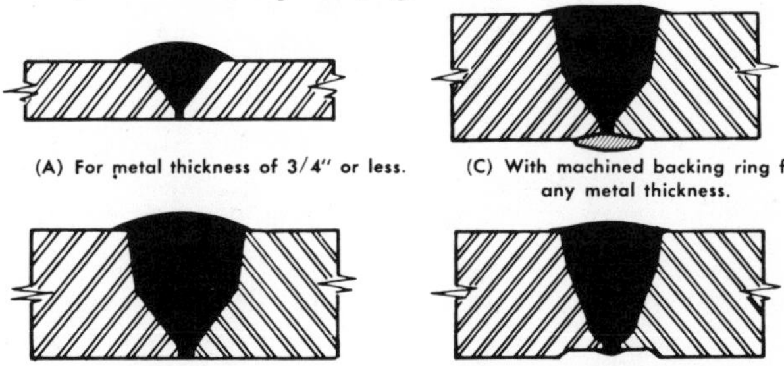

(A) For metal thickness of 3/4″ or less.

(C) With machined backing ring for any metal thickness.

(B) For metal thickness greater than 3/4″.

(D) For all metal thicknesses with inert arc first pass.

Courtesy *Crane Co.*

Fig. 31. Typical butt welds used to join pipe.

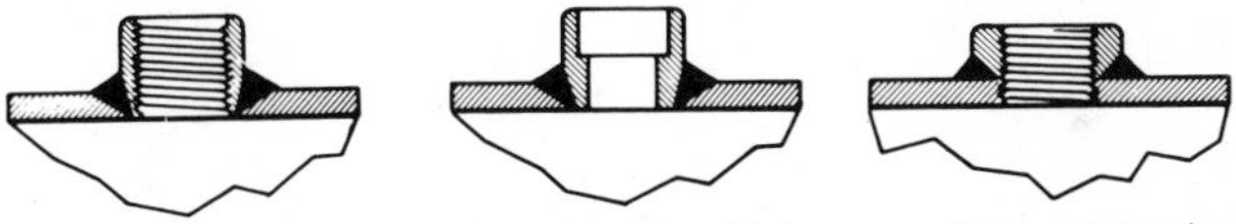

(A) Threaded coupling.

(B) Coupling bored for socket welding.

(C) Threaded pad.

Courtesy *Crane Co.*

Fig. 32. Welded couplings and pads.

Welded Nozzles

In the fabrication of pipe with welded nozzles, two types are utilized—the *standard type,* which is not reinforced; and the *reinforced type,* which provides reinforcement of the nozzle weld. Fig. 33A shows a 90° standard type, while Fig. 33B shows a 90° reinforced nozzle. Angular types with standard and reinforced welds are shown in Figs. 33C and D, respectively.

Reinforcing of welded nozzles is done as required by the customer and in accordance with ASMF Boiler and Pressure Vessel Code or the ASA Pressure Piping Code when compliance with these codes is mandatory. The ring type of reinforcement is furnished in the absence

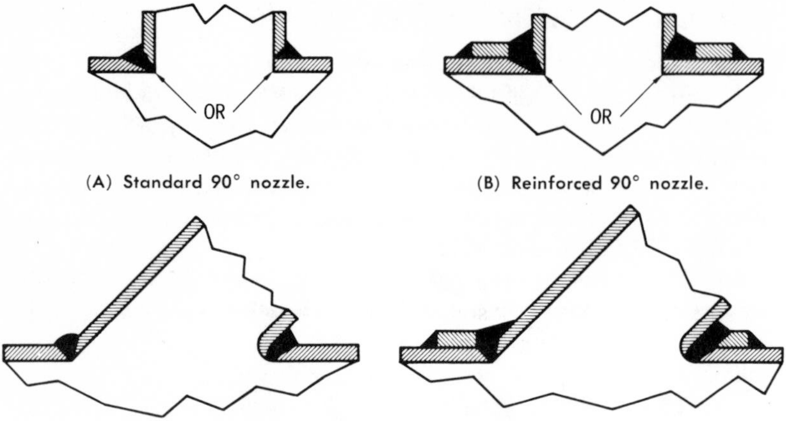
OR
OR

(A) Standard 90° nozzle.

(B) Reinforced 90° nozzle.

(C) Standard angular nozzle.

(D) Reinforced angular nozzle.

Courtesy *Crane Co.*

Fig. 33. Typical angular and 90° nozzle welds.

of definite specifications for fabricated piping with reinforced welded nozzles.

Miter Weld

A miter weld (Fig. 34) may be used for fabricating pipe in the larger sizes or for unusual conditions. The use of pipe bends or welding elbows is recommended in preference to miter welds, where possible. However, where space or design limitations require, or when suitable welding elbows or bends are not available, miter welding can be used. Such welds are made up of various combinations of segments, angles, etc., and must be designated for each condition encountered.

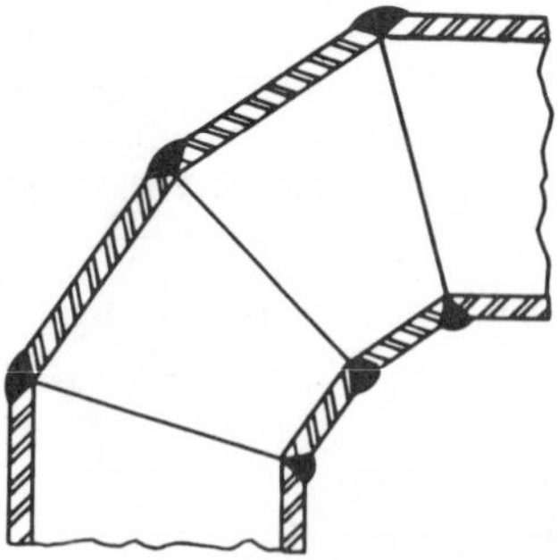

Fig. 34. A miter-type weld.

Courtesy *Crane Co.*

MICRO-WIRE PIPE WELDING

(Courtesy Hobart Brothers Co.)

The piping industry, a major user of welding, is always looking for ways to reduce costs. A method developed by Hobart Brothers Company, called the *Micro-wire* welding process, was introduced to meet these demands. It was promptly investigated by the piping industry and, after extensive laboratory work and field experiments, was used in actual production in pipe welding.

Micro-wire welding is a gas-shielded, arc-welding process which uses a small-diameter, consumable electrode wire that is continuously fed into the arc. Metal is transferred through the gas-protected arc column to the work. The process may be either automatic or semi-automatic, the latter method being the more popular. The diagram of the process in Fig. 35 shows the electrode wire and the shielding gas being directed into the arc by a lightweight, manually controlled welding gun.

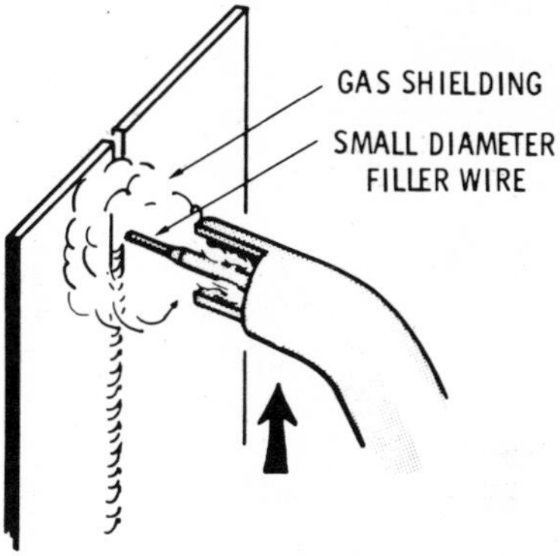
GAS SHIELDING
SMALL DIAMETER FILLER WIRE

Fig. 35. Micro-wire gas-shielded metal arc.

Courtesy *Hobart Brothers Co.*

Advantages

Many distinct advantages of *Micro-wire* over manual coated-electrode welding are claimed by the manufacturer. Some of these are:

1. Reduce welding time per pipe joint.
2. Elimination of flux and slag, thus reducing the cleaning time per weld joint.
3. Fewer passes required per joint.
4. Fewer starts and stops per weld pass.
5. Usable on almost all sizes of pipe and pipe-wall thickness.

6. High-quality, "low-hydrogen" type of weld metal.
7. Smooth weld surface produced inside the pipe.
8. Back-up rings normally not required.
9. Training period for welding operators shorter than for manual coated-electrode welding operators.
10. Weld appearance superior to that produced with manual coated electrodes.
11. No electrode stub loss.
12. Distortion and warpage reduced, especially on thin-wall pipe.
13. Initial or root pass thicker and stronger, allowing faster "move-up."
14. Usable for hot tapping.

There are many other minor advantages claimed for *Micro-wire* welding, such as smoothness of the exterior surface, ease of painting, and reduced warpage.

Piping Categories

The pipe industry is divided into four distinct groups. Each has its own specific requirements and is governed by different conditions and specifications. These four groups are:

1. Subcritical piping.
2. Pressure or power piping.
3. Cross-country transmission-line piping.
4. Water-supply piping.

Subcritical piping consists mostly of small-diameter pipe applications, such as domestic water supply, sanitary systems, and air conditioning. There is no code nor qualification procedures which cover subcritical piping.

Pressure and power piping is used for steam generating, refining, and chemical processing. Welding for this type of piping is governed by the ASA Pressure Piping Code, which is based on Section IX of the ASME Pressure Vessel Code. When properly applied, *Micro-wire* welding satisfies these requirements.

Cross-country transmission piping carries fuel gas and/or petroleum products over long distances. Welding of such piping (usually of large diameter) is governed by the American Petroleum Institute Standard 1104 Code. *Micro-wire* welding qualifies for this type of piping also, when properly applied.

Water transmission piping is not governed by any code except that specified by the owner, which is usually a local government code. This piping is of large diameter, often as large as 96′.

Typical companies which use the *Micro-wire* process for welding pipe are as follows:

Cross-country pipeline contractors
Mechanical contractors
Utility companies
Heating, ventilating, and air-conditioning contractors
Refineries
Chemical plants
Food processing industries
Beverage industries

Application

Micro-wire welding features versatility and economy. Welding is possible in all positions, the arc is visible to the operator, there is no slag to remove, and the weld surface is smooth. The process is used primarily for welding low- and medium-carbon steels, and low-alloy, high-strength steels of thin to medium gauge. Welding can be performed on extremely thin materials (20-gauge) and will successfully bridge wide gaps. The *Micro-wire* process has produced highly satisfactory welds in products of the automotive and household appliance industries, and in the fabrication of metal furniture, fuel tanks, and pressure vessels. Other applications include electrical equipment, industrial machinery, pipe joining, salvage operations, and highway, bridge, and structural fabrication.

Equipment

Fig. 36 shows the more important components of a *Micro-wire* welding outfit. Included are the shielding-gas system and controls, the welding machine (power source), the wire-feeding mechanism and controls, the electrode wire, and the welding gun and cable assembly.

Various shielding gases can be used. Carbon dioxide CO_2 gas is the least expensive and the most popular. Mixtures of CO_2 and argon and mixtures of argon and oxygen, are also sometimes used. The CO_2 gas should be of *welding grade*, meaning that it should have a low moisture content, indicated by the *dew point temperature*.

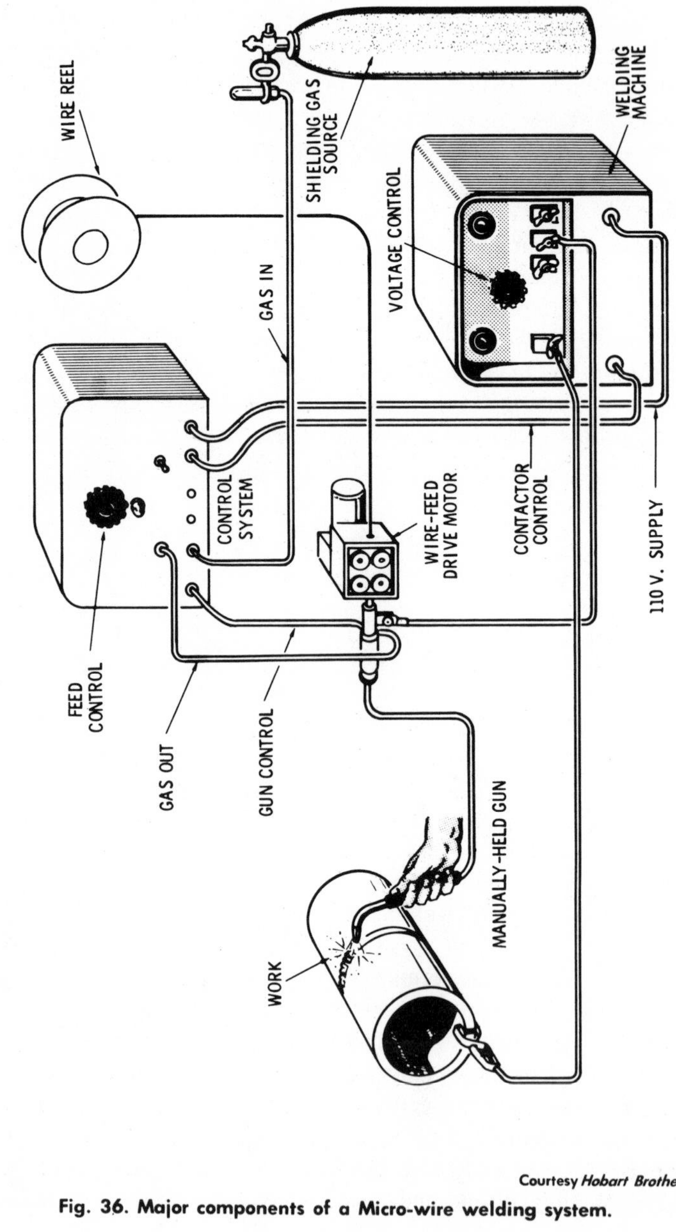
WIRE REEL
SHIELDING GAS SOURCE
WELDING MACHINE
GAS IN
VOLTAGE CONTROL
CONTROL SYSTEM
WIRE-FEED DRIVE MOTOR
CONTACTOR CONTROL
110 V. SUPPLY
FEED CONTROL
GAS OUT
GUN CONTROL
MANUALLY-HELD GUN
WORK

Courtesy *Hobart Brothers Co.*

Fig. 36. Major components of a Micro-wire welding system.

A dew-point temperature of −40°F or lower is recommended. The gas-flow rate is very important. A pressure-reducing regulator and a flowmeter are required on the gas cylinder. For *Micro-wire* welding, the flow rate is from 15 to 20 cubic feet per hour as measured and set by the flowmeter. When welding outside, or when air currents disturb the gas shield, a higher gas-flow rate is necessary, sometimes for 30 to 35 cubic feet per hour. To maintain the higher rate, two or more gas cylinders are manifolded together. When high flow rates are used, it is sometimes necessary to heat the gas by means of a gas heater.

A specially designed *welding machine* (power source) is used for *Micro-wire* welding. It is a constant-voltage type. It can be a DC rectifier or generator, with the generator motor—or engine-driven. The output welding power of a constant-voltage machine has essentially the same voltage, no matter what the welding current may be. The output voltage is regulated by a rheostat on the welding machine. *Micro-wire* welding uses an arc voltage of from 16 to 24 volts; the open-circuit voltage (while not welding) and the arc voltage (while welding) are read on the welding machine's volt-meter. There is no current control on a constant-voltage type machine and, for this reason, it *cannot be* used for manual electrode welding. The welding-current output is determined by the wire feeder.

The specially designed *wire-feeding mechanism* and the constant-voltage welding machine constitute the heart of the *Micro-wire* welding process. There is a fixed relationship between the rate of electrode wire burn-off and the welding current. In other words, at a given wire-feed rate, the welding machine will produce the current required to maintain the arc. Thus, the electrode wire-feed rate determines the welding current. Wire-feed speed (thus, the welding current) is set by the wire-feed speed control on the wire feeder. The welding current, which can be read on the welding machine's ammeter, ranges from 50 to 300 amperes for *Micro-wire* welding. Reverse polarity is used with direct current.

The electrode wire is, normally, 0.030″, 0.035″, or 0.045″ in diameter. It is solid and bare, except for a very thin copper coating, or preservative, on the surface to prevent rusting. It contains dioxidizers which help to clean the weld metal and to produce sound, solid welds. The composition of the electrode wire must be matched to the base metal being welded.

The *welding gun and cable assembly* is used to carry the electrode wire, the welding current, and the shielding gas from the wire feeder to the arc area. The operator directs the arc and controls the weld with the welding gun. The distance from the tip of the gun to the work is called the *stickout.* This stickout distance is controlled by the operator, but the arc length is automatically controlled by the voltage setting on the welding machine. The stickout can be varied somewhat, and is useful in controlling the weld. Too much stickout reduces the gas shield and must be avoided. A stickout (tip-to-work) distance of 1/4″ to 3/8″ is best. The gun nozzle must be cleaned regularly to assure efficient gas shielding. Special antisplatter conpounds are used frequently to prevent a spatter build-up on the nozzle.

The equipment required for automatic *Mirco-wire* welding is essentially the same as for the semiautomatic process, except that fixturing is provided and the welding gun is replaced by an automatic welding torch.

Fig. 37 shows a typical *Micro-wire* semiautomatic wire-feed welding machine used where a power source is available. Fig. 38 shows a generator-type welder rated at 350 amperes using a water-cooled gas engine for in-the-field-use.

HOBART
Micro-wire

Courtesy *Hobart Brothers Co.*

Fig. 37. A portable Micro-wire welding system driven by an electric motor.

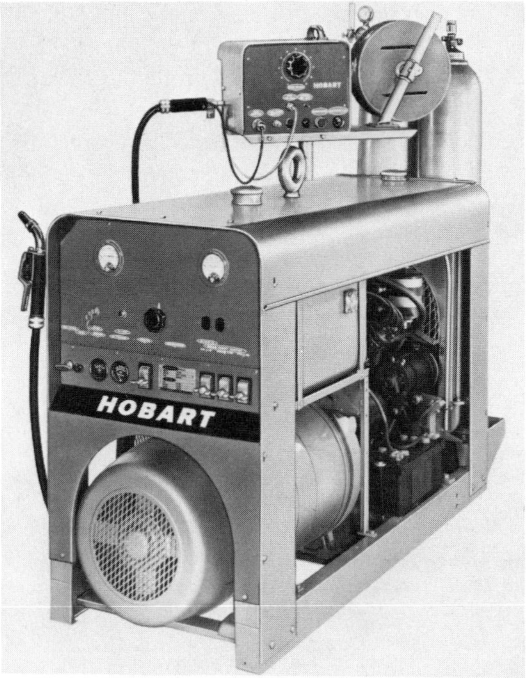
HOBART

Courtesy *Hobart Brothers Co.*

Fig. 38. A Micro-wire welding system powered by a gasoline engine.

CHAPTER 2

Planning a Heating System

This chapter has been included because heating installations are often of concern to those persons engaged in the plumbing trade. In fact, plumbing and heating are allied trades.

Heating is generally required in nearly every section of the United States, and in most areas is required for most of the year. In some instances, the same equipment is used for summer cooling as well as heating during the cooler seasons. When both heating and cooling are combined in an integrated unit, the functions required for both must be considered when planning for the heating system.

The first step in determining the size of a heating system is to accurately determine the heat loss of the building. Improved construction methods and building materials outmode the rule-of-thumb method formerly used to determine heat loss. To avoid oversizing the heating plant, with consequent increase in installation costs, every effort should be made to obtain as accurate a heat-loss determination as possible.

When the outside temperature is lower than the inside temperature, heat is lost through floors, ceilings, walls, doors, and windows. Windows and doors allow some infiltration of air through cracks. Thus, heat is lost to the outdoors in two ways—by infiltration around doors and windows, and by transmission through the building materials.

HEAT LOSS PER SQUARE FOOT

(STEP 1)

To properly determine the amount of heat lost through a particular type of construction, it is necessary to work with a specific value known as a *coefficient of heat transmission.* This value is designated as the *"U" factor,* which represents the time rate of heat flow (expressed in Btu/hr) for one square foot of surface with a temperature difference of one degree Fahrenheit between the air on one side and the air on the other. "U" factors have been determined for a wide variety of wall, floor, ceiling, window, and door construction, These are available from the American Society of Heating & Ventilating Engineers as well as many manufacturers of heating equipment.

The location of the walls, ceilings, windows and skylights, floors, and doors are listed on a tabulation sheet (Fig. 1) along with the type of construction. Walls may be located according to their exposure (North, West, etc.), by the room, or both.

NOTE: Heat loss is calculated for all surfaces exposed to the outdoors, to unheated spaces, or to partially heated spaces.

List the "U" factor for each entry on the sheet,

NOTE: It is common practice in calculating the transmission heat loss of a door to use the "U" factor for a comparable size single-pane glass window. This is because accuracy in selecting the proper "U" factor for the actual door construction varies widely. By using the glass factor, the heat-loss calculation will be on the safe side.

Design Temperature

The heat-loss calculation also depends on the desired inside air temperature. 70°F is usually taken to be adequate, but some building

STEP 1—heat loss per square foot

OUTSIDE DESIGN TEMPERATURE -10°F INSIDE AIR TEMPERATURE 70°F

	LOCATION	CONSTRUCTION	Transmission Coefficient "U"		Design Temp. Difference °F.		Heat Loss Per Sq. Ft.* "HL"
Walls	a 1st floor	wood siding, bldg. paper, wood sheathing, studding,	0.13	×	80	=	10.4
	b	gypsum lath, 1" mineral wool (blanket form)		×		=	
	c basement	above frost line: 8" hollow concrete block (gravel aggregate)	0.10	×	80	=	8.0
	d	below frost line: same	0.10	×	18	=	1.8
Floors	a basement	4" concrete	0.10	×	18	=	1.8
	b			×		=	
	c			×		=	
Ceilings (Roofs)	a 1st floor	roof & ceiling: roof asphalt shingles on wood sheathing - no insulation	0.11	×	80	=	8.8
	b	ceiling - gypsum lath, plastered, single wood attic floor		×		=	
	c	1" mineral wool between ceiling joists		×		=	
Windows and Doors	a 1st floor	windows: double-hung wood with storm sash	0.45	×	80	=	36.0
	b	doors: storm doors for each entrance	0.45	×	80	=	36.0
	c basement	steel casement windows - no storm sash	1.13	×	80	=	90.4
	d			×		=	

Courtesy *Dunham-Bush, Inc.*

Fig. 1. A sample tabulation sheet used to calculate the heat loss per square foot of exposed area.

owners may require a higher temperature. The outside temperature depends on geographical location. The coldest outside temperature expected during a normal heating season is known as the *outside design temperature.* It is not necessary to install a heating system capable of meeting the lowest temperature on record because such a temperature may never be reached again, or even approached except in rare instances. In design practice, a temperature is selected that will normally be reached during extreme cold weather. Outside design temperatures for the major cities in the United States are listed in Table 1.

NOTE: Local weather authorities should be consulted for temperatures in localities other than those listed. Design temperatures for cities separated by only a few miles or at different altitudes may vary widely.

Design Temperature Difference

The design temperature difference is the variation between the outside design temperature and the temperature desired inside the building. For example, if the rooms are to be heated to 70°F and the outside design temperature is 10° below zero, the design temperature difference is 80°.

NOTE: If a room wall is adjacent to a partially heated space, such as a garage heated to 40°F, the garage temperature would be used as the outside design temperature in calculating the heat loss of that particular room wall.

Heat-Loss Determination

List the design temperature difference on the tabulation sheet and multiply this difference by the "U" factor to determine the heat loss per square foot.

Exposure

"U" factors are calculated on a wind velocity of 15 mph. Higher velocities may be disregarded in determining the heat loss by transmission, but they must be taken into account when calculating the heat loss from infiltration.

Table 1. Winter Outside Design Temperatures for Major Cities in the United States.

State	City	Outside Design Temperature Commonly Used	State	City	Outside Design Temperature Commonly Used
Alabama	Birmingham	10	Missouri	Kansas City	—10
Arizona	Tucson	25		St. Louis	0
Arkansas	Little Rock	5	Montana	Billings	—25
California	San Francisco	35		Helena	—20
	Los Angeles	35		Miles City	—35
Colorado	Denver	—10	Nebraska	Lincoln	—10
Connecticut	New Haven	0		Valentine	—25
Dist. of Columbia	Washington	0	Nevada	Reno	— 5
Florida	Jacksonville	25	New Hampshire	Concord	—15
	Key West	45	New Jersey	Atlantic City	5
Georgia	Atlanta	10		Trenton	0
	Savannah	20	New Mexico	Albuquerque	0
Idaho	Boise	—10	New York	Albany	—10
Illinois	Cairo	0		Buffalo	— 5
	Chicago	—10		New York City	0
Indiana	Indianapolis	—10	North Carolina	Asheville	0
Iowa	Des Moines	—15		Charlotte	10
	Sioux City	—20	North Dakota	Bismarck	—30
Kansas	Topeka	—10	Ohio	Akron	0
Kentucky	Louisville	0		Cincinnati	0
Louisiana	New Orleans	20		Columbus	—10
Maine	Portland	— 5	Oklahoma	Oklahoma City	0
Maryland	Baltimore	0	Oregon	Portland	10
Massachusetts	Boston	0	Pennsylvania	Erie	— 5
Michigian	Detroit	—10		Harrisburg	0
	Escanaba	—15		Philadelphia	0
	Sault Ste. Marie	—20	Rhode Island	Providence	0
Minnesota	Duluth	—20	South Carolina	Charleston	15
	Minneapolis	—20	South Dakota	Huron	—20
Mississippi	Vicksburg	10	Tennessee	Knoxville	0

Courtesy *Dunham-Bush, Inc.*

Table 1. Winter Outside Design Temperatures for Major Cities in the United States (Continued).

State	City	Outside Design Temperature Commonly Used	State	City	Outside Design Temperature Commonly Used
Texas	Abilene	15	Washington	Seattle	15
	Austin	20		Spokane	—15
	Brownsville	30	West Virginia	Parkersburg	—10
	Corpus Christi	20	Wisconsin	Green Bay	—20
	Houston	20		Madison	—15
	Dallas	0		Milwaukee	—15
Utah	Salt Lake City	—10	Wyoming	Cheyenne	—15
Vermont	Burlington	—10			
Virginia	Lynchburg	5			
	Norfolk	15			

Courtesy *Dunham-Bush, Inc.*

Basements

Basements, unless used as a living area, normally have enough piping to heat the space. If they are occupied, however, radiation should be installed. Heat loss should be calculated to determine the amount of radiation required.

Heat in a basement is generally lost by transmission and by infiltration through doors and windows. Basement walls and floors also have a heat loss which is dependent on the ground temperature. The "U" factor for a basement wall below grade is but only 0.10; therefore an assumed ground temperature will not greatly affect the tabulated transmission heat loss. Ground temperatures do vary in accordance with the geographical location, however, so that the heat loss through the below-grade floor should be considered. A practical method is as follows:

Consider the basement wall as divided into two parts by a line at about the "mean" frost line, The wall above this line should be considered as being exposed to the outdoors, so the heat loss for this portion is calculated in the normal way. Everything below this line, including the basement floor, may be considered as being affected by a ground temperature selected from Table 2.

AREA CALCULATION
(STEP 2)

After the transmission heat loss per square foot has been determined it is necessary to find the total area of each surface.

Table 2. Ground Temperatures Below the Frost Line.

State	City	Ground Temperature Commonly Used	State	City	Ground Temperature Commonly Used
Alabama	Birimingham	66	Nevada	Reno	52
Arizona	Tucson	60	New Hampshire	Concord	47
Arkansas	Little Rock	65	New Jersey	Atlantic City	57
California	San Francisco	62	New Mexico	Albuquerque	57
	Los Angeles	67	New York	Albany	48
Colorado	Denver	48		New York City	52
Connecticut	New Haven	52	North Carolina	Greensboro	62
Dist. of Columbia	Washington	57	North Dakota	Bismarck	42
Florida	Jacksonville	70	Ohio	Cleveland	52
	Key West	78		Cincinnati	57
Georgia	Atlanta	65	Oklahoma	Oklahoma City	62
Idaho	Boise	52	Oregon	Portland	52
Illinois	Cairo	60	Pennsylvania	Pittsburgh	52
	Chicago	52		Philadelphia	52
	Peoria	55	Rhode Island	Providence	52
Indiana	Indianapolis	55	South Carolina	Greenville	67
Iowa	Des Moines	52	South Dakota	Huron	47
Kentucky	Louisville	57	Tennessee	Knoxville	61
Louisiana	New Orleans	72	Texas	Abliene	62
Maine	Portland	45		Dallas	67
Maryland	Baltimore	57		Corpus Christi	72
Massachusetts	Boston	48	Utah	Salt Lake City	52
Michigan	Detroit	48	Vermont	Burlington	46
Minnesota	Duluth	41	Virginia	Richmond	57
	Minneapolis	44	Washington	Seattle	52
Mississippi	Vicksburg	67	West Virginia	Parkersburg	52
Missouri	Kansas City	57	Wisconsin	Green Bay	44
Montana	Billings	42		Madison	47
Nebraska	Lincoln	52	Wyoming	Cheyenne	42

Courtesy *Dunham-Bush, Inc.*

Rooms

List each room, using two lines for large rooms that have various sizes of windows and doors. List the length, width, and ceiling height of each room.

1. Determine the floor area by multiplying the room length by the width (Column 1 × 2, Fig. 2).

2. Calculate the ceiling area in the same manner as the floor.
3. Calculate the total outside wall area by multiplying the length and width of each room by the ceiling height (Column 1 × Column 3, and Column 2 × Column 3, Fig. 2).

STEP 2—area calculations

Room	1	2	3	4	5	6	7	8	9	10	11
	Room Size						Windows and Doors				
	L	W	H	Floor Area	Ceiling Area	Gross Wall Area	No.	W	H	Area	Net Wall Area
living rm.	12	4	8^6		298 (1)	317	D 1	3	6^8	20	252
	20	11^4	8^6				W 1	9	5	45	
	10	1^6	8^6								
dining rm.	9^8	9	8^6		87	159	W 3	3	5	45	
kitchen	7	8^6	8^6		76	72	D 1	3	6^8	20	
	2^6	6^6	8^6								
bath	7	5	8^6		35	43	W 1	3	3	6	37
bedroom #1.	13^4	11^4	8^6		151	210	W 3	3	5	45	165
bedroom #2.	10^8	9^7	8^6		102	172	W 2	3	5	30	142
recreation rm.	18	13	8^6	234							
wall above frost		31	3^6			109	W 2	3	1^6	9	100
wall below frost		31	3^6			109					109

(1) ceiling area includes area over stairway living rm. proper and area over bay window

Courtesy *Dunham-Bush, Inc.*

Fig. 2. A sample tabulation sheet used to calculate area.

Windows and Doors

List the number of windows of each type and their sizes. Window sizes are based on the outside measurements of the sash. See Fig. 3. Calculate the area of each window.

List the number of doors and their sizes. Door sizes are generally based on the actual size of the swinging door. Calculate the area of each door. Calculate the net wall area by subtracting the total area of the windows and doors (Column 10) from the gross wall area (Column 6).

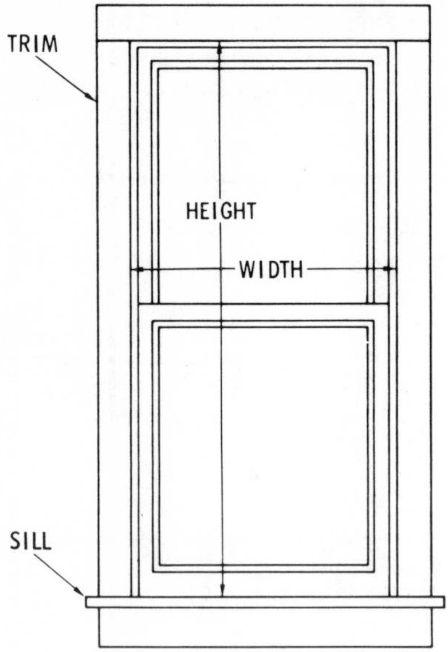
TRIM
HEIGHT
WIDTH
SILL

Fig. 3. Heat-loss calculations for windows are based on the outside measurements of the sash.

HEAT LOSS FROM TRANSMISSION

(STEP 3)

Calculate the heat loss through walls by multiplying the total wall area (last column in Fig. 2) by "HL" (last column in Fig. 1). Calculate the heat loss through the floors by multiplying the floor area (Column 4, Fig. 2) by "HL." Calculate the heat loss through the ceilings by multiplying the ceiling area (Column 5, Fig. 2) by "HL." Finally, determine the heat loss through the windows and doors by multiplying the window and door areas (Column 10, Fig. 2) by "HL." These various losses are listed on a form similar to the one in Fig. 4.

INFILTRATION HEAT LOSS

One of two methods may be used for calculating the heat loss due to infiltration around windows and doors—the *air-change method* or the *crack* method.

Air-Change Method

The air-change method may be used with some accuracy for calculating heat loss of residential construction. The amount of air leakage

Room	12	13	14	15	16	17	18	19
	STEP 3—heat loss from transmission							
	Walls		Floors		Ceilings (Roofs)		Windows and Doors	
	'HL'	Heat Loss	'HL'	Heat Loss	'HL'	Heat Loss	'HL'	Heat Loss
living rm	104	2624			8.8	2552	36.0	2340
dining rm	10.4	1182			8.8	766	36.0	1620
kitchen	10.4	544			8.8	667	36.0	720
bath	10.4	380			8.8	308	36.0	216
bedroom #1	10.4	1712			8.8	1330	36.0	1620
bedroom #2	10.4	1478			8.8	899	36.0	1080
recreation rm			1.8	421				
wall above grade	8.0	796			90.4	814		
wall below grade	8.0	195						

Courtesy *Dunham-Bush, Inc.*

Fig. 4. Sample tabulation sheet used in calculating the heat loss by transmission.

is estimated by assuming a certain number of air changes per hour. This method does not consider wind velocities.

Crack Method

The crack method is based on actual air leakage through the cracks around windows and doors according to wind velocities which may be expected in the particular area. See Table 3. The crack method is recommended as being more accurate for determining the infiltration heat loss of a building.

Calculate the lineal feet of crack. The length of cracks for a double-hung window is three times the width plus two times the height. For a pivoted metal sash, the length of the crack is the total perimeter of the ventilating sections.

Table 3. Typical Window Infiltration Chart

(cubic feet per foot of crack per hour)

Window Type	Remarks	Wind Velocity (miles per hour)					
		5	10	15	20	25	30
Unlocked woodsash windows, double-hung	Perimeter of frame in wood-frame building	2	5	10	16	22	28
	Caulked perimeter in masonry wall	1	3	4	5	6	7
	Non-caulked frame perimeter in masonry wall	2	3	12	18	25	34
	Poorly fitted window, no weather stripping with above hairline cracks including wood-frame leakage	25	65	102	140	180	240
Metal windows, double-hung	Locked without weatherstripping	18	42	68	92	120	148
	Unlocked without weatherstripping	18	45	72	98	130	164
	Unlocked but weatherstripped	4	16	30	42	56	72

It is not necessary to use the total amount of crack when calculating infiltration heat loss because the wind does not blow through the cracks on all sides of a building at the same time. For instance, in a building without partitions, the air enters through cracks on the windward side and must leave through cracks on the opposite side. For this type of building, take one-half the total lineal feet of crack. For a room with only one exposed wall, use all the crack; but in no case take less than half of the crack. List the leakage per foot of crack (Table 3). This is the cubic feet of air per hour per foot of crack.

NOTE: Room exposure should be considered in selecting wind velocities. Prevailing winter winds in most sections of the country are north, northwest, and west, and will have an effect on the rooms on these sides of the building.

Crack thickness and the clearance determine the leakage through a double-hung window. The difference between the width of the window sash and the width of the track formed by the window guides is the clearance. When calculating the infiltration loss of a double-hung wood window, use the unlocked condition.

For a well-fitted door, use the leakage values for a poorly fitted window. Double this value if the door is poorly fitted; if the door is weatherstripped, use one-half this value. For a commercial building, where the door may be frequently opened, use three times this value.

Calculate the infiltration heat loss by using the following formula:

$$L \times B \times \text{temp. diff.} \times .018 \times \text{Btu/hr infiltration heat loss}$$

where,

L is the lineal feet of crack (Column 25, Fig. 5),

B is the air leakage in cubic feet per hour per foot of crack (Column 26, Fig. 5).

	20	21	22	23	24	25	26	27	28	29
	STEP 4—heat loss from infiltration									
	AIR CHANGE METHOD					CRACK METHOD				
Room	No. Exp. Walls	No. Air Chgs.	Room Volume	Design Temp. Diff. x .018	Heat Loss	Lineal ft. of Crack	Leakage per Ft. (Cu. ft. /Hr.)	Design Temp. Diff. x .018	Heat Loss	Total Heat Loss† (Btu /hr.)
Living rm.			×	×	=	D 19	× 70 (2)	× 1.44	= 1915	12,818
			×	×	=	W 48	× 49 (2)	× 1.44	= 3387	
			×	×	=		×	×	=	
dining rm.			×	×	=	32	× 24	× 1.44	= 1106	4674
kitchen			×	×	=	19	× 56 (1)	× 1.44	= 1532	3463
			×	×	=		×	×	=	
bath			×	×	=	10	× 24	× 1.44	= 346	1250
bedroom #1			×	×	=	32	× 24	× 1.44	= 1106	5768
bedroom #2			×	×	=	16	× 24	× 1.44	= 553	4010
recreation rm.			×	×	=	18	× 57	× 1.44	= 1348	3574
			×	×	=		×	×	=	
			×	×	=					
			×	×	=					
			×	×	=					
			×	×	=					
			×	×	=					
			×	×	=					

(1) ½ the leakage of a poorly fitted window (because of storm door)
(2) door leakage based on ½ the leakage of a poorly fitted window plus 25%.
Windows leakage based on that of average fitted windows plus 25%
25% allows for opening of doors and for fireplace.

Courtesy *Dunham-Bush, Inc.*

Fig. 5. A sample tabulation sheet used to calculate the heat loss by infiltration.

Example: What is the Btu/hr infiltration heat loss through the crack of a 3 × 5 ft. poorly fitted window, nonweatherstripped, having a 3/32″ crack and a 3/32″ clearance? Wind velocity is expected to average 10 mph. Outside design temperature selected is 0°F and the inside air temperature will be 70°F.

Answer: According to Table 3, the air leakage through a window of this type will be 65 cubic feet per hour per foot of crack. Therefore, B equals 65. Lineal feet of crack (L) will be (2 × 5) plus (3 × 3), or 19 feet. The design temperature difference is 70°. Subtracting in the formula:

$$65 \times 19 \times 70 \times 0.018 = 1556 \text{ Btu/hr}$$

The total heat loss can now be determined by adding Columns 13, 15, 17, and 19 of Fig. 4, and Column 28 of Fig. 5. The total heat loss is placed in Column 29 of Fig. 5. After the heat loss calculations have been computed, there are often factors to be considered. The heating capacity of the radiation placed in a room is (on the average) the same as that in Column 29 of Fig. 5. However, for some installations it will be necessary to make allowances for certain conditions after the heat loss has been calculated. For example, the radiation for bathrooms should be approximately 20% higher than the figured heat loss to make certain that comfort conditions exist for bathing, etc. High ceilings are another example. Since warm air is lighter than cool air, it rises and accumulates at the ceiling, resulting in a temperature differential between the ceiling and floor. The term applied to this condition is *stratification.* For ceilings 12 feet and under, this temperature difference is almost negligible. However, if the ceiling is higher, the difference in temperature may be considerable, depending on the type of heating units installed. Cast-iron radiators without shields, or a warm-air heating system, if not carefully designed, are most likely to cause a stratification condition. Unit heaters, if properly selected and installed, and convectors reduce stratification by causing air movement. A good solution is the installation of baseboard radiation, as the outlet air is warm, not hot, and the air flow is near the floor.

NOTE: In practice, it is common to add 20% to the heating capacity of the units to be installed for each foot in ceiling height from 12 feet to 24 feet. If a ceiling is over 25 feet high, 25% can be added to the heating capacity of the units to be installed.

Some buildings, such as churches, lodges, gymnasiums, large meeting halls, etc., may only need to be heated on certain days. In extremely cold weather, the building may become very cold, requiring extra heating to raise the room temperatures to a comfortable level in a short time. A common practice in this type of building is to add an approximate 25% to the radiation heating capacity if specific calculations are not made for a determination of the actual radiation requirements to provide a quick rise in temperature.

EDR DETERMINATION

Hot-water system radiation to be installed is generally selected on the basis of the Btu/hr capacity of each unit. With steam, the radiation to be used is chosen on the basis of the square-feet EDR capacity of each unit. EDR (Equivalent Direct Radiation) originally was used in rating direct radiation (free-standing, cast-iron radiators, and pipe coils) on the basis of the square-foot area of the heating surface. Radiators, however, varied in heat output per square foot of surface depending on section depth, width, and height. As a result, a standard was brought into being—that which is now used—240 Btu/hr output per square foot of heating surface. This standard is in use for rating the radiation used with steam.

A square foot of EDR equals 240 Btu/hr. As an example: If it requires 14,400 Btu/hr to heat a room, the required steam radiation would be 60 square-feet of EDR capacity.

RADIATION SELECTION

Convector Radiation

A well-designed convector (Fig. 6) supplies heat by a gentle air motion and a mild radiant effect for proper comfort conditions. The absence of intense radiated heat, such as occurs with radiators, permits the occupants to be comfortable even when close to the convectors. The unit may be equipped with a damper which regulates the air flow.

A convector is composed of two major parts—the heating element and the enclosure. The heating element consists of nonferrous fins attached to copper tubes. The size of the element, location of the outlet, and the enclosure height determine the convector capacity. The enclo-

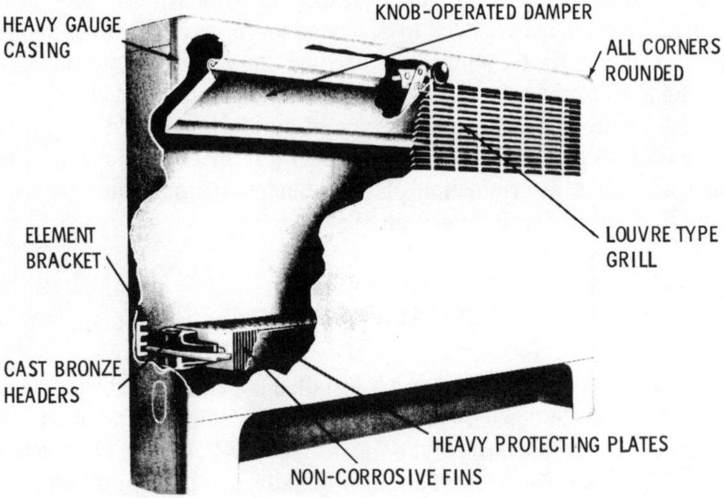
HEAVY GAUGE CASING
KNOB-OPERATED DAMPER
ALL CORNERS ROUNDED
ELEMENT BRACKET
LOUVRE TYPE GRILL
CAST BRONZE HEADERS
HEAVY PROTECTING PLATES
NON-CORROSIVE FINS

Courtesy *Dunham-Bush, Inc.*

Fig. 6. A convector-type radiator.

sure is actually a flue or chimney which directs air through the heating element and out into the room.

Factors to consider when selecting convectors are:

1. Limitations involved by the building construction.
2. Space available, and the cost and appearance of the convectors.
3. Interior decoration, occupancy, and type of building.

A convector should be located so as to offset probable air currents which could result in discomfort. The total number of units required is determined by placing them according to exposure requirements. Taking into consideration the probable air currents, radiation subdivisions should be made so as to locate the convectors in high heat-loss areas. Distribute the radiation so as to maintain a uniform temperature of both the air and the wall surfaces. At the bottom of stair wells and under very high windows where a downdraft effect can be very pronounced, front outlet convectors are preferable to top outlet and sloping types. Fig. 7 shows a typical wall-hung convector for use with a hot-water system.

The piping layout and the type of cabinet determine the manner in which the convector heating elements are connected. The supply connections to the heating element may be made at the bottom, top, or end of the inlet header. The return connections are made at the bottom

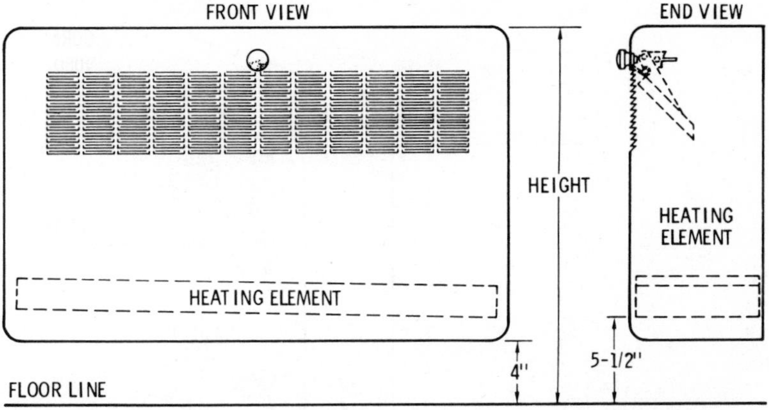
FRONT VIEW
END VIEW
HEIGHT
HEATING ELEMENT
HEATING ELEMENT
4"
5-1/2"
FLOOR LINE

Courtesy *Dunham-Bush, Inc.*

Fig. 7. A typical wall-hung convector showing the suggested installation measurements.

or end of the opposite header. Fig. 8 shows a typical convector connection, and Fig. 9 the typical piping connections for steam systems.

Piping connections for hot-water systems are illustrated in Fig. 10. An air chamber and vent should be part of each convector installation. If it is not desirable for the air valve to protrude through the side of the convector, the air chamber may be turned and a section of copper tubing connected between the air chamber and air valve. The valve can then be located below the heating element, near the inlet air opening, and will be easily accessible.

Baseboard Radiation

Baseboard radiation has both the advantages of convected and radiant heat (Fig. 11). Fig. 11 shows the cross section of a typical flush-back type, while Fig. 12 shows a recessed type.

A baseboard convector consists of two major parts—the baseboard enclosure and the heating element or elements. The enclosure is generally constructed of sheet steel with openings so placed as to be inconspicuous.These openings permit the air to pass over the heating element and into the space to be heated. The heating element is a finned tube usually made of copper with aluminum fins or steel with steel fins.

This type of radiation is generally located along the outside wall, where it may counteract the cooler air currents resulting from heat loss.

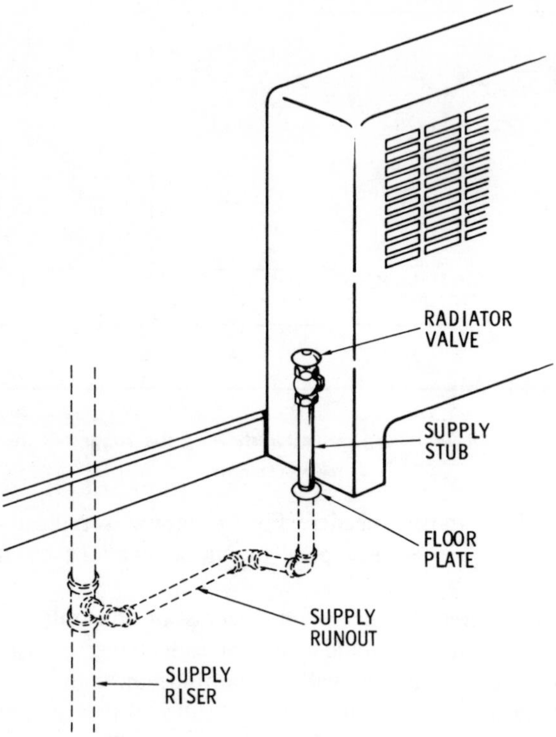
RADIATOR
VALVE
SUPPLY
STUB
FLOOR
PLATE
SUPPLY
RUNOUT
SUPPLY
RISER

Courtesy *Dunham-Bush, Inc.*

Fig. 8. Typical connection to a convector in a steam heating system.

Generally, the ceiling-to-floor temperature is very uniform with this type of radiation, varying less than 4 degrees as compared to the usual 12 degrees with other types of radiation. Baseboard radiation is particularly recommended for rooms having wide windows with low sills.

Fin-Tube Radiation

Fin-tube radiation (Fig. 13) is designed to meet the demand for a heating unit which is shallow in depth, low in height, attractive in appearance, and high in heating capacity. It is adaptable for almost any type of building, but particularly useful in schools, industrial plants, and public and commercial buildings. A series of units may be used to make one long attractive enclosure, and may be used with either hot water or steam.

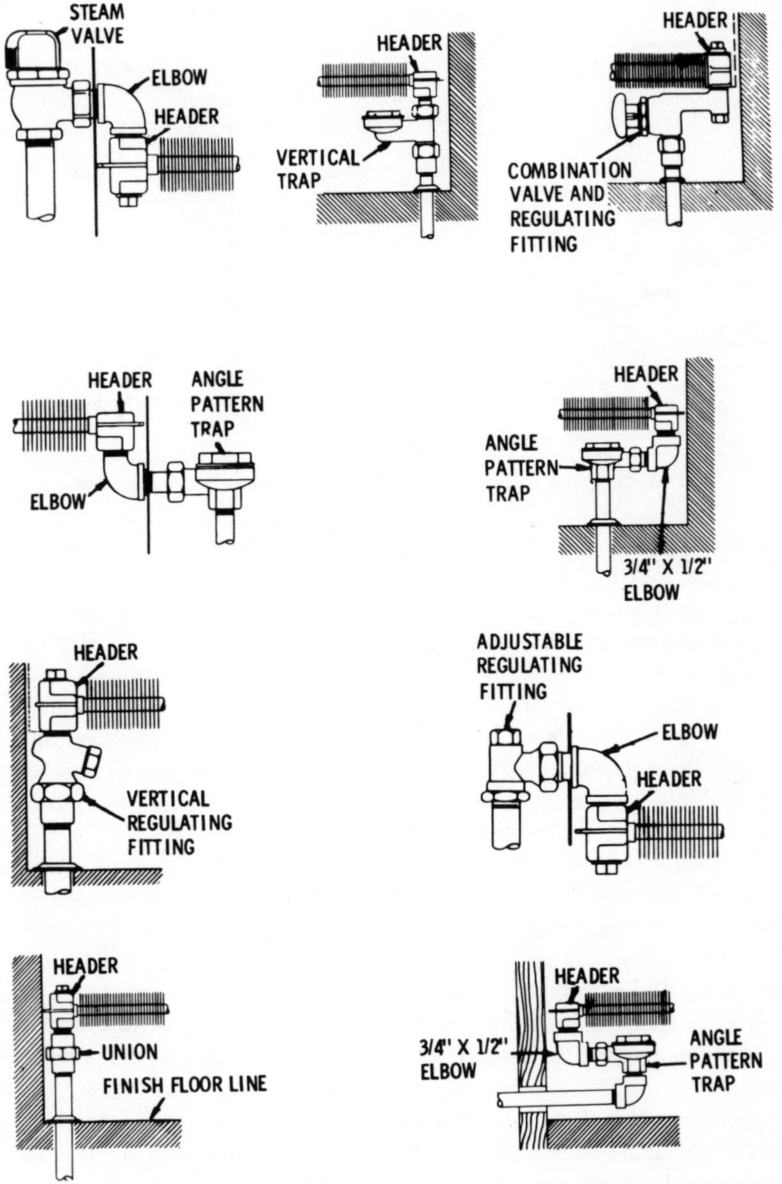
STEAM VALVE
ELBOW
HEADER
HEADER
VERTICAL TRAP
HEADER
COMBINATION VALVE AND REGULATING FITTING
HEADER
ANGLE PATTERN TRAP
ELBOW
HEADER
ANGLE PATTERN TRAP
3/4" X 1/2" ELBOW
HEADER
VERTICAL REGULATING FITTING
ADJUSTABLE REGULATING FITTING
ELBOW
HEADER
HEADER
UNION
FINISH FLOOR LINE
HEADER
3/4" X 1/2" ELBOW
ANGLE PATTERN TRAP

Courtesy *Dunham-Bush, Inc.*

Fig. 9. Convector piping connections in a steam heating system.

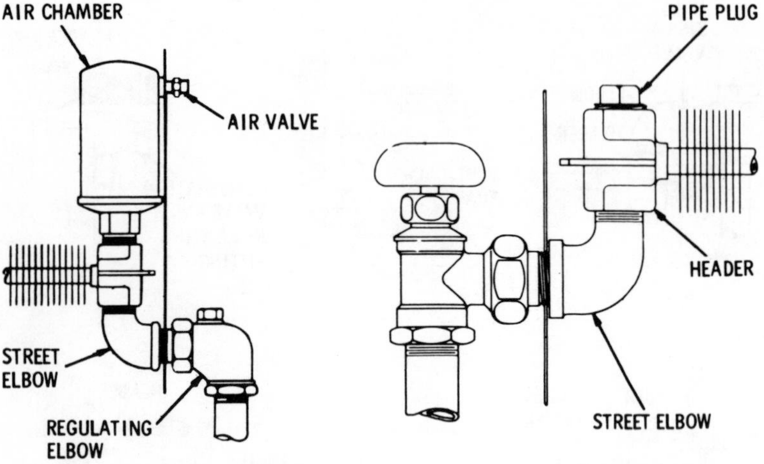
AIR CHAMBER
AIR VALVE
STREET ELBOW
REGULATING ELBOW
PIPE PLUG
HEADER
STREET ELBOW

Courtesy *Dunham-Bush, Inc.*

Fig. 10. Convector piping connections in a hot-water heating system.

This type of radiation should be mounted close to the floor (the height from the floor to the bottom of the cover should not be less than four inches in order to permit an unrestricted air flow). Capacity tables in manufacturers' information folders and brochures are usually based on this position as the one which provides maximum distribution of heat. If, however, this type of radiation is mounted too high, cold air may pocket at the floor level. Therefore, it should never be mounted any higher than immediately below the line of the window sill, except for special jobs such as clerestory windows, dormers, and skylights where they are used to eliminate downdrafts.

Appearance is generally a controlling factor in the selection of the type of radiation. The tiered type shown in Fig. 14 is often used in schools and other public buildings. The front outlet in the cover avoids the danger of smudging the walls. When a front outlet might be blocked by furniture or equipment, a sloping top or flat-topped cover should be used. In applications where the covers may be damaged, replaceable covers such as the flat-topped or expanded metal type may be preferable.

The type of wall construction and the extent of cold drafts may be a deciding factor on the kind of cover which should be installed. On some masonry walls, for example, the lack of a nailing strip may make the mounting of covers which fasten to the walls costly. Backs may be

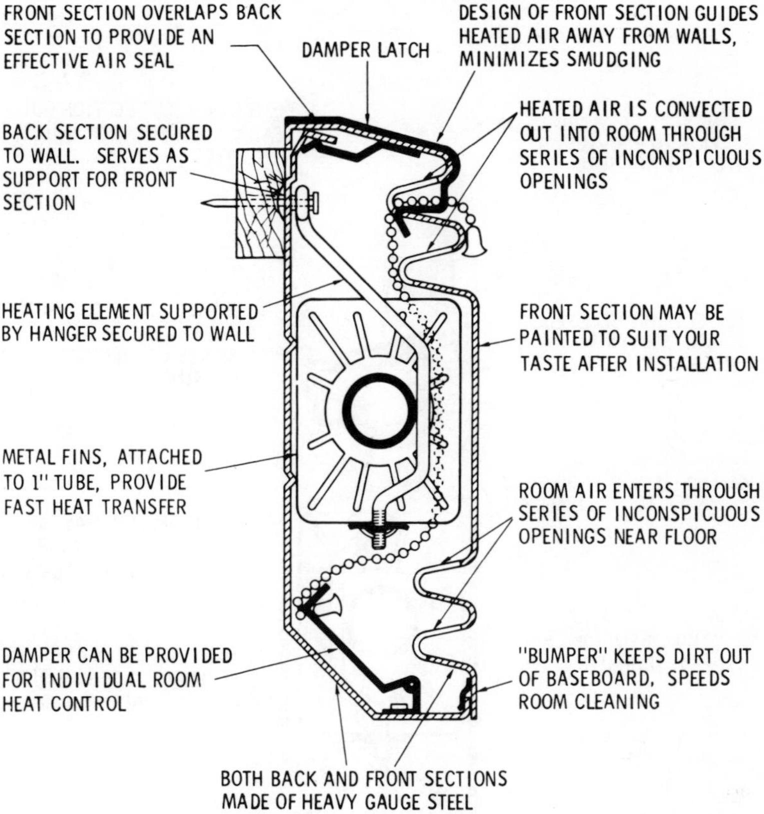
FRONT SECTION OVERLAPS BACK SECTION TO PROVIDE AN EFFECTIVE AIR SEAL
DAMPER LATCH
DESIGN OF FRONT SECTION GUIDES HEATED AIR AWAY FROM WALLS, MINIMIZES SMUDGING
BACK SECTION SECURED TO WALL. SERVES AS SUPPORT FOR FRONT SECTION
HEATED AIR IS CONVECTED OUT INTO ROOM THROUGH SERIES OF INCONSPICUOUS OPENINGS
HEATING ELEMENT SUPPORTED BY HANGER SECURED TO WALL
FRONT SECTION MAY BE PAINTED TO SUIT YOUR TASTE AFTER INSTALLATION
METAL FINS, ATTACHED TO 1" TUBE, PROVIDE FAST HEAT TRANSFER
ROOM AIR ENTERS THROUGH SERIES OF INCONSPICUOUS OPENINGS NEAR FLOOR
DAMPER CAN BE PROVIDED FOR INDIVIDUAL ROOM HEAT CONTROL
"BUMPER" KEEPS DIRT OUT OF BASEBOARD, SPEEDS ROOM CLEANING
BOTH BACK AND FRONT SECTIONS MADE OF HEAVY GAUGE STEEL

Courtesy *Dunham-Bush, Inc.*

Fig. 11. Cross section of a typical flash-back type of baseboard radiator.

necessary in some instances to prevent leakage of air into furred spaces. In other cases, it may be necessary to place an insulating board in back of the units.

BOILER AND PUMP SIZES

After all heating calculations have been tabulated, and after the type and size of radiation and boiler have been determined, the pumps and/or units can be selected from manufacturers' brochures and information literature.

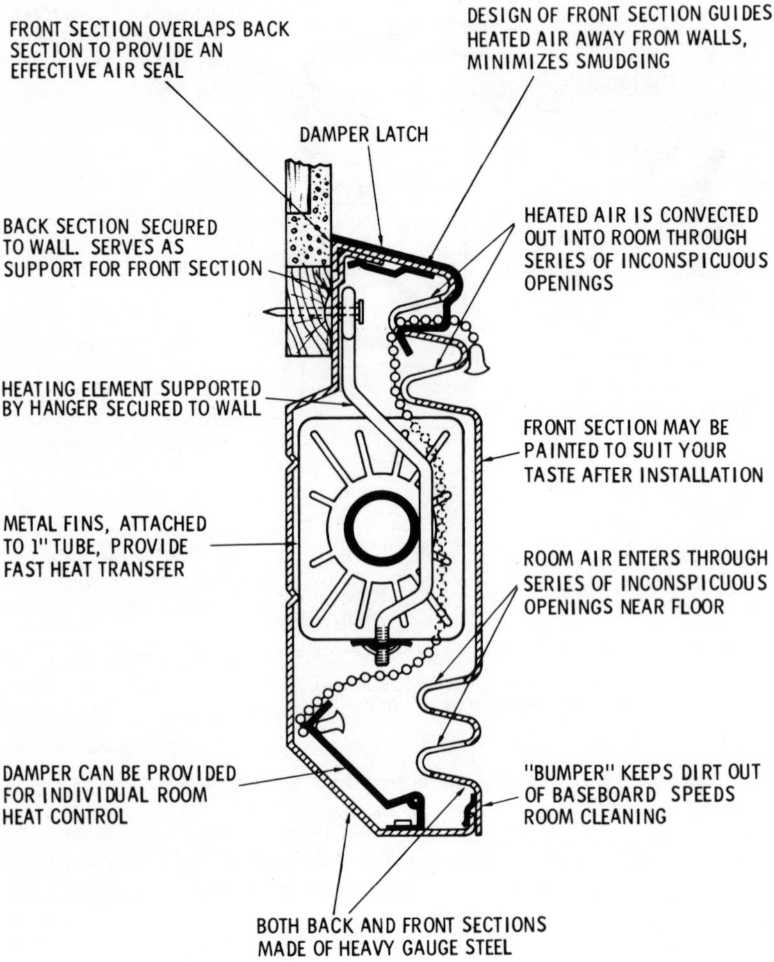
FRONT SECTION OVERLAPS BACK SECTION TO PROVIDE AN EFFECTIVE AIR SEAL
DESIGN OF FRONT SECTION GUIDES HEATED AIR AWAY FROM WALLS, MINIMIZES SMUDGING
DAMPER LATCH
BACK SECTION SECURED TO WALL. SERVES AS SUPPORT FOR FRONT SECTION
HEATED AIR IS CONVECTED OUT INTO ROOM THROUGH SERIES OF INCONSPICUOUS OPENINGS
HEATING ELEMENT SUPPORTED BY HANGER SECURED TO WALL
FRONT SECTION MAY BE PAINTED TO SUIT YOUR TASTE AFTER INSTALLATION
METAL FINS, ATTACHED TO 1" TUBE, PROVIDE FAST HEAT TRANSFER
ROOM AIR ENTERS THROUGH SERIES OF INCONSPICUOUS OPENINGS NEAR FLOOR
DAMPER CAN BE PROVIDED FOR INDIVIDUAL ROOM HEAT CONTROL
"BUMPER" KEEPS DIRT OUT OF BASEBOARD SPEEDS ROOM CLEANING
BOTH BACK AND FRONT SECTIONS MADE OF HEAVY GAUGE STEEL

Courtesy Dunham-Bush, Inc.

Fig. 12. Cross section of a typical reversed type of baseboard radiator.

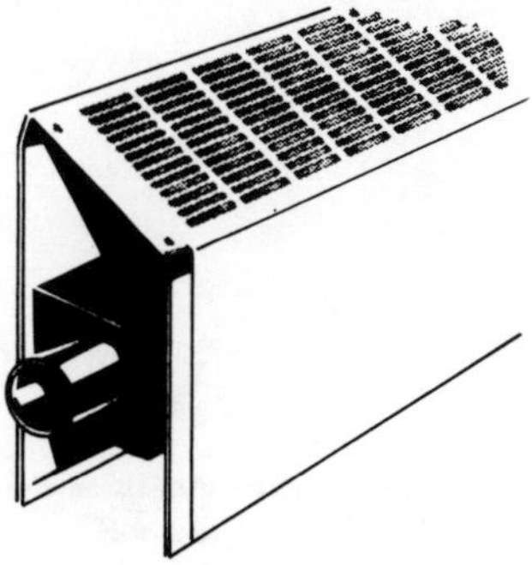

(A) One-tier assembly.

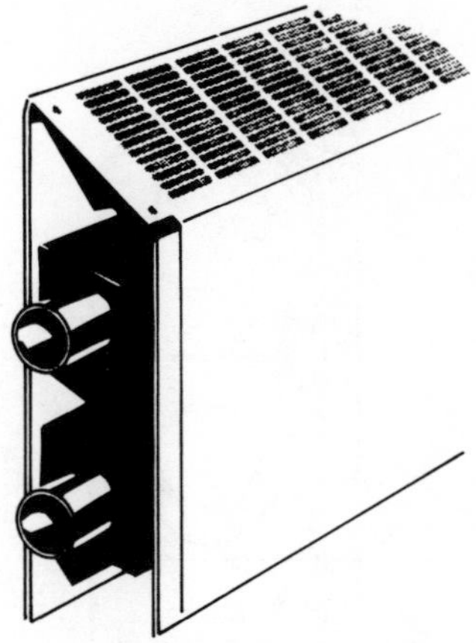

(B) Two-tier assembly.

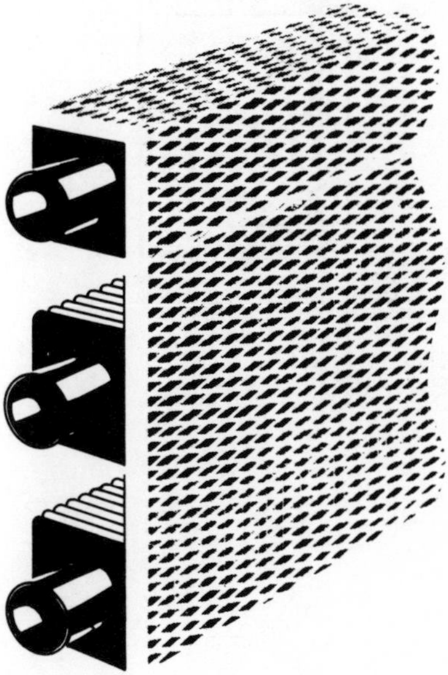

(C) Three-tier assembly.

Fig. 13. Typical covers for fin tube radiators.

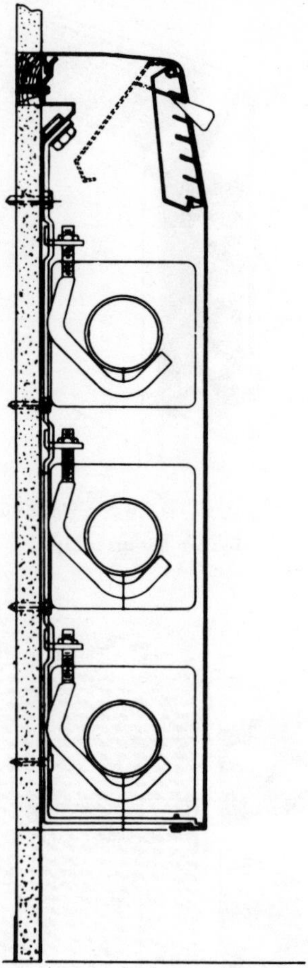

Fig. 14. Cross section of a three-tier fin-tube wall radiator.

CHAPTER 3

Steam Heating Systems

The widespread use of steam for space heating points up the long recognized fact that steam, as a heating medium, has numerous basic characteristics which can be advantageously employed. Some of the most important advantages are discussed in the following paragraphs.

Steam's ability to give off heat.—Properties of saturated steam are shown in steam tables and given much information regarding the temperature and the heat (energy) contained in one pound of steam for any pressure. As an example, to change one pound of water from 212°F. into steam at the same temperature of 212°F. at atmospheric pressure (14.7 psi) requires a heat content of 1150.4 Btu which is made up of 180.1 Btu of *sensible heat,* (the heat required to raise one pound of water from 32°F. to 212°F.) and 970.3 Btu of *latent heat.* The latent heat is the heat added to change the one pound of water from 212°F. into steam at 212°. This stored up latent heat is required to transform the water into steam and it reappears as heat when the process is reversed to condense the steam into water.

Because of this basic fact, the high latent heat of vaporization of a pound of steam permits a large quantity of heat to be transmitted efficiently from the boiler to the heating unit or radiator with little change in temperature.

Steam promotes its own circulation through piping.—Steam will flow naturally from a higher pressure as generated in the boiler to a lower pressure (existing in the steam mains). Circulation or flow is caused by the lowering of the steam pressure along the steam supply mains and in the heating units due to pipe friction and to the condensing process of steam as it gives up heat to the space being heated. This action is shown in Fig. 1. Because of this fact, the natural flow of steam does not require a pump such as needed for hot water heating, or a fan as employed in warm air heating.

Steam heats more readily.—Steam circulates through a heating system faster than other fluid mediums. This can be important where fast pickup of the space temperature is desired. It will also cool down more rapidly when circulation is stopped. This is an important consideration in Spring and Fall when comfort conditions can be adversely affected by long heating-up or slow cooling-down periods.

Flexibility of Steam Heating.—Other advantages in using steam as a heating medium can be found in the easy adaptability to meet unusual conditions of heat requirement with a minimum of attention and maintenance. Here are some examples:

1. Temporary heat during construction is easily provided without undue risks and danger of freeze-ups.
2. Additional heating units can be added to existing systems without making basic changes to the system design.

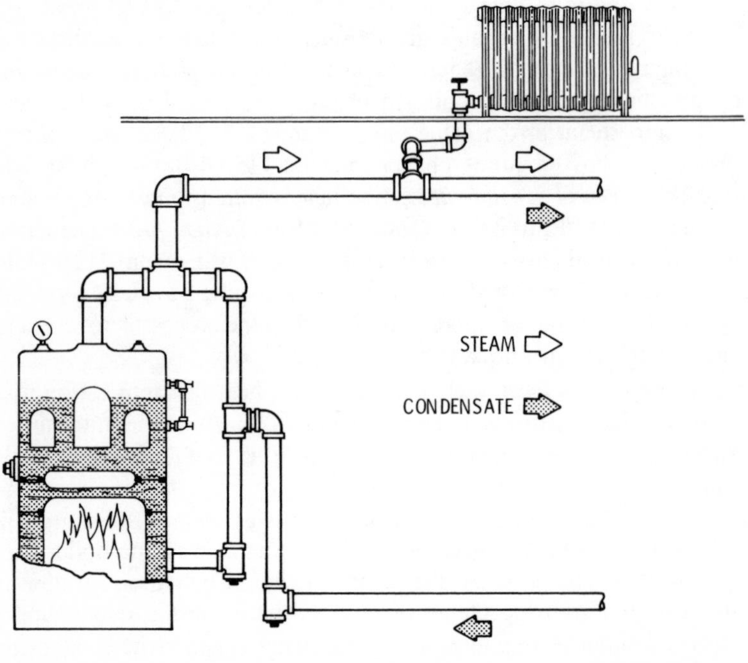
STEAM
CONDENSATE

Fig. 1. Steam condensing as it gives up heat to the space being heated.

3. Increased heat output from heating units can be easily accomplished by increasing the steam pressure the proper amount.
4. Steam heating systems are not prone to leak; however, leaks that may occur in the system piping, pipe fittings, or equipment, cause less damage than leaks in a system using hot water. A cubic foot of steam condenses into a relatively small quantity of water. In many cases, a small leak does not cause any accumulation of water at the location of the leak; instead, it evaporates into the air and causes no damage.
5. Repair or replacement of system components such as valves, traps, heating units (radiators), and similar equipment, can be made by simply closing off the steam supply. It is not necessary to drain the system and to spend additional time to re-establish circulation. There is less need to worry about freezing since the water in a steam heating system is mainly in the boiler. Boiler water can easily be protected from freezing during shutdowns, during new construction, repairs, or replacement of parts, by installing an aquastat below the boiler water level to control water temperature.
6. Steam is a flexible medium when used in combination processes and heating applications. These often require different pressures which are easily obtained. In addition, exhaust steam, when available, can be utilized to the fullest advantage.
7. Steam heating systems are considered to be "lifetime" in-vestments. Many highly efficient systems are in operation today after more than 50 years of service. Fig. 2 shows a steam boiler with the water at normal operating level, and a hot water boiler with the system completely filled with water.

Steam is easy to distribute and control in a heating system.—The distribution of steam to heating units or radiators is easy to accomplish with distribution orifices located at the steam inlet to the unit. Metering orifices can also be employed when used with proper controls to maintain steam pressure in accordance with the flow characteristic of the metering orifices. These orifices can be either a fixed type or a variable type. The controls for steam systems are simple and effective. They include those used to control space temperatures by the application of "ON-OFF" valves. Modulating controls can also be applied which respond to indoor-outdoor temperature conditions to control the quantity of steam flowing to orificed radiators.

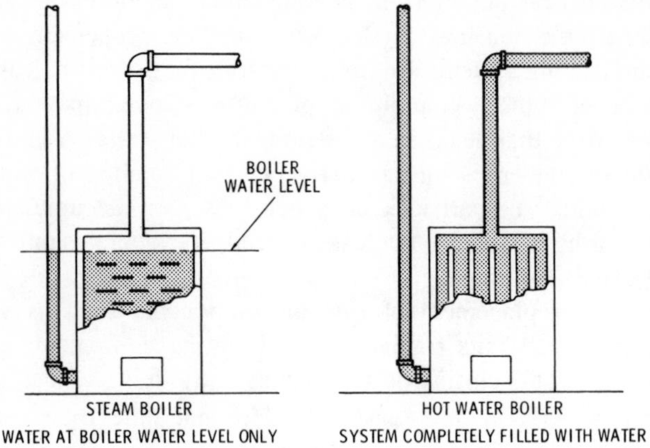
BOILER
WATER LEVEL
STEAM BOILER
WATER AT BOILER WATER LEVEL ONLY
HOT WATER BOILER
SYSTEM COMPLETELY FILLED WITH WATER

Fig. 2. Water content in boiler.

Where steam can be used.—Steam can be used as the heating medium for all types of heating units such as convectors, wall fin-tube radiation, cast-iron radiators, unit heaters, unit ventilators, heating and ventilating units, all types of coils in ventilating and air conditioning systems, and steam absorption units used in air conditioning. Steam can be used for all types of systems applicable to a variety of building designs ranging from residences to large industrial, commercial, multi-story apartment groups, offices, churches, or schools.

Recommended steam applications.—The following applications for the use of steam are recommended to provide trouble-free and efficient heating systems:

1. Where there is more than one job to do, such as providing comfort heating as well as steam for processes in:
   industrial plants
   hospitals
   institutions
   restaurants
   dry cleaning plants
   laundries
2. Where outdoor air is heated for ventilation (especially in cold climates) as in:
   factories
   school classrooms
   auditoriums

buildings with central air conditioning or ventilating systems
gymnasiums

3. Where there is a surplus of steam from processes which can be used for air cooling or water chilling.
4. Where the heating medium must travel a great distance from the boiler to the heating units, as found in:
   high multi-story buildings
   long, rambling buildings
   scattered buildings
   supplied from a central station
5. Where intermittent changes in heat loads are required, as in:
   schools
   office buildings
   churches
6. Where central heat control or individual room control of temperature is important as found in:
   schools
   hospitals
   office buildings
   hotels and motels
7. Where there is a chance of freezing, in cold climates or where sub-freezing air is handled
8. Where there may be additions or alterations of space or change in occupancy in a building.
9. Where extra heat is needed as in buildings with large or frequently used doors, as in:
   department stores
   garages
   shipping departments or warehouses
   airplane hangers.

ONE PIPE STEAM HEATING SYSTEMS

All steam heating systems can be said to be of two basic types, they are either one pipe systems or two pipe systems. The one pipe system uses the same pipe to deliver steam to a unit and to return the condensate from the unit. The unit (radiation, unit heater, etc.) has only one piping connection, the steam enters the unit and the condensate returns through the one piping connection, at the same time.

One pipe systems which are correctly designed and installed require a minimum of mechanical equipment. Open type air vents as shown in Fig. 3A and B are used on the end of mains and on each heating unit. If the system is correctly designed and installed the condensate will flow (return) back to the boiler by gravity. The advantage of the one pipe system is that it is very dependable and the initial cost is low. Fig. 4 shows a typical modern one pipe installation of a simple up-feed gravity system.

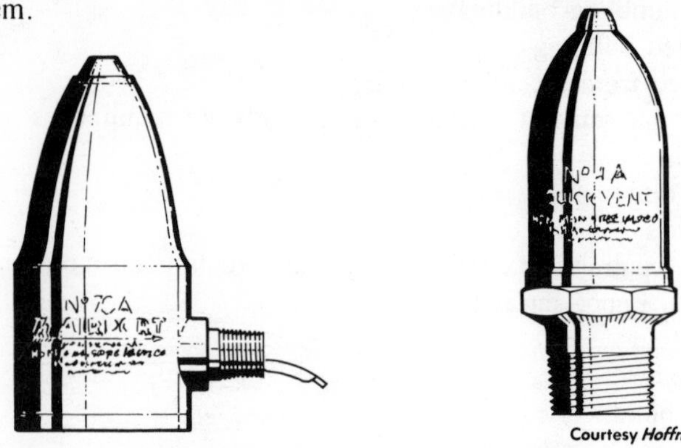

Courtesy *Hoffman Specialty ITT*

(A) An open type radiator air vent. **(B) An end of main vent**

Fig. 3. Typical one pipe system vents

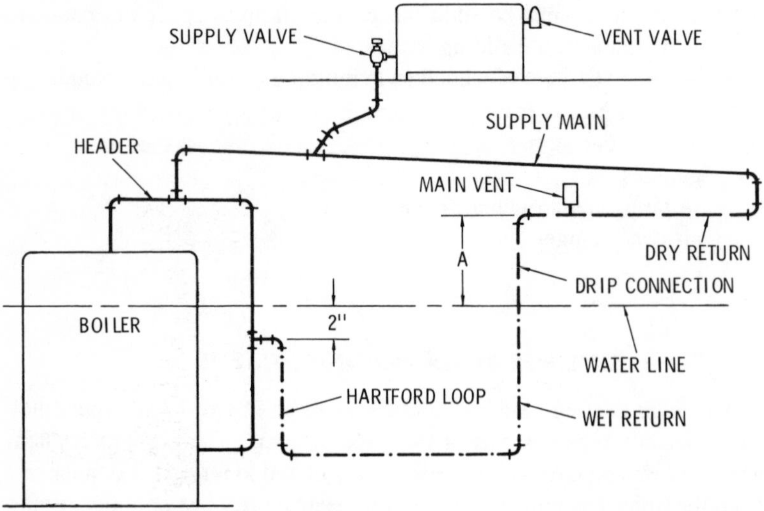
SUPPLY VALVE
VENT VALVE
SUPPLY MAIN
HEADER
MAIN VENT
A
DRY RETURN
DRIP CONNECTION
BOILER
2"
WATER LINE
HARTFORD LOOP
WET RETURN

Courtesy *Hoffman Specialty ITT*

Fig. 4. A basic one-pipe up-feed gravity steam system.

The Steam Boiler

The boiler in most cases will be fired automatically and will be equipped with controls for maintaining the system pressure. It should be equipped with an automatic water feeder and a low water cut-off. It should have the required safety devices for proper combustion of the type of fuel used.

The Steam Supply Main

The steam supply main conducts the steam from the boiler to the various radiators in the system. Condensate flows back from the radiators to the drip connection through the steam supply main. As shown in Fig. 4 the condensate flow is in the same direction as the steam flow. This type system is called a parallel flow system.

Radiator Valves

The steam enters each radiator through an angle pattern radiator supply valve installed at the bottom inlet tapping of the radiator. The valve should be an angle pattern valve because steam must enter the heating unit and condensate must leave the unit through the same valve, and at the same time. A straightway horizontal pattern valve should not be used because it has a natural obstruction to the free flow of steam and condensate. A radiator supply valve on a single pipe system can only be used to turn the radiator OFF completely or ON completely. It can not be used as a throttling valve due to the noise that would be created. Furthermore, if a number of valves in a system were throttled, steam condensate could not readily return to the boiler. The correct type valve for a one pipe steam system is shown in Fig. 5.

Fig. 5. A typical radiator supply valve.

Courtesy *Hoffman Specialty ITT*

The Hartford Loop

The Hartford Loop is a pressure balancing loop which introduces full boiler pressure on the return side of the boiler. This pressure prevents *reversed circulation,* or water leaving the boiler via the return piping. Some engineers in the past have attempted to stop reversed circulation by installing a check valve in the return piping. This method is unsatisfactory, indeed unsafe, because dirt, corrosion, or scale can lodge under the check valve and prevent it from closing, permitting reversed circulation to occur. The Hartford loop is the only safe method of preventing reversed circulation. It is actually two loops, the first is an equalizer line and the second is an extension from the steam main, through the return and back to a connection at the equalizer loop, as shown in Fig. 4.

In order for the Hartford loop to be effective, the loop must be at least the full size of the return main, and the horizontal nipple at the point where the return connects should be 2 inches below the boiler water line. It is important to use a close nipple to construct the Hartford loop; if a long nipple is used at this point and the water line of the boiler becomes low, water hammer noise will occur.

End of Main Vent Valves

Air must be eliminated from the piping and radiation of a one pipe steam system before steam can enter the piping or the radiators. Air which is present in the piping or in the radiators will block the flow of steam. An end of main vent should be installed on large one pipe systems to assure quick venting of air from the horizontal steam main. Venting of the air will open a path and assure prompt distribution of steam to vertical runs and risers supplying steam to radiators. An end of main vent is made with a single port and has a much larger venting capacity than a radiator vent. A non vacuum or open vent type main vent is shown in Fig. 6. Main vents are never constructed for adjustable, variable venting because once all the air has been expelled from the steam main, the vent port is closed and has no further function until the next boiler firing or heating-up cycle occurs.

A vacuum type of end of main vent is shown in Fig. 7. It has all the features of a smaller vacuum type radiator vent. A vacuum type vent has all the features of an ordinary air vent, plus it prevents the return of air into the system through the vent valve.

Fig. 6. A float-type End-of-Main air vent valve.

Courtesy *Hoffman Specialty ITT*

Fig. 7. A float-type Vacuum End-of-Main air vent valve.

Courtesy *Hoffman Specialty ITT*

End of main vents should always be installed near the end of the steam main. It is *not* good practice to install a main vent on the last fitting at the end of the steam main.

The correct way to install the end of main vent is shown in Fig. 8. When this vent is incorrectly installed as shown in Fig. 9 the float can be damaged by water surge which can create a very high pressure. Although the surge pressure lasts only a fraction of a second, it can cause the float to be collapsed.

Radiator Vent Valves

Air which is present in radiators will prevent the steam in the supply main from entering the radiator. Each radiator must be equipped with an air vent valve installed in the vent tapping located on the end opposite the supply valve. Radiator vent valves are made in two basic types:

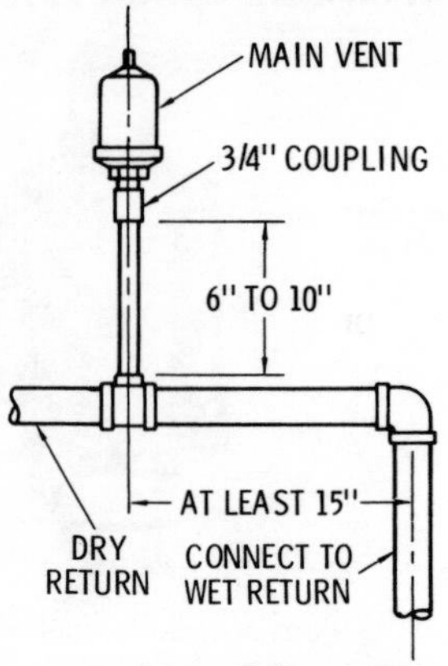

MAIN VENT
3/4" COUPLING
6" TO 10"
AT LEAST 15"
DRY RETURN
CONNECT TO WET RETURN

Courtesy *Hoffman Specialty ITT*

Fig. 8. The correct way to install end-of-main air vents.

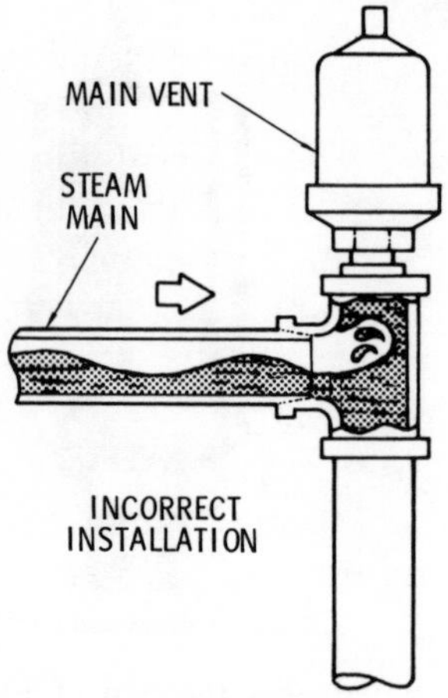

MAIN VENT
STEAM MAIN
INCORRECT INSTALLATION

Courtesy *Hoffman Specialty ITT*

Fig. 9. Incorrect installation of an end-of-main air vent.

1. The non-vacuum or open vent type which is available with a single non-adjustable port, or with an adjustable port which is used for proportional venting.
2. The vacuum type which has an adjustable port for proportional venting.

The functions of a vent valve in a heating system are:

1. To permit air to be vented (pushed out) so that the steam can occupy the space in the piping and heating units.
2. To close when steam contacts the valve, thus preventing the steam from escaping through the vent port.

3. To close when water contacts the valve and thus prevent the water from escaping through the vent port.
4. To re-open when the steam temperature has dropped sufficiently, or the water has drained away, thus permitting the air venting process to continue.
5. Vacuum vent *only* the vacuum type vent prevents the return of air, through the vent valve, into the system.

Proportional Venting

When the steam starts through a piping system which has more than one radiator, it is desirable for the steam to be able to enter all radiators at approximately the same time. Because there will be a difference in size (EDR-Equivalent Direct Radiation) of the radiators, due to the space being heated by each unit, single port vent valves should not be used. Instead, adjustable port vent valves should be installed on each unit of radiation and the vents adjusted to the size of each unit. When the vents are properly adjusted steam will enter each unit of radiation at the same time during each firing cycle. An adjustable port vent valve is shown in Fig. 10. A single port (non-adjustable) vent valve is shown in Fig. 11.

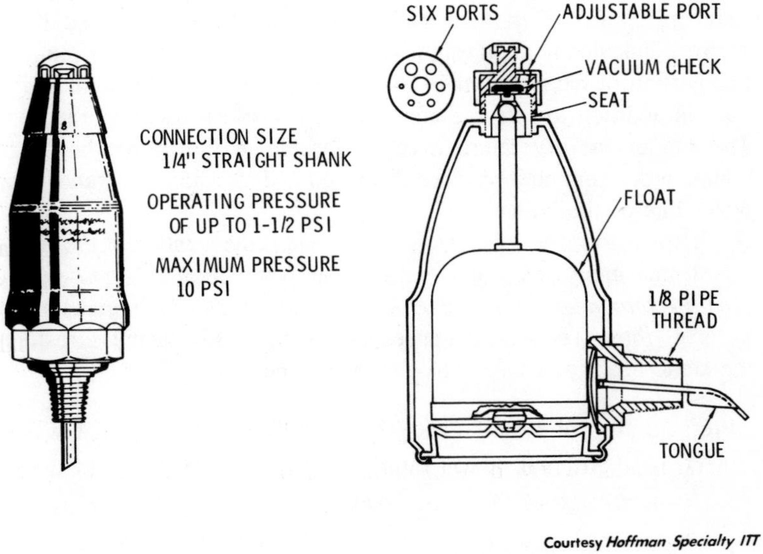
CONNECTION SIZE
1/4" STRAIGHT SHANK
OPERATING PRESSURE
OF UP TO 1-1/2 PSI
MAXIMUM PRESSURE
10 PSI
SIX PORTS
ADJUSTABLE PORT
VACUUM CHECK
SEAT
FLOAT
1/8 PIPE
THREAD
TONGUE

Courtesy *Hoffman Specialty ITT*

Fig. 10. An adjustable port vent valve.

Fig. 11. A single-port (non-adjustable) air valve.

Courtesy *Hoffman Specialty ITT*

We have followed the path of the steam through the supply main to each unit of radiation and to the end of the supply main. At the end of the supply main the steam will revert to the water from which it came and will become condensate. This condensate will be joined by the condensate which formed in the various units of radiation and will return to the boiler through the wet return piping of the system. Referring again to Fig. 4, the section of piping between the end of the supply main and the main vent is called the *dry return.* The dry return is that portion of the return main which is located above the boiler water line. In addition to carrying condensate, it also carries steam and air. The end of the dry return must be located at the proper height to maintain the minimum required distance for dimension "A" above the water line of the boiler.

Dimension "A" is shown in Fig. 4 and is the location of the end of the steam supply main above the boiler water line. *A small system having a total heat loss of not more than 100,000 Btu/hr.* is sized on the basis of 1/8psi. The three distances shown in Fig. 12, the pressure drop, the static head, and the safety factor will be:

Pressure drop of system (1/8 psi)	= 3-1/2″ of water
Static head (friction of wet return)	= 3-1/2″ of water
Safety factor (twice the static head)	= 7″ of water
Total distance	= 14″ of water

For a small system it is standard practice to make the minimum distance for Dimension "A" not less than 18″.

For a larger system; assume that the piping is sized for a total pressure drop of 1/2 psi. The three distances would then be as follows:

Pressure drop of system (1/2 psi)	= 14″ of water
Static head (friction of wet return)	= 4″ of water
Safety factor (twice the static head)	= 8″ of water
Total distance	= 26″ of water

It is standard practice for a system based on 1/2 psi pressure drop to make the minimum distance for Dimension "A" not less than 28 inches.

The Wet Return

The portion of the return piping which carries the condensation back to the boiler and is installed below the level of the boiler water line is called the wet return. It is completely filled with water and does not carry air or steam. Since water will seek its own level when the system is first filled or is cold, the water in the wet return will be at the same level as the water in the boiler.

Types of One Pipe Systems

There are several different types of one pipe heating systems, determined by the piping arrangement. In a gravity system the condensate returns directly to the boiler. The proper slope or pitch must be given to the steam supply and dry return mains in a gravity system in order to insure the proper flow of steam, air, and condensate. The proper pitch is not less than one inch in 20 feet in the direction of the gravity flow of the condensate. The wet return requires no pitch. Some typical examples of different piping arrangements and types are shown and explained in the following paragraphs.

The Counter Flow System—In a counter flow system the steam flows in the opposite direction to the flow of condensate. This type of system is limited to small residences, especially those with unexcavated basements. The sizing of the steam main depends on the load requirements, it must be one pipe size larger than the steam main required on other types of one pipe systems, and it must pitch upwards from the boiler towards the end of the main. A slope of one inch in 10 feet is required to facilitate proper condensate return to the boiler. The end of the steam supply main, dimension "A" in Fig. 12, must be of sufficient height above the boiler water line to return the condensate to the boiler. A counter flow system is shown in Fig. 13.

Parallel Flow Systems—In a parallel flow system, steam and condensate flow in the same direction in the horizontal steam or return mains. A parallel flow system with a wet return is shown in Fig. 14. A parallel flow system using a dry return is shown in Fig. 15. The end of the dry return main drops below the boiler water line and becomes a wet return. In both the dry and wet return systems the condensate returns to the boiler through a Hartford loop. In both the dry and wet return systems dimension "A" in Fig. 12 must be maintained.

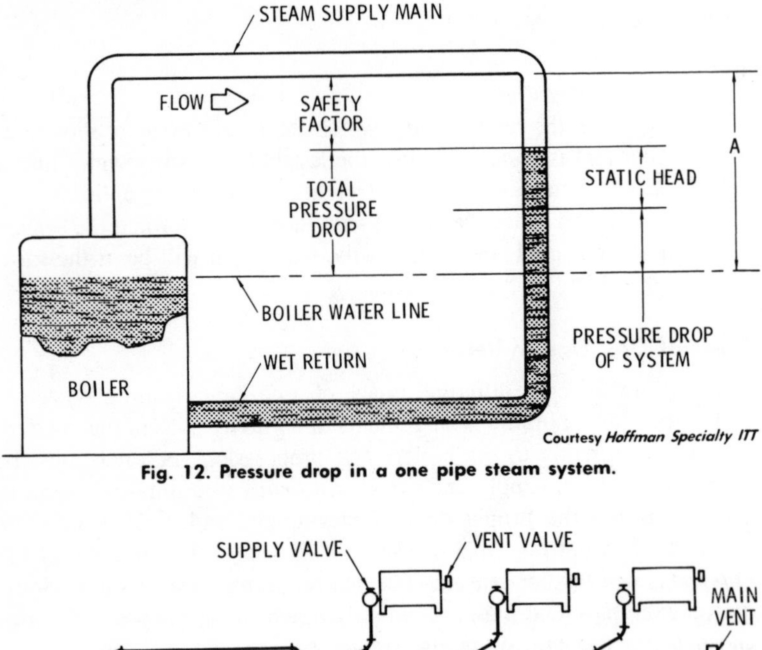
STEAM SUPPLY MAIN
FLOW
SAFETY FACTOR
TOTAL PRESSURE DROP
STATIC HEAD
A
BOILER WATER LINE
PRESSURE DROP OF SYSTEM
WET RETURN
BOILER

Courtesy *Hoffman Specialty ITT*

Fig. 12. Pressure drop in a one pipe steam system.

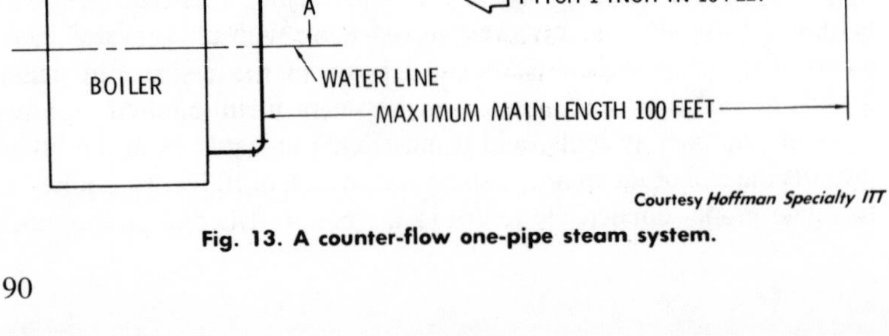
SUPPLY VALVE
VENT VALVE
MAIN VENT
PITCH 1 INCH IN 10 FEET
A
WATER LINE
BOILER
MAXIMUM MAIN LENGTH 100 FEET

Courtesy *Hoffman Specialty ITT*

Fig. 13. A counter-flow one-pipe steam system.

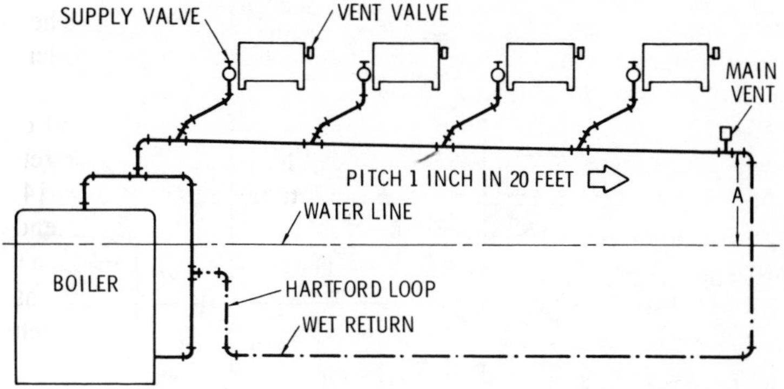
SUPPLY VALVE
VENT VALVE
MAIN VENT
PITCH 1 INCH IN 20 FEET
WATER LINE
A
BOILER
HARTFORD LOOP
WET RETURN

Courtesy *Hoffman Specialty ITT*

Fig. 14. A parallel flow system using a wet return.

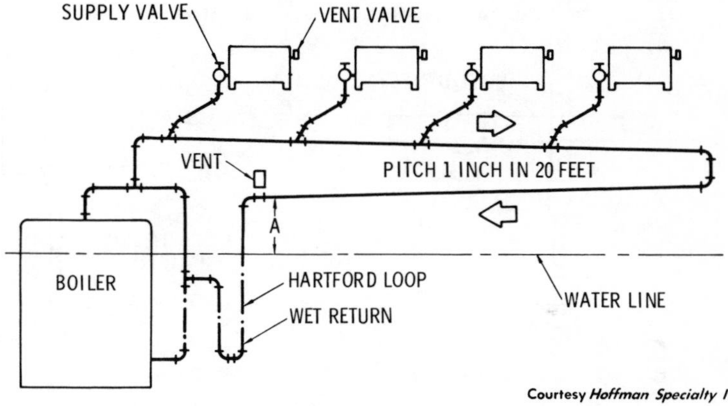
SUPPLY VALVE
VENT VALVE
VENT
PITCH 1 INCH IN 20 FEET
A
BOILER
HARTFORD LOOP
WET RETURN
WATER LINE

Courtesy *Hoffman Specialty ITT*

Fig. 15. A parallel flow system using a dry return.

Parallel Flow Up-Feed Systems—The parallel flow up-feed system is designed for buildings having more than one floor. As shown in Fig. 16, steam is distributed to the units of radiation through a basement main from which two up-feed risers supply the second and third floors. The up-feed risers are dripped back to the wet return. Connecting the risers, as shown in Fig. 17, is advisable in order to keep the horizontal main free of condensate accumulation and assure unobstructed flow of steam. The up-feed branch connection to the first floor radiators are not dripped. The details of these connections are shown in Fig. 18. If the up-feed run-out to a riser is not to be dripped to the wet return, the horizontal pipe must be increased one pipe size and a greater pitch

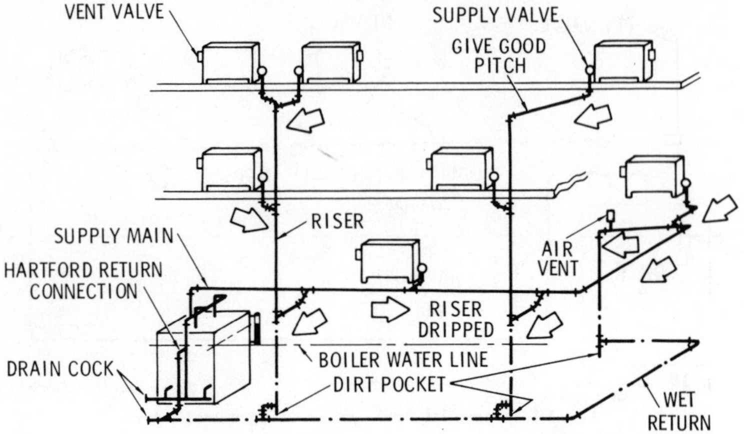
VENT VALVE
SUPPLY VALVE
GIVE GOOD PITCH
RISER
SUPPLY MAIN
HARTFORD RETURN CONNECTION
AIR VENT
RISER DRIPPED
BOILER WATER LINE
DRAIN COCK
DIRT POCKET
WET RETURN

Courtesy *Hoffman Specialty ITT*

Fig. 16. A parallel flow up-feed system.

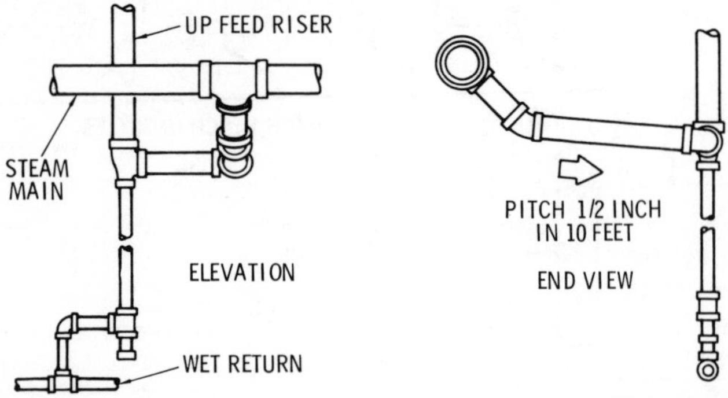
UP FEED RISER
STEAM MAIN
ELEVATION
WET RETURN
PITCH 1/2 INCH IN 10 FEET
END VIEW

Courtesy *Hoffman Specialty ITT*

Fig. 17. Correct installation of up-feed risers in a parallel flow up-feed system.

provided as shown in Fig. 19. Two methods of connecting a branch or runout to an up-feed riser are shown in Fig. 20. The 45° connection provides less obstruction to the free flow of steam to the riser which must also carry the reverse flow of condensate from the riser.

Parallel Flow Down-Feed System—A one pipe system with the distribution main installed above the radiators, as shown in Fig. 21 is called a down-feed system. In this type system the steam and condensate flow the same way in the down-feed risers. The main steam supply riser should be installed directly from the boiler to the overhead supply main as shown in the illustration. The downfeed runout connections are taken from the bottom of the horizontal supply main thus assuring the

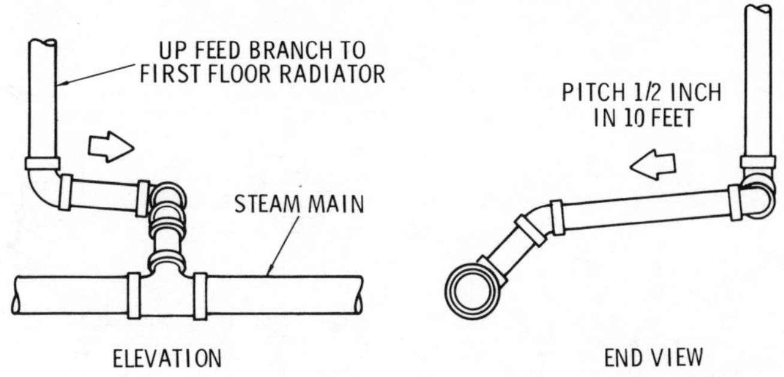
UP FEED BRANCH TO
FIRST FLOOR RADIATOR
STEAM MAIN
ELEVATION
PITCH 1/2 INCH
IN 10 FEET
END VIEW

Courtesy *Hoffman Specialty ITT*

Fig. 18. Correct installation of an up-feed branch connection which is not dripped to the wet return.

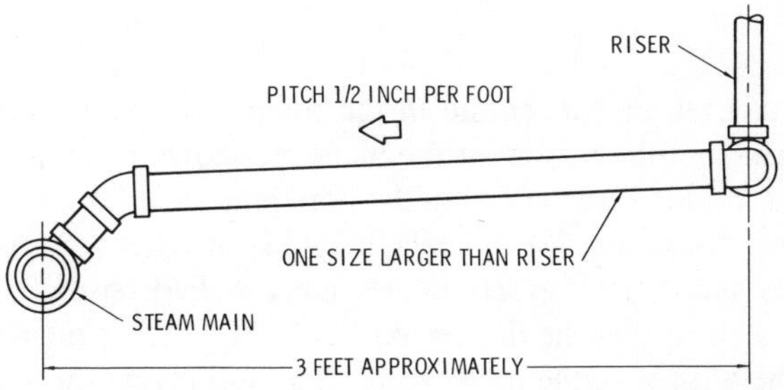
RISER
PITCH 1/2 INCH PER FOOT
ONE SIZE LARGER THAN RISER
STEAM MAIN
3 FEET APPROXIMATELY

Courtesy *Hoffman Specialty ITT*

Fig. 19. Correct installation of an up-feed riser which is not dripped to the wet return.

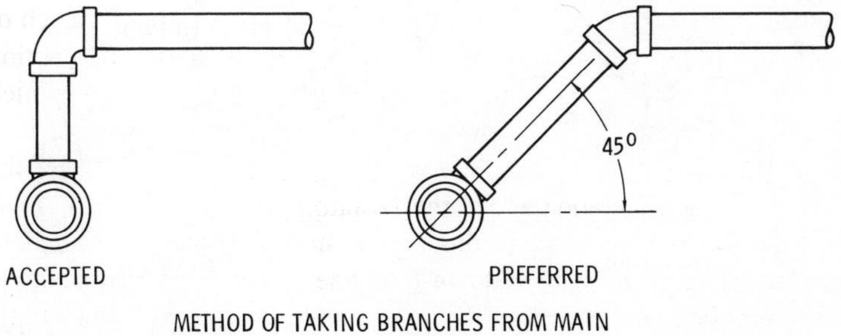
45°
ACCEPTED
PREFERRED
METHOD OF TAKING BRANCHES FROM MAIN

Courtesy *Hoffman Specialty ITT*

Fig. 20. The preferred method of taking branch connections from a main.

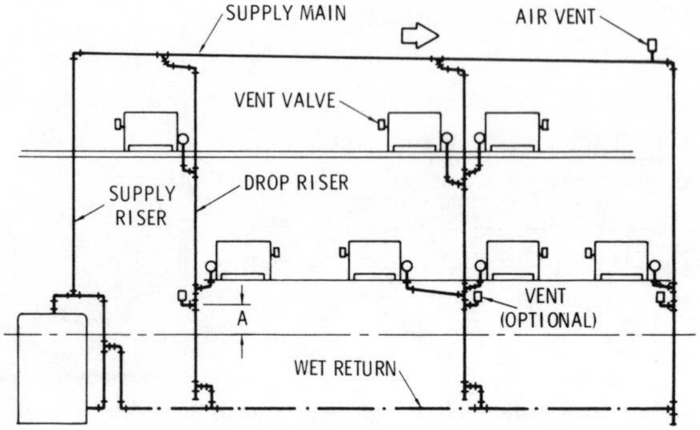
SUPPLY MAIN
AIR VENT
VENT VALVE
SUPPLY RISER
DROP RISER
A
VENT (OPTIONAL)
WET RETURN

Courtesy *Hoffman Specialty ITT*

Fig. 21. A parallel flow down-feed system.

least accumulation of condensate in the main. The piping details for these runouts from the bottom of the main are shown in Fig. 22. If the main vent is installed on the end of the horizontal main, only one main vent will be required. Main vents may be installed on the risers (optional) as shown in Fig. 21; if they are installed on the riser they must be located so that the dimension "A" (Fig. 12) is maintained at the proper height above the boiler water line, to prevent the vent valve from being closed by water rising to a sufficient height to raise the vent valve float.

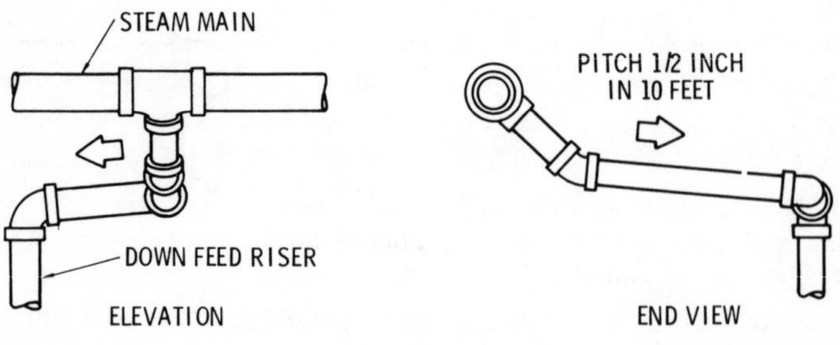
STEAM MAIN
DOWN FEED RISER
ELEVATION
PITCH 1/2 INCH IN 10 FEET
END VIEW

Courtesy *Hoffman Specialty ITT*

Fig. 22. All down-feed run-out connections should be taken from the bottom of the main.

Returning The Condensate To The Boiler By Mechanical Means

A condensate pump must be used to return the condensate to the boiler when there is insufficient height to maintain dimension "A" at its minimum above the boiler water line.

When a condensate pump is used, the boiler pressure, the end of the steam main pressure, or the boiler water line elevation, have no bearing on the height of the end of the steam main, as long as it is above the maximum water level in the condensate pump receiver. The return main begins at the discharge of a float and thermostatic trap as shown in Fig. 22. This trap is sized to handle the entire maximum condensate load of the system if it is a single main, or the connected load of each individual main. It functions to discharge air and condensate accumulations into the return, and to close against the passage of steam. The return main for this type of system is referred to as a "no pressure return" because the open vent on the condensate pump receiver maintains it at atmospheric pressure. The piping must be uniformly pitched to the pump receiver without pockets which will trap air and prevent the gravity flow of condensate to the receiver. The pump discharge is connected directly to the boiler return opening *without* the use of a Hartford loop. A Hartford loop connection can cause noise when used with a pumped discharge of condensate. It is not good practice to install the discharge from a float and thermostatic trap directly into, or close to, the pump receiver because the "flash" of steam vapor may be sufficient to affect the pump operation. The term "flash," as used here, means the rapid passing into steam, of water at a high temperature when the pressure it is under is reduced. For best pump operation the condensate temperature in the receiver should not exceed 200°, for this reason the return piping should not be insulated when a pump is used to return the condensate to the boiler. Often construction conditions exist which will not permit the condensate to be returned by gravity flow to the condensate receiver if the receiver is installed above the floor as shown in Fig. 23. In this case an underground receiver can be installed. The return main can also be installed underground to permit the condensate to return to the receiving tank by gravity flow. The condensate pump could then be mounted on the receiver, above ground, to pump the condensate back to the boiler. An underground receiver with the pump mounted on top is shown in Fig. 24.

When a horizontal main is installed without uniform pitch, usually due to improper or insufficient support, it can cause a water pocket due

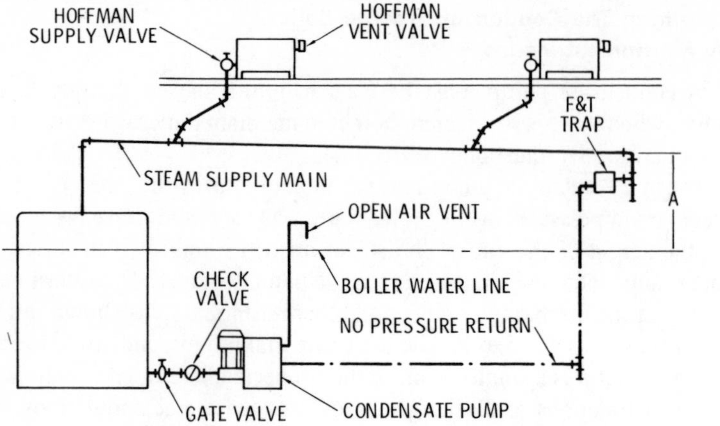
HOFFMAN
SUPPLY VALVE
HOFFMAN
VENT VALVE
F&T
TRAP
STEAM SUPPLY MAIN
A
OPEN AIR VENT
CHECK
VALVE
BOILER WATER LINE
NO PRESSURE RETURN
GATE VALVE
CONDENSATE PUMP

Courtesy *Hoffman Specialty ITT*

Fig. 23. An above ground condensate receiver and pump.

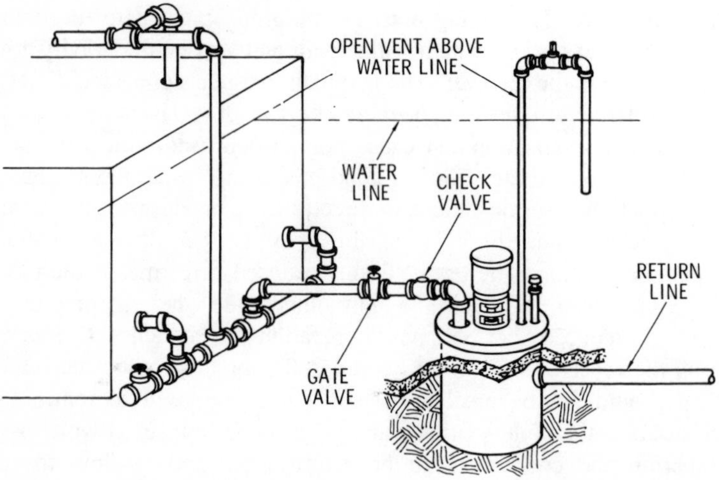
OPEN VENT ABOVE
WATER LINE
WATER
LINE
CHECK
VALVE
RETURN
LINE
GATE
VALVE

Courtesy *Hoffman Specialty ITT*

Fig. 24. A condensate receiver installed underground. The condensate pump is mounted on top of the receiver.

to a sag in the piping. Fig. 25 shows such a condition and it is evident that the free flow of steam and air is impossible until the condition is corrected. Such a sag can cause noise when steam reaches the water pocket on a cold start. Also, depending on its location, it can be responsible for water hammer noise and, sometimes, destructive damage to valves and vents.

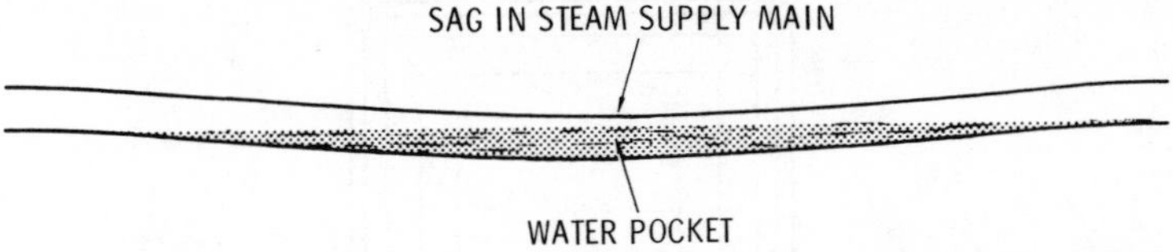
SAG IN STEAM SUPPLY MAIN
WATER POCKET

Courtesy *Hoffman Specialty ITT*

Fig. 25. The cause of noise in a sagging pipe.

There are times when it is necessary to install a steam main around a construction obstruction such as a steel beam or girder. The details of the proper method of piping around such an obstruction are shown in Fig. 26. Steam and air will flow above the obstruction and the condensate will flow below.

Another type of obstruction is a doorway which interferes with the uniform pitch of a dry return which must be located above the floor. Fig. 27 shows the correct method of piping around the doorway. When the piping is installed as shown the condensate will flow in the piping below the doorway and air will flow through the air loop over the top.

A radiator which does not set level because of a sagging floor, as shown in Fig. 28, can be the cause of noise and poor heating. Condensate remains trapped in the lower end of the radiator, thus setting up an obstruction to the steam which is trying to enter the radiator. Water hammer can result from this condition. The radiator

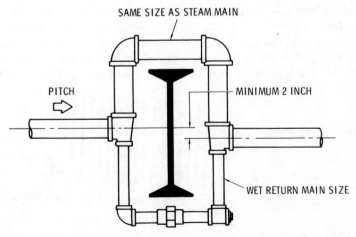

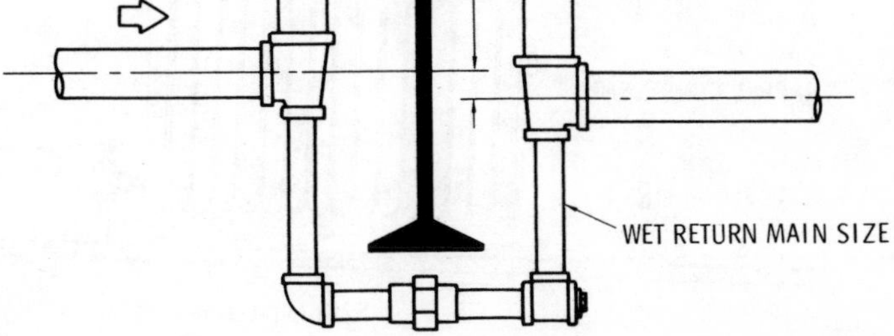
SAME SIZE AS STEAM MAIN
PITCH
MINIMUM 2 INCH
WET RETURN MAIN SIZE

Courtesy *Hoffman Specialty ITT*

Fig. 26. Correct method of piping around a beam.

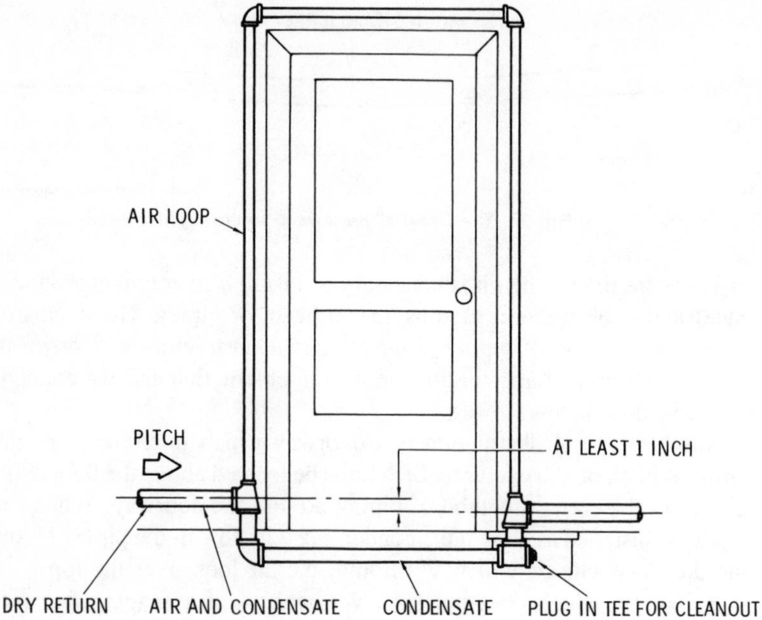
AIR LOOP
PITCH
AT LEAST 1 INCH
DRY RETURN
AIR AND CONDENSATE
CONDENSATE
PLUG IN TEE FOR CLEANOUT

Courtesy *Hoffman Specialty ITT*

Fig. 27. The correct method for piping steam and condensate around a doorway.

should be set as level as possible. On a one pipe system, a slight pitch of the radiator towards the supply end will help eliminate problems.

There are times when it may be desirable to add another radiator to an existing one pipe steam heating system. Fig. 29 shows the "wrong" and "right" method for adding the additional radiator. Incorrect installation will cause the supply pipe from the existing radiator to fill with

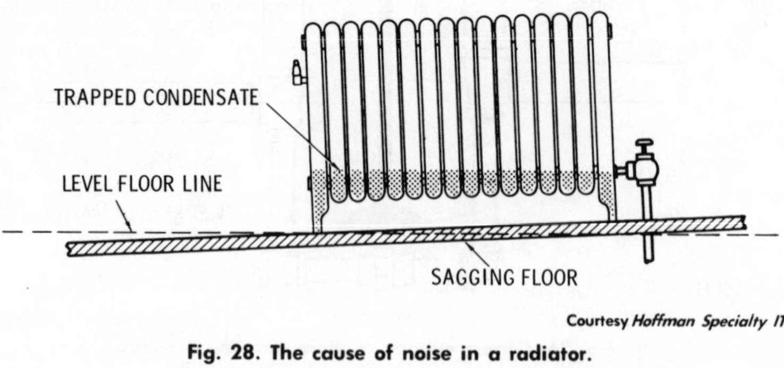
TRAPPED CONDENSATE
LEVEL FLOOR LINE
SAGGING FLOOR

Courtesy *Hoffman Specialty ITT*

Fig. 28. The cause of noise in a radiator.

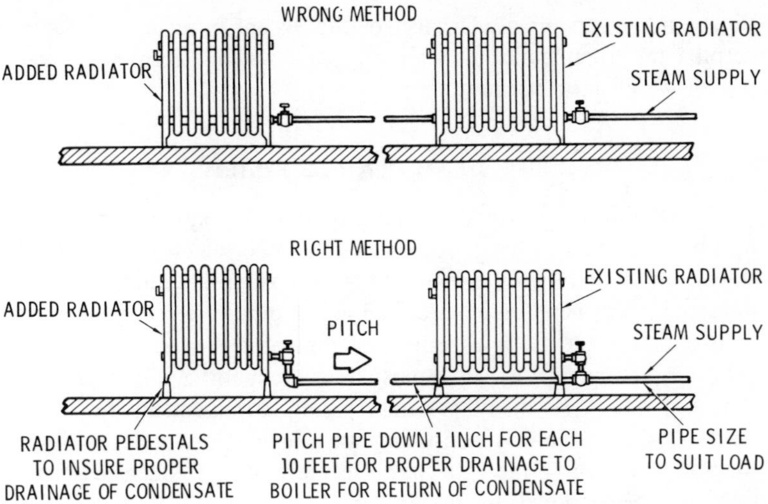
WRONG METHOD
ADDED RADIATOR
EXISTING RADIATOR
STEAM SUPPLY
RIGHT METHOD
ADDED RADIATOR
PITCH
EXISTING RADIATOR
STEAM SUPPLY
RADIATOR PEDESTALS
TO INSURE PROPER
DRAINAGE OF CONDENSATE
PITCH PIPE DOWN 1 INCH FOR EACH
10 FEET FOR PROPER DRAINAGE TO
BOILER FOR RETURN OF CONDENSATE
PIPE SIZE
TO SUIT LOAD

Courtesy *Hoffman Specialty ITT*

Fig. 29. The correct way to add a radiator to a system.

condensate and thus block the passage of steam to the added radiator. The right method allows the steam from the supply pipe to go directly to the additional radiator. The new steam supply piping must be pitched as shown in Fig. 29 so that condensate can flow back to the return system.

Piping Fundamentals for One Pipe Systems

There are essential factors which must be considered in the design and installation of the piping for a one pipe steam system. To insure the proper flow of steam, air, and condensate, the steam supply main and the dry return must be installed with a uniform pitch. Fig. 30 shows the proper method of reducing the size of the pipe when installed horizontally. Use an eccentric reducing coupling instead of a regular (concen-

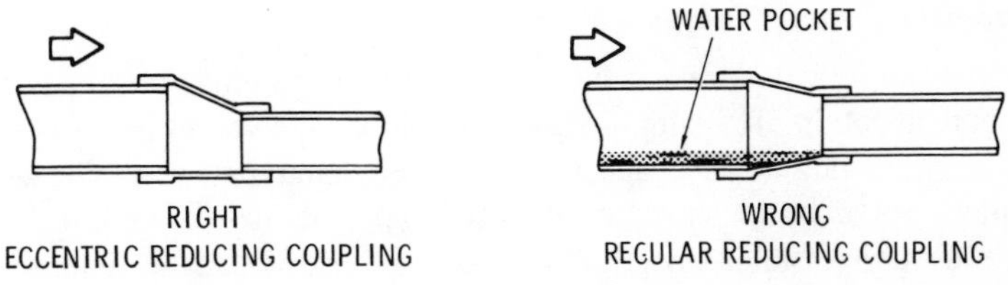
WATER POCKET
RIGHT
ECCENTRIC REDUCING COUPLING
WRONG
REGULAR REDUCING COUPLING

Courtesy *Hoffman Specialty ITT*

Fig. 30. How to reduce horizontal piping.

tric) reducing coupling. The use of the eccentric reducing coupling permits the continuance of uniform pitch without forming a water pocket which restricts flow.

TWO PIPE STEAM HEATING SYSTEMS

The definition of a two pipe heating system is—a system in which the heating units have two piping connections, one which is used for the steam supply and the other is used for the condensate return. Some of the same components which are used in one pipe systems are also used in two pipe systems. Two pipe systems are designed to operate at pressures ranging from sub-atmospheric (vacuum) to high pressure. Although they use many practical piping arrangements to provide up-flow or down-flow systems, they are conveniently classified by the method of returning the condensate to the boiler. Condensate can be returned to the boiler by gravity or by use of any one of several mechanical return means.

The Steam Boiler

Boilers are manufactured as either a cast-iron sectional boiler or as a steel boiler. The modern boiler is automatically fired, using coal, oil, or gas as the fuel. Suitable pressure controls, safety firing devices, and low water cut-off valves should be installed on the boiler. Package boilers, furnished complete with all equipment factory assembled and tested, are used extensively in larger installations.

Heating Units

A variety of heating units such as cast-iron radiation, wall fin-tube radiation, convectors, unit ventilators, cabinet heaters, and unit heaters are used in two pipe heating systems.

Radiator Supply Valves

The supply valve for a radiator in a two pipe system is installed at the inlet connection. The supply valves are globe type valves and are made in angle, straight way, and right and left hand patterns. The angle pattern valve is the most widely used type, being adaptable to most heating unit installations. Radiator supply valves can be obtained in either the modulating (adjustable flow) or non modulating type. A supply valve which uses stem packing is shown in Fig. 31. A packless

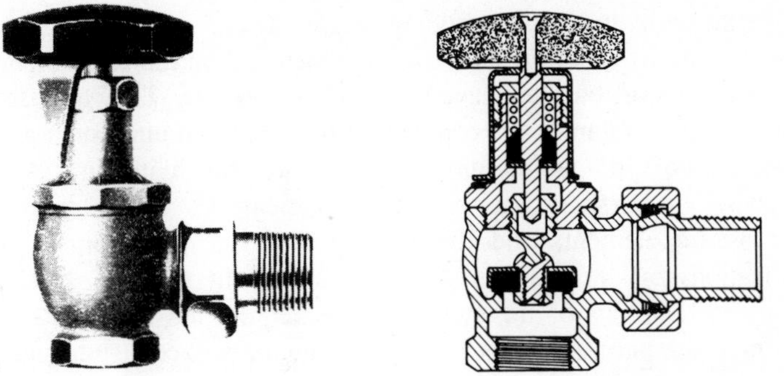

Courtesy *Hoffman Specialty ITT*

Fig. 31. A modulating type supply valve having a spring packed stem.

type supply valve is shown in Fig. 32. The packless type valve is a modulating type valve also; leakage around the stem is prevented by the metal bellows. A packless type valve is particularly suited for use in two pipe vacuum systems since the construction of this valve prevents air leakage into the system and steam leakage from the system.

Thermostatic Steam Traps

Thermostatic steam traps are the most commonly used traps in two pipe steam heating systems. They are of simple construction, small in physical size and weight with ample capacity for the usual heating system pressure. Thermostatic traps open in response to pressure as well as temperature to discharge condensate and air. When steam reaches the thermostatic element the trap closes to prevent the passage of steam into the return.

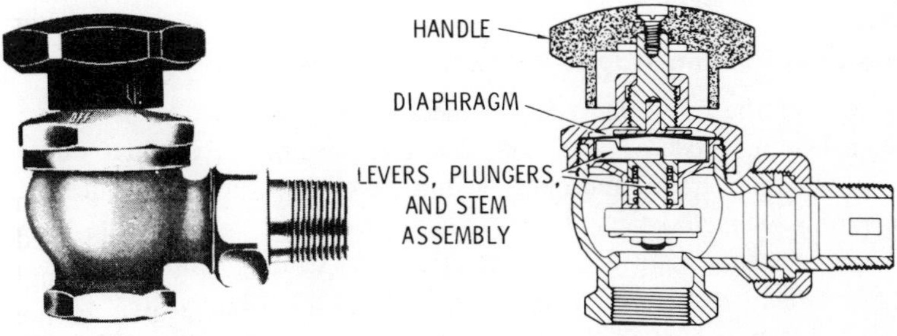
HANDLE
DIAPHRAGM
LEVERS, PLUNGERS, AND STEM ASSEMBLY

Courtesy *Hoffman Specialty ITT*

Fig. 32. A packless type modulating supply valve.

Thermostatic traps are made in angle, swivel, straight-way, and vertical patterns. Thermostatic traps are installed on the return connection of a radiator, but they have many other uses also. They are used as drip traps and to handle the condensate from fan coil units such as unit heaters. Two types of thermostatic traps are shown in Fig. 33. The cut-away views show the operating elements of these traps. The thermostatic elements and the seats are replaceable. Some typical applications for thermostatic traps are shown in Fig. 34. Note that certain installations of thermostatic traps require the installation of a cooling leg to permit the condensate accumulation to cool sufficiently to open the trap and discharge the condensate into the return. An installation requiring the use of a cooling leg is shown in Fig. 34.

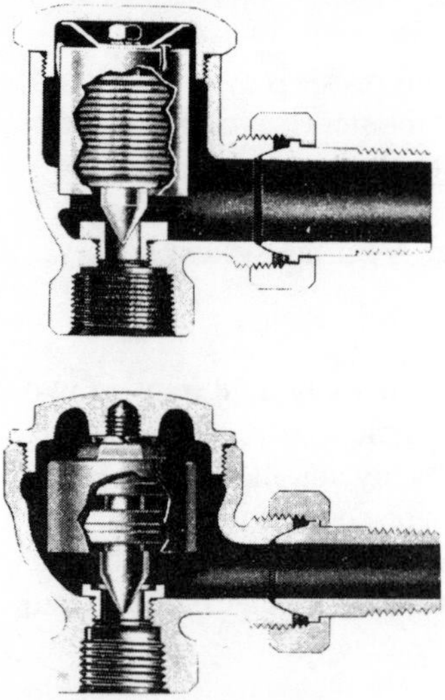

Fig. 33. Cut-away view of two types of thermostatic traps.

Courtesy *Hoffman Specialty ITT*

Mechanical Steam Traps

Some of the different types of mechanical traps used in two pipe steam heating systems are: float and thermostatic traps, float traps, inverted bucket traps, and open or upright bucket traps. The different operating characteristics of these traps makes them applicable to a variety of steam main drips, riser drips, and heating units.

Float and Thermostatic Traps—A float and thermostatic trap, called an F & T trap, is opened by the action of the condensate collecting in the trap and raising the float. As the float raises, it opens the discharge port and permits the condensate to enter the return piping. The float causes a

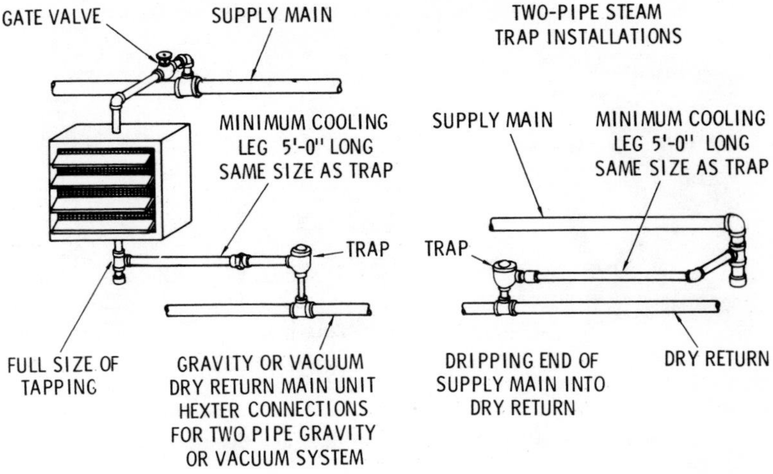
GATE VALVE
SUPPLY MAIN
TWO-PIPE STEAM
TRAP INSTALLATIONS
MINIMUM COOLING
LEG 5'-0" LONG
SAME SIZE AS TRAP
SUPPLY MAIN
MINIMUM COOLING
LEG 5'-0" LONG
SAME SIZE AS TRAP
TRAP
TRAP
FULL SIZE OF
TAPPING
GRAVITY OR VACUUM
DRY RETURN MAIN UNIT
HEXTER CONNECTIONS
FOR TWO PIPE GRAVITY
OR VACUUM SYSTEM
DRIPPING END OF
SUPPLY MAIN INTO
DRY RETURN
DRY RETURN

Courtesy *Hoffman Specialty ITT*

Fig. 34. Two typical installations of thermostatic traps showing the use of a cooling leg.

throttling action of the valve in the seat port; therefore, the discharge of the condensate is continuous. When air or condensate are present in the trap at a temperature below its designed closing pressure the thermostatic air by-pass remains open. The thermostatic air by-pass closes as steam enters the trap. Cooling legs are not required with an F & T trap installation because the trap will discharge condensate at any temperature up to a temperature very close to the saturated steam temperature corresponding to the pressure at the trap inlet. An F & T trap is shown in Fig. 35. F & T traps are used on unit heaters, unit ventilators, to drip the ends of steam mains, and to drip the heels of up-feed steam risers and down-feed steam risers.

Float Traps—Float traps are similar in appearance to an F & T trap. Float traps depend entirely upon the liquid level in the trap for operation. The temperature of the condensate does not affect the operation of the trap. When sufficient water or condensate enters the trap, the float rises and lifts the valve pin off the valve seat sufficiently to allow the liquid to be discharged into the return piping. The sealing of the pin on its seat against leakage does not present a problem because the steam pressure in the trap is always higher than the pressure in the return. An external by-pass, using a thermostatic trap, can be installed around a float trap, if needed. Fig. 36 illustrates some typical applications for Float & Thermostatic Traps and Float Traps.

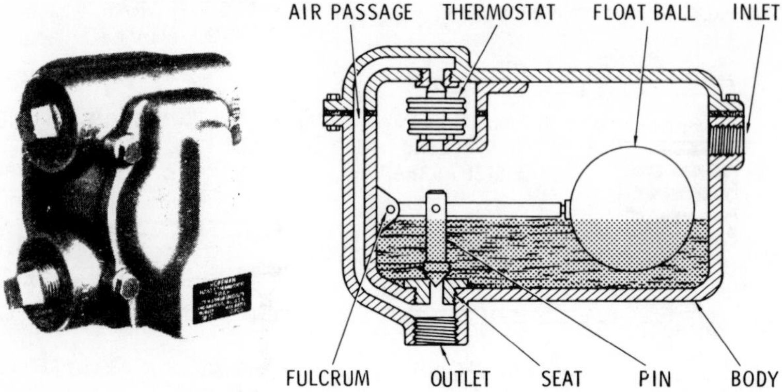
AIR PASSAGE
THERMOSTAT
FLOAT BALL
INLET
FULCRUM
OUTLET
SEAT
PIN
BODY

Courtesy *Hoffman Specialty ITT*

Fig. 35. Illustrating the action of a float and thermostatic trap.

Inverted Bucket Trap—Bucket traps are used to drain condensate and air from equipment such as unit heaters, steam coils, steam heated vats, pressing machinery, and to lift condensate to overhead return mains. On a new installation the trap must be primed by pouring water through the plug hole on top of the trap. The operation of an inverted bucket trap is shown in Fig. 37.

TYPES OF TWO PIPE SYSTEMS

Fig. 38 shows a typical gravity installation. A continuous flow type float and thermostatic trap is shown ahead of the meter with a float-type air vent at the trap discharge. This vent discharges all the air from the system because some meter constructions will not handle air.

Fig. 39 shows the installation where an intermittent type of bucket trap is used. A receiver must always be used with an intermittent operating type trap so as not to overload the meter when discharging. The float-type vent is also used on the receiver.

Fig. 40 shows the installation of a float and thermostatic trap ahead of the meter for a vacuum installation. The by-pass air vent line around the meter is required for this vacuum installation.

Fig. 41 shows a vacuum installation without the use of a master trap. This can be used when other traps in the system are performing without leakage. The air by-pass piping is also required.

The following table gives Dimension "A" for a variety of meter sizes for installation as shown by Figs. 39 through 41.

METER CAPACITY LB. PER HR.	250	500	750	1500	3000	6500	12000
			DIMENSION A—INCHES				
FIG. 39	4	4	8	8	8	12	12
FIG. 40	12	12	15	15	15	18	18
FIG. 41	12	12	15	15	15	18	18

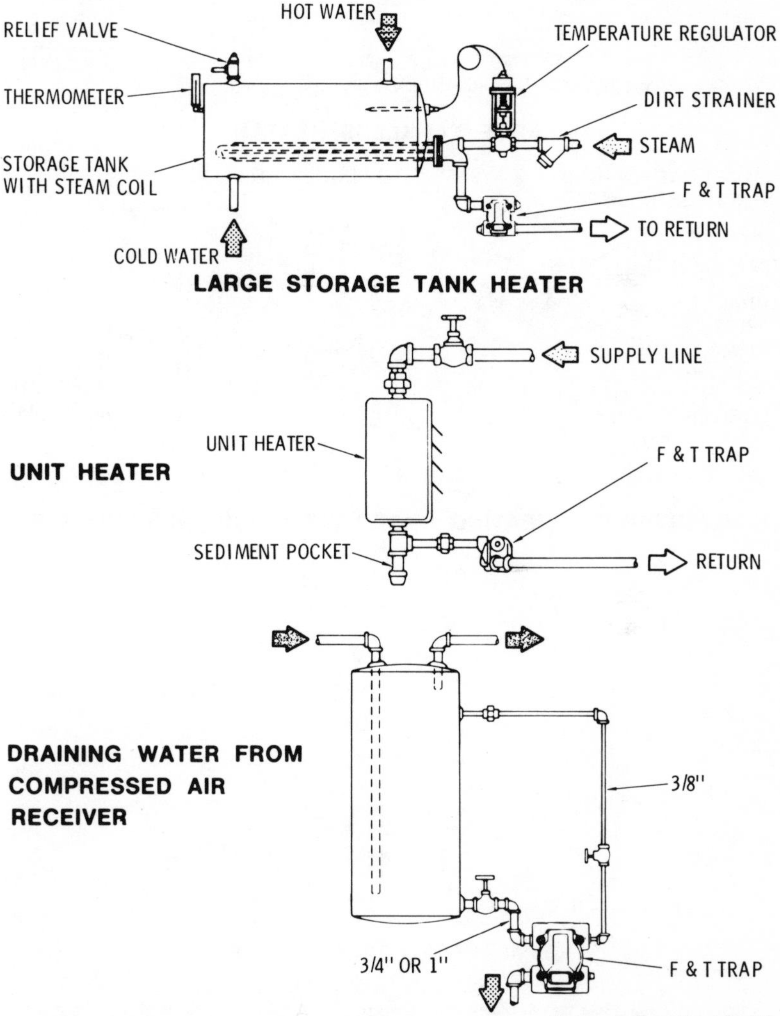
HOT WATER
RELIEF VALVE
TEMPERATURE REGULATOR
THERMOMETER
DIRT STRAINER
STEAM
STORAGE TANK
WITH STEAM COIL
F & T TRAP
TO RETURN
COLD WATER
LARGE STORAGE TANK HEATER
SUPPLY LINE
UNIT HEATER
UNIT HEATER
F & T TRAP
SEDIMENT POCKET
RETURN
DRAINING WATER FROM
COMPRESSED AIR
RECEIVER
3/8''
3/4'' OR 1''
F & T TRAP

Courtesy *Hoffman Specialty ITT*

Fig. 36. Some typical applications of float and thermostat traps and float traps.

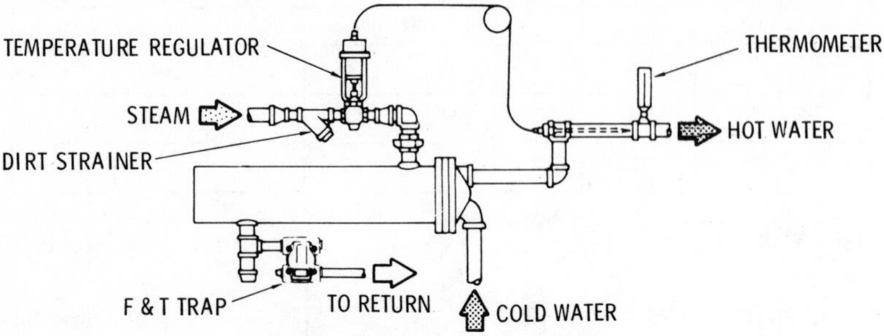
TEMPERATURE REGULATOR
THERMOMETER
STEAM
HOT WATER
DIRT STRAINER
F & T TRAP
TO RETURN
COLD WATER

LARGE TANKLESS HEATER

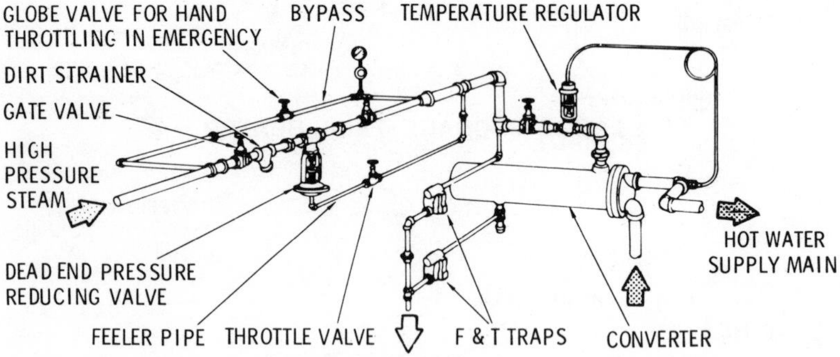
GLOBE VALVE FOR HAND THROTTLING IN EMERGENCY
BYPASS
TEMPERATURE REGULATOR
DIRT STRAINER
GATE VALVE
HIGH PRESSURE STEAM
DEAD END PRESSURE REDUCING VALVE
HOT WATER SUPPLY MAIN
FEELER PIPE
THROTTLE VALVE
F & T TRAPS
CONVERTER

CONVERTER FOR HEATING WATER WITH HIGH PRESSURE STEAM

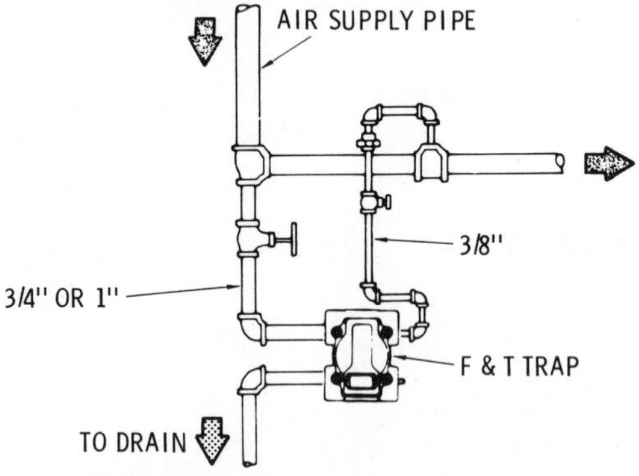
AIR SUPPLY PIPE
3/8"
3/4" OR 1"
F & T TRAP
TO DRAIN

COMPRESSED AIR SUPPLY PIPE

Courtesy *Hoffman Specialty ITT*

OPERATION

PRIMING: The inverted bucket of trap must be submerged to operate. On new installsions, trap should be primed by pouring water thru the plug hole on cover.

FIG. A

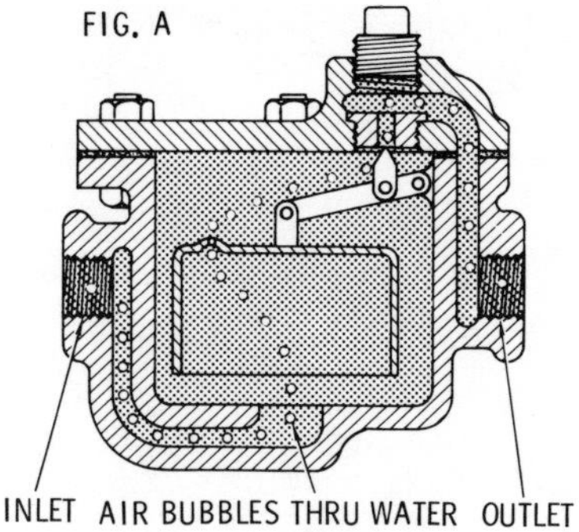
INLET
AIR BUBBLES THRU WATER
OUTLET

START-UP: Figure A shows the position of the bucket when steam is turned on. Excess condensate and air flow thru the valve seat.

FIG. B

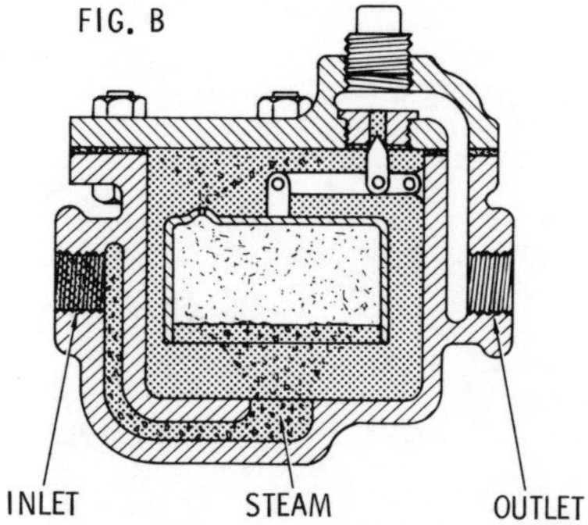
INLET
STEAM
OUTLET

VALVE CLOSES: Figure B shows the position of the bucket when condensate ceases to flow and steam enters the trap, and accumulates in the bucket. The steam pressure will push out more condensate until enough water has been displaced by steam in the bucket and it floats, holding the valve seat closed.

FIG. C

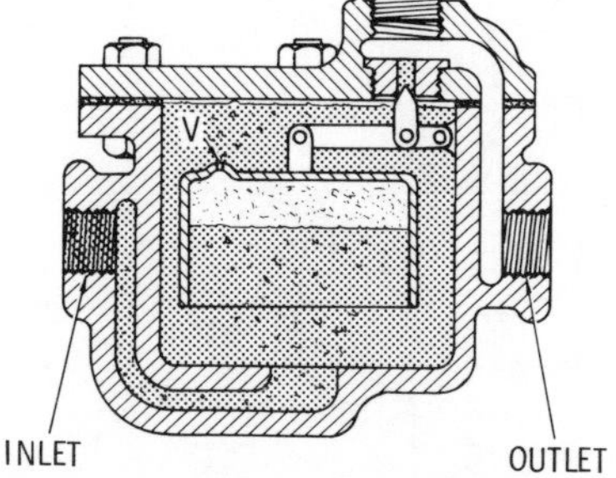
V
INLET
OUTLET

STEAM CONDENSES: Figure C steam slowly condensed inside the bucket and entrained air gradually escapes through the vent in the bucket (V), rising to the top of the trap. When steam condenses the water level rises inside the bucket causing the bucket to lose bouyancy

FIG. D

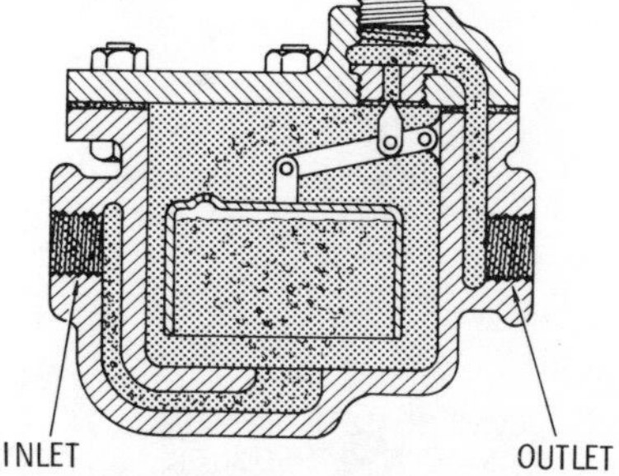
INLET
OUTLET

VALVE OPENS: Figure D when the weight of the bucket times the lever arm ratio equals the pressure on the seat port, the bucket drops. Condensate and air are forced from the trap by steam pressure until sufficient steam has displaced the water in the bucket causing the bucket to become buoyant.

Courtesy *Hoffman Specialty ITT*

Fig. 37. An inverted bucket trap.

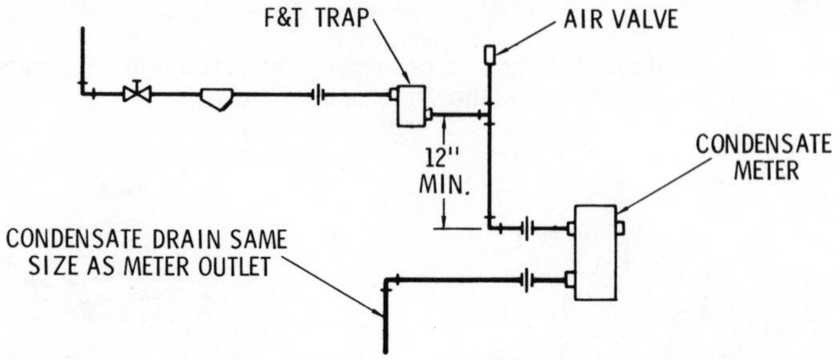
F&T TRAP
AIR VALVE
12" MIN.
CONDENSATE METER
CONDENSATE DRAIN SAME SIZE AS METER OUTLET

Courtesy *Hoffman Specialty ITT*

Fig. 38. A typical gravity installation.

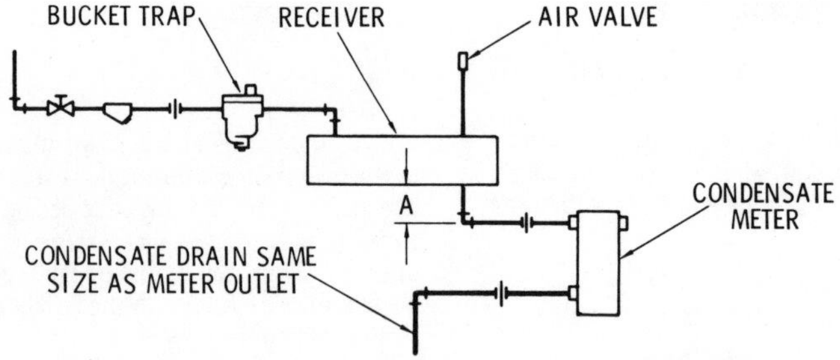
BUCKET TRAP
RECEIVER
AIR VALVE
A
CONDENSATE METER
CONDENSATE DRAIN SAME SIZE AS METER OUTLET

Courtesy *Hoffman Specialty ITT*

Fig. 39. Installation using an intermittent type of bucket trap.

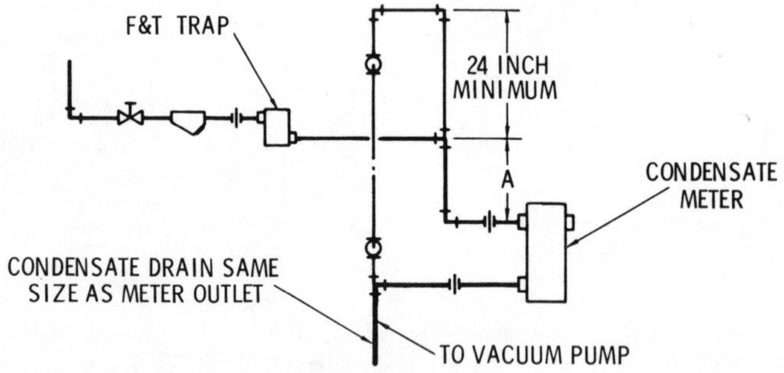
F&T TRAP
24 INCH MINIMUM
A
CONDENSATE METER
CONDENSATE DRAIN SAME SIZE AS METER OUTLET
TO VACUUM PUMP

Courtesy *Hoffman Specialty ITT*

Fig. 40. Installation of a float and thermostatic trap ahead of the meter for a vacuum installation.

Local codes in most cities do not permit the discharge of condensate into the sewer at temperatures above a specific amount. This is usually between 140°F and 150°F. In order to assure control of this maximum

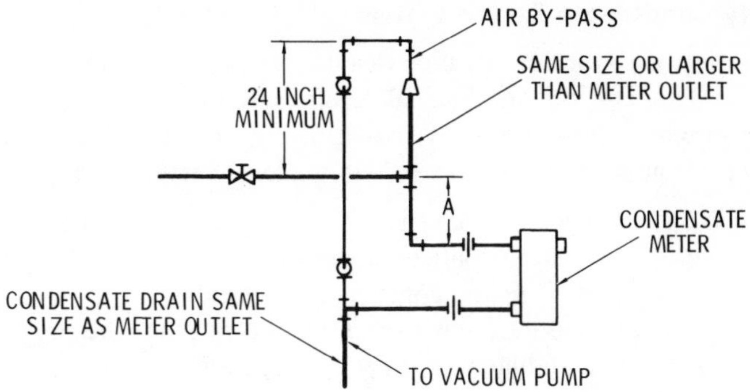
AIR BY-PASS
SAME SIZE OR LARGER THAN METER OUTLET
24 INCH MINIMUM
A
CONDENSATE METER
CONDENSATE DRAIN SAME SIZE AS METER OUTLET
TO VACUUM PUMP

Courtesy *Hoffman Specialty ITT*

Fig. 41. A vacuum installation without the use of a master trap.

temperature requirement and to extract all the heat possible from the condensate before discharging it to the sewer, a method of salvaging this heat is required.

Fig. 42 shows a method of discharging the condensate through an economizer to preheat domestic water before it goes to the primary water heater. If the condensate is pumped, it should first be discharged into a vented receiver from which it can flow by gravity through the economizer, or preheater, and the flow meter.

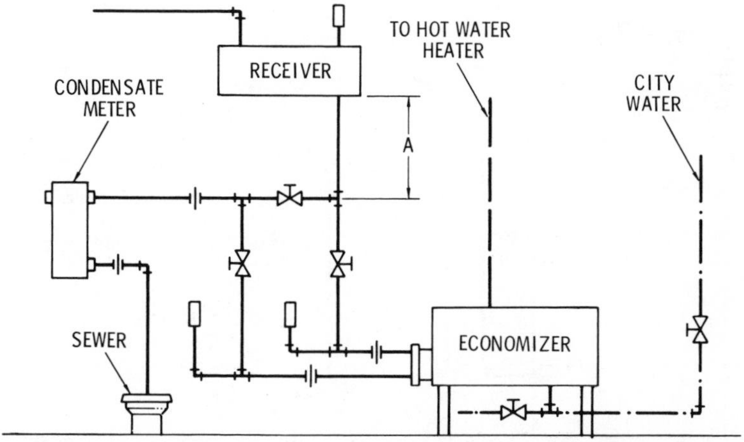
RECEIVER
TO HOT WATER HEATER
CONDENSATE METER
CITY WATER
A
SEWER
ECONOMIZER

Courtesy *Hoffman Specialty ITT*

Fig. 42. Discharging the condensate through an economizer to preheat domestic water before it goes to the primary water heater.

Gravity Condensate Return System

The two basic types of two-pipe steam heating systems are the gravity return system and the mechanical return system. The diagram of a simple gravity return system is shown in Fig. 43. The definition and function of the system components are explained referring to Fig. 43.

(A) The steam supply header is the horizontal piping connecting to one or more steam outlets at the top of the boiler. The steam supply main or mains connect to the header.

(B) The steam supply main carries the steam from the boiler and header to the radiators or other type heating units in the system. As shown in Fig. 43 an up-feed runout carries steam to the radiator through a supply valve connected to the inlet connection. A thermostatic trap is connected to the outlet connection on the radiator. The return runout piping carries the trap discharge to the dry return main.

(C) Part of the return main is called a *dry return* to distinguish it from the *wet return.* As shown in Fig. 43, the *dry return* is located above the boiler water line. The dry return carries the air and condensate discharge from the radiator thermostatic trap and from the F & T trap which drips the end of the steam supply main.

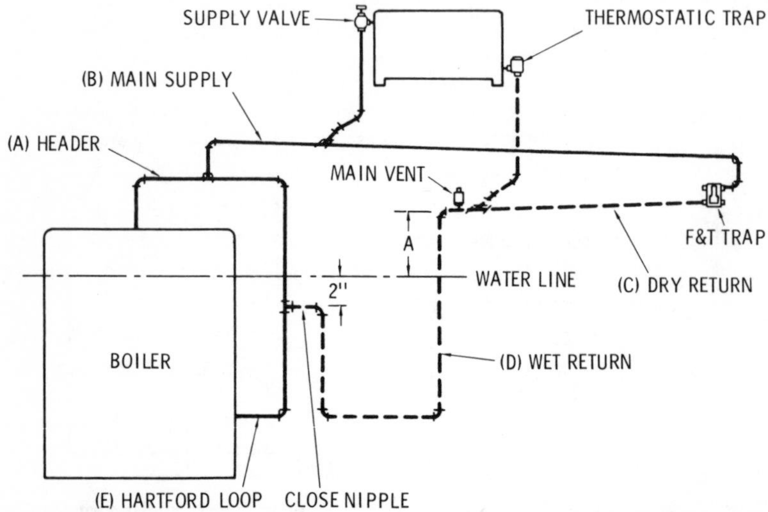
SUPPLY VALVE
THERMOSTATIC TRAP
(B) MAIN SUPPLY
(A) HEADER
MAIN VENT
F&T TRAP
A
2"
WATER LINE
(C) DRY RETURN
BOILER
(D) WET RETURN
(E) HARTFORD LOOP
CLOSE NIPPLE

Courtesy *Hoffman Specialty ITT*

Fig. 43. The main components in a two pipe gravity return system.

(D) The *wet return* is that portion of the return main which carries condensate back to the boiler and which is installed below the boiler waterline. Again referring to Fig. 43, it will be seen that the wet return is completely filled with water.

(E) The purpose of the Hartford loop is to prevent the water from leaving the boiler through the return piping. The construction details of a Hartford loop are explained in the one pipe section of this chapter.

To sum up the basic differences between a one pipe gravity return system and a two pipe gravity return system, the two pipe system requires a separate return line. Because the supply main is used as a steam supply line *only* in a two pipe system, a thermostatic or a float and thermostatic trap must be installed on the end of the steam supply main.

Mechanical Condensate Return System

As steam heating systems increase in size, requiring higher steam pressures in order to obtain steam circulation, some means other than gravity must be used to return the steam condensate to the boiler. The use of a condensate pump for this purpose has become the accepted method. The condensate from the return piping system flows by gravity into a tank or receiver. A condensate pump (or pumps) mounted on or adjacent to the receiver pumps the accumulated condensate into the boiler return piping. The return main must be graded uniformly to a slope of 1/4 inch in 10 feet when carrying the condensate to the receiver to insure that condensate and air will be separated in the piping. The air can then be vented into the atmosphere through the receiver vent connection. If it is necessary, either because of a job condition or to obtain uniform pitch to the receiver inlet, the condensate receiver can be located underground or in a pit. Fig. 44 shows a receiver filling with condensate. As the water level rises, the float also rises until the point is reached where the float actuates the float switch. The float switch energizes the pump and the water level drops as the condensate is pumped into the boiler return piping (Fig. 45). When the float, dropping with the water level, reaches its low position, the pump switch is turned off, and the receiver starts to accumulate condensate, beginning the cycle over again.

VACUUM SYSTEMS

A vacuum system is one which operates with a steam pressure that is less than atmospheric pressure. The object of such a system is to take

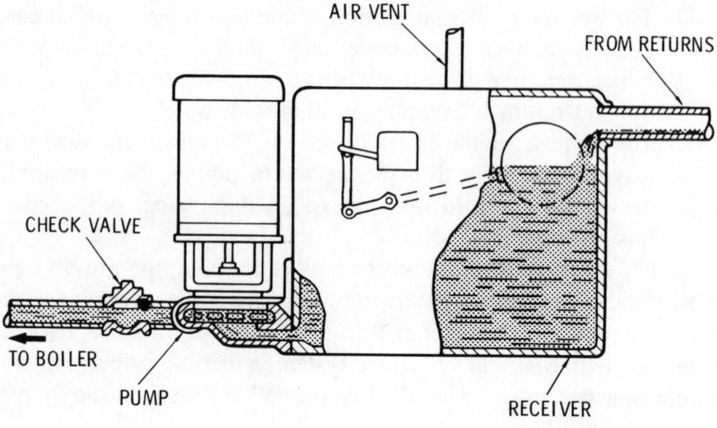
AIR VENT
FROM RETURNS
CHECK VALVE
TO BOILER
PUMP
RECEIVER

Courtesy *Hoffman Specialty ITT*

Fig. 44. A condensate receiver filling with water.

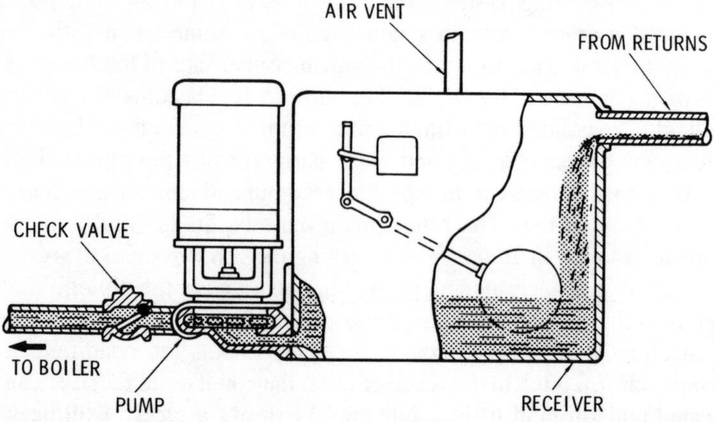
AIR VENT
FROM RETURNS
CHECK VALVE
TO BOILER
PUMP
RECEIVER

Courtesy *Hoffman Specialty ITT*

Fig. 45. Water level in a receiver dropping as condensate is pumped in the boiler return piping.

advantage of the low working temperature of the steam at this low pressure, giving a mild form of heat similar to that obtained with hot-water heating systems.

Vacuum Return-Line System

A vacuum return-line system is similar to a condensate return system except that a vacuum pump is installed to provide a low vacuum in the return line to return the condensate to the boiler. This is shown in Fig. 46. Pressure in the steam supply lines and in the radiators is maintained

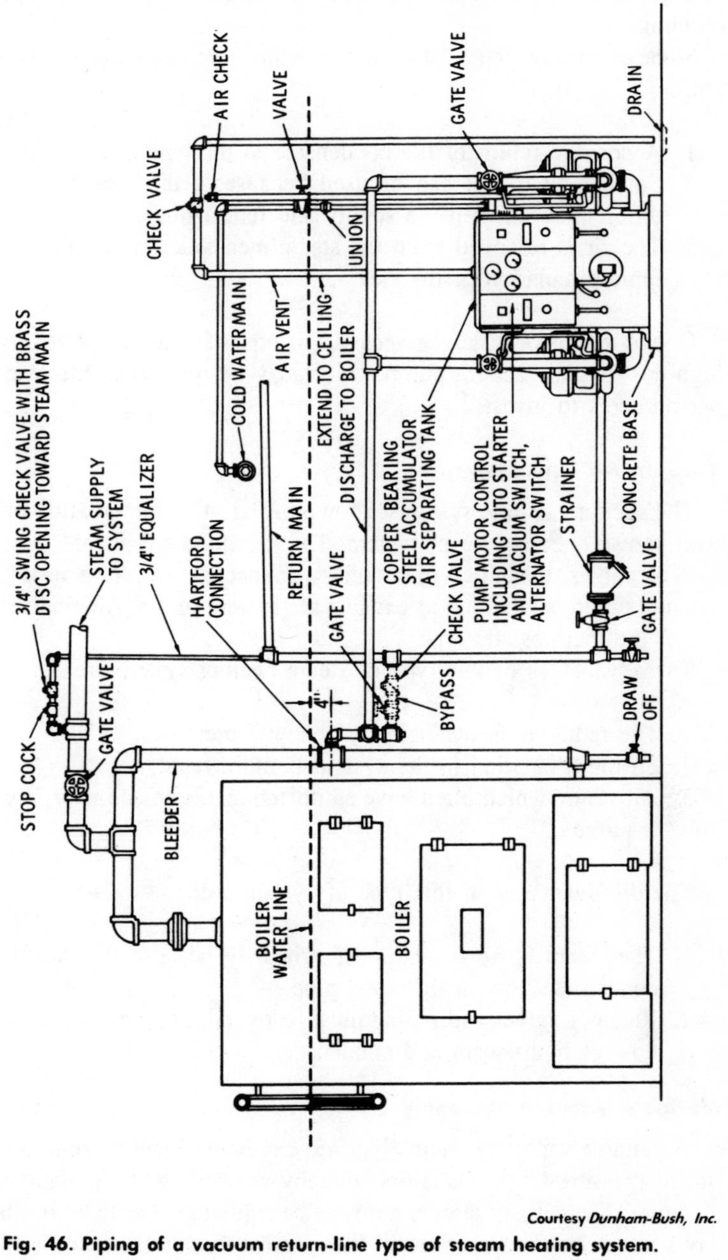
3/4" SWING CHECK VALVE WITH BRASS DISC OPENING TOWARD STEAM MAIN
STEAM SUPPLY TO SYSTEM
3/4" EQUALIZER
HARTFORD CONNECTION
STOP COCK
GATE VALVE
BLEEDER
COLD WATER MAIN
RETURN MAIN
AIR VENT
CHECK VALVE
AIR CHECK
VALVE
UNION
EXTEND TO CEILING
GATE VALVE
4"
DISCHARGE TO BOILER
COPPER BEARING STEEL ACCUMULATOR AIR SEPARATING TANK
CHECK VALVE
BYPASS
PUMP MOTOR CONTROL INCLUDING AUTO STARTER AND VACUUM SWITCH, ALTERNATOR SWITCH
GATE VALVE
STRAINER
DRAW OFF
CONCRETE BASE
GATE VALVE
DRAIN
BOILER WATER LINE
BOILER

Courtesy *Dunham-Bush, Inc.*

Fig. 46. Piping of a vacuum return-line type of steam heating system.

in a positive manner. This system is adaptable to any size or type of building.

Some of the advantages of a vacuum return-line steam heating system are:

1. A positive return of the condensate to the boiler.
2. Smaller pipe sizes can be used because of the greater pressure differential between the supply and return lines.
3. The air is removed from the steam mechanically, resulting in a rapid circulation of the steam.

One of the major disadvantages of this type of steam is the relatively high cost of the vacuum pump which may be more than the average person cares to invest.

Vacuum Air-Line System

The vacuum air-line systems, shown in Fig. 47, is a variation of the well-known one pipe steam system. The radiator air vents are replaced with air valves, the outlets of which are connected to a return air line. A vacuum pump is included to exhaust the system of air. Air-line valves are of the thermostatic type.

The advantages of this type of steam heating system are:

1. The radiators heat efficiently at lower pressures.
2. Steam circulation in the system is more rapid.
3. Air vents, which often have an objectionable noise factor, are not required.

The disadvantages of this type of system are:

1. The steam is noisy at times, primarily because the steam and condensate flow in the same pipe.
2. Radiator valves and piping must be oversize to accommodate the flow of both steam and condensate.

Variable Vacuum System

A variable vacuum system (Fig. 48) can better regulate and vary the steam pressure in the radiators, thereby controlling the radiator temperature. The radiator temperature can be regulated so as to be similar to hot-water radiator temperatures. Steam pressure can range from a high

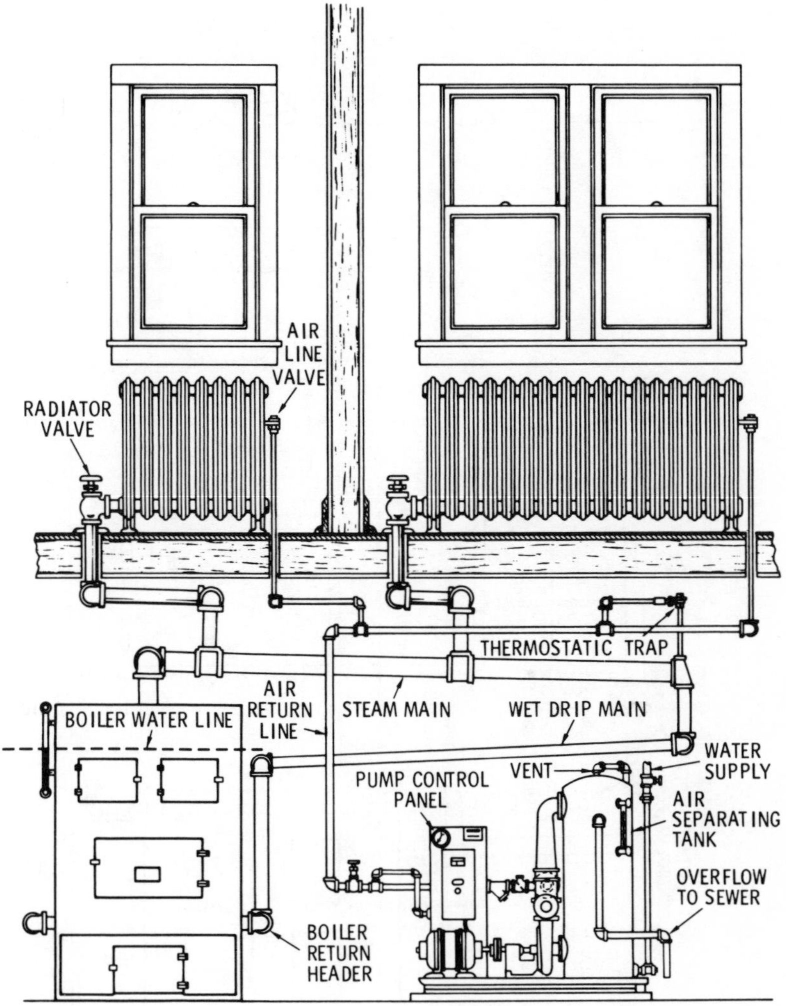
AIR
LINE
VALVE
RADIATOR
VALVE
THERMOSTATIC TRAP
AIR
RETURN
LINE
STEAM MAIN
WET DRIP MAIN
BOILER WATER LINE
WATER
SUPPLY
VENT
PUMP CONTROL
PANEL
AIR
SEPARATING
TANK
OVERFLOW
TO SEWER
BOILER
RETURN
HEADER

Courtesy *Dunham-Bush, Inc.*

Fig. 47. An air-line pump installed on a one pipe steam system to create a vacuum air-line system.

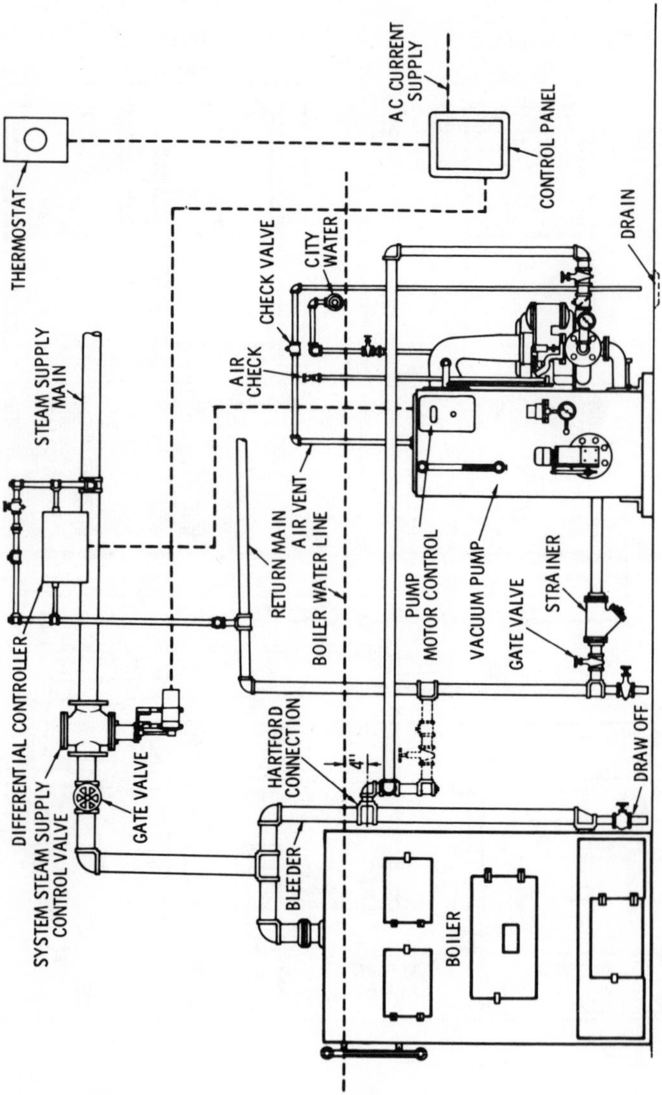
THERMOSTAT
AC CURRENT SUPPLY
CONTROL PANEL
DRAIN
STEAM SUPPLY MAIN
AIR CHECK
CHECK VALVE
CITY WATER
DIFFERENTIAL CONTROLLER
RETURN MAIN
AIR VENT
BOILER WATER LINE
PUMP MOTOR CONTROL
VACUUM PUMP
STRAINER
GATE VALVE
SYSTEM STEAM SUPPLY CONTROL VALVE
GATE VALVE
HARTFORD CONNECTION
4"
BLEEDER
BOILER
DRAW OFF

Fig. 48. A typical variable-vacuum steam heating system installation.

vacuum (low absolute pressure) to a positive value above atmospheric pressure. A continuous controlled steam supply can be maintained by regulating its generation or by regulating its admission to the steam mains. The vacuum-pump controller regulates and maintains a slight differential between the supply piping and the return piping-the return is always slightly lower to speed the steam circulation when operating at positive pressures.

If lift connections are used, they are restricted to a single lift between the accumulator tank and the vacuum pump. It is recommended that the piping be installed so that all condensate flows by gravity to the accumulator tank.

The principal advantages of a variable-vacuum steam heating system are:

1. A uniform floor-to-ceiling temperature can be maintained.
2. The types of radiation can be varied with maximum efficiency expected.
3. The system can be operated at much higher temperatures resulting in more efficiency during extreme cold weather.
4. Heat distribution can be balanced.
5. The system responds rapidly to temperature changes.
6. The steam supply is continuous, creating a better indoor climate without extreme temperature fluctuations.

The main disadvantage of this system is the relatively high cost of the vacuum pump for small heating systems. This may not make it competitive in price with other types of installations.

CHAPTER 4

Steam Heating Systems Design

Steam heating systems are identified by one or more of the following combinations which relate to the features of the system: 1. piping arrangement, 2. pressure range and, 3. methods employed for the return of condensate.

1. When identified by piping arrangement they can be either one pipe systems or two pipe systems. They can be either up-feed or down-feed depending on the direction of steam flow in the risers. They can be described as a wet return system or a dry return system depending on the location of the condensate return main above or below the water line of the boiler or of the condensate receiver.
2. Systems are called vapor systems when they operate at pressures ranging from low pressure to vacuum without the use of a vacuum pump. It is a vacuum system when it operates over this same pressure range using a vacuum pump.
   A system is a low pressure system when it employs pressure ranging from 0 to 15 psig. It is a high pressure system when it operates at pressure above 15 psig.
3. Systems are described as gravity systems when condensate returns directly to the boiler by gravity against the pressure causing the steam flow. When the condensate cannot be returned by gravity, mechanical means must be used to return the condensate accumulation to the boiler. The return mains can be at atmospheric

pressure for non-vacuum systems or under vacuum when a vacuum pump is used. In either system, using mechanical means for condensate return, the condensate must flow by gravity to the receiver.

DESIGN PRESSURE DROPS

Steam distribution in a heating system depends on the pressure drop selected for the system. The steam supply piping should be sized so that the pressure drop for the developed length of branches from the same supply are as nearly uniform as possible for the same supply pressure. The pressure drop is expressed in a unit length of 100 feet of equivalent length of run.

The pressure drop in common use for steam piping is as follows:

Return piping is always sized for the same pressure drop as selected for the supply piping.

SYSTEM PRESSURE PSIG	PRESSURE DROP per 100 ft.	TOTAL SYSTEM PRESSURE DROP un STEAM SUPPLY PIPING
Vacuum Systems	2—4 oz.	1—2 PSI
1	2 oz.	1—4 oz.
2	2 oz.	8 oz.
5	4 oz.	1½ PSI
10	8 oz.	3 PSI
15	1 PSI	4 PSI

Arranging the system piping so that the total distance of the supply from the boiler to the radiator is the same as the distance of the return from the radiator back to the boiler, gives the desirable resistance to and from the radiator. It is important that the piping be sized to handle the full design load conditions.

A heating system operates at less than half of its design load conditions during an average winter. However, the pick-up load required to raise the temperature of the metal in the piping up to the steam temperature and to raise the indoor temperature up to design conditions requires a large amount of heat even during moderate winter outdoor temperature. This increased pick-up value is usually considered to be 143 percent of the design load.

SIZING BOILER CONNECTIONS AND HEADER PIPING

Steam boilers for heating systems have one or more outlets. When there is more than one, all the outlets should be used. The vertical pipe to the supply main or the header should be the same size as the boiler outlet tapping and never reduced except at the supply main or header. The header should be sized on the basis of the maximum load that must be carried by any part of it. The sizing can be done from the same capacity table as used for sizing the steam mains. Increasing the size of the header to a larger pipe size than required can be undesirable when the pick-up load during the heating-up period is considered.

The return header is the same size as the required return main at the boiler. If the boiler has more than one return tapping, both should be used and sized the same as the return main.

DESIGNING A ONE PIPE STEAM SYSTEM

The piping for a one pipe gravity air vent system in which the total equivalent length of run does not exceed 200 ft. can be sized from Tables 3, 5, and 7. Instructions for their use is as follows:

1. Where steam and condensate flow in the same direction in the steam main, dripped run-outs to up-feed risers, and down-feed risers, use Table 3, Col. 2 (1 oz. per 100 ft. pressure drop).
2. Where run-outs to risers are not dripped and steam and condensate flow in the opposite directions, use Table 5, Col. F. Use the same for run-outs to radiators.
3. For up-feed risers carrying condensate back from the radiators, use Table 5, Col. D.
4. For down-feed systems, the main riser between the boiler and the overhead attic main, which does not carry condensate from the radiators, use Table 5, Col. B.
5. Size radiators supply valves and vertical connections from Table 5, Col. E.
6. Use Table 7, Col. O for sizing dry return mains.
7. Use Table 7, Col. N for sizing wet return mains.

Additional notes on the installation of the piping for a one-pipe system are as follows.

1. Supply and dry-return mains should not be pitched less than 4″ in 10 ft.
2. Horizontal run-outs to risers and radiators should not be less than 4″ per foot. Where this pitch cannot be obtained for run-outs over 8 ft. in length, increase the pipe one size larger.

3. It is not desirable to have a supply main smaller than 2″.
4. All supply mains, run-outs to risers, or risers should be dripped where necessary.
5. Where supply mains are decreased in size they should be dripped or eccentric reducer coupling should be used as shown by Fig. 19B.

The following six (6) steps are required for a complete design of a one pipe steam system.

1. Calculate the heat loss for each room or space. Use ASHRAE GUIDE, I=B=R or other available publications. Heat loss calculations are not performed for this example.

2. Decide on type and size of heating units for each room. Consult ASHRAE GUIDE and manufacturer's catalogs for specific information.
   A. Convert the Btu per hour heat loss for each room to Equivalent Direct Radiation—1 sq. ft. EDR=240 Btu per hour.
   B. Locate heating unit to suit space and wall exposure. This location is usually on exposed outside walls and under or near windows so as to replace heat losses where they occur-

NOTE: Example shows size only of required heat output for a unit and not a specific make of unit.

3. Make a piping layout to obtain the most direct piping arrangement for supplying steam from the boiler to the most remote radiator, and for the return of condensate. Usually a double circuit main has the advantage of quick and uniform delivery of steam to the heating units. Fig. 1 shows the piping layout for the two circuit system.

   A. Measure actual length of main and risers from the boiler to most remote radiator to establish maximum pipe length.

Section ABCG	=	103 ft.
Riser (1)	=	20 ft.
	Total	123 ft.

   B. Determine total equivalent length of piping by making proper allowance for fittings—See Table 6.

Main
2—21/2″ elbows = 10 ft.
3—21/2″ side outlet tees = 33 ft.
2—2″ elbows = 8.6 ft.
Riser
1—2″ side outlet tee = 8 ft.
1—1/2″ side outlet tee = 7 ft.
3—1/2″ elbows = 10.5 ft.
Total 77.1 ft.
Total Equivalent Length = 123 ft. + 77 ft. = 200 ft.

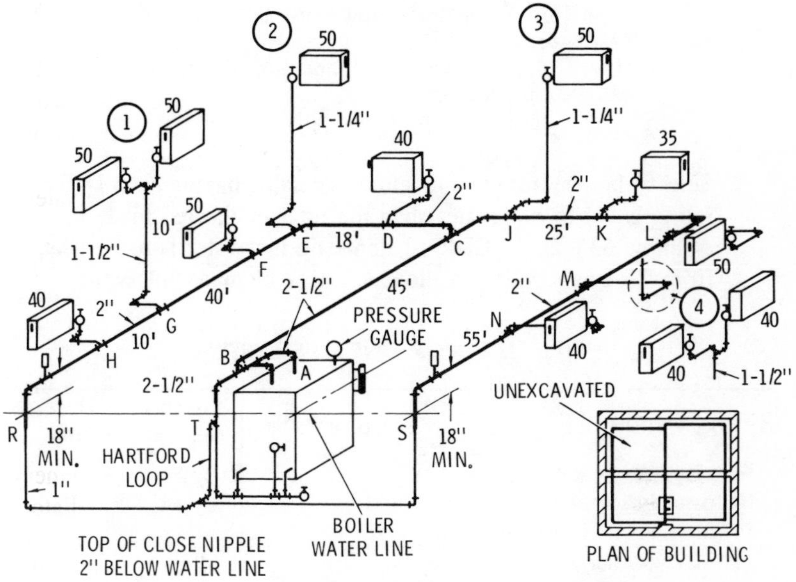
1
2
3
4
50
50
50
50
50
50
40
40
40
40
40
40
35
1-1/4"
1-1/4"
1-1/2"
1-1/2"
2"
2"
2"
2"
2-1/2"
2-1/2"
10'
10'
18'
25'
40'
45'
55'
E
D
C
J
K
L
F
G
H
B
A
M
N
R
T
S
PRESSURE GAUGE
18" MIN.
18" MIN.
HARTFORD LOOP
1"
UNEXCAVATED
TOP OF CLOSE NIPPLE 2" BELOW WATER LINE
BOILER WATER LINE
PLAN OF BUILDING

Courtesy *Hoffman Specialty ITT*

Fig. 1. A piping layout for a two circuit system.

4. Determine system pressure drop and boiler pressure:
   A. For Gravity system where equivalent length is less than 200 ft. the total pressure drop should not exceed 2 oz. For equivalent length greater than 200 ft. the total pressure drop should not exceed 4 oz. Because the total equivalent length of this example is not more than 200 ft., the piping will be sized on the basis of a total pressure drop of 2 oz.
   B. Boiler pressure should not be less than twice the total pressure drop.

The boiler pressure will be maintained at a minimum practical control limit of 1/4 PSI.

5. Size supply mains, risers, return piping.
   A. Divide total system pressure drop by equivalent length of run to obtain pressure drop for 100 ft. length.
   B. Consult Table 3, Col. 2 for supply main sizes. The minimum horizontal supply main should not be less than 2″.
   C. Consult Table 7, Col. N for wet return main sizes. Size returns on basis of same pressure drop for 100 ft. length as that selected for supply main.
   D. Consult Table 5 for sizing the following pipe connections:
      (1) Use Col. F for horizontal run-out to up-feed riser or to radiator.
      (2) Use Col. E for vertical connection to radiator and radiator supply valve size.
      (3) Use Col. D for up-feed supply riser.

6. Select a boiler, from manufacturer's catalog having a net I=B=R rating equal to or greater than the total connected load; in this example 535 sq. ft. EDR. If domestic hot water is to be heated from this steam boiler no allowance need be made unless there are

PIPE SIZES FOR ONE PIPE SYSTEM

DESCRIPTION	SECTION	LOAD—Sq. Ft. EDR	PIPE SIZE
Horizontal Run-out to radiator		35	1¼″
Horizontal Run-out to radiator		40	1¼″
Horizontal Run-out to radiator		50	1¼″
Vertical Connection Radiator Supply Valve		35	1¼″
Vertical Connection Radiator Supply Valve		40	1¼″
Vertical Connection Radiator Supply Valve		50	1¼″
Up-feed Riser	(1)	100	1½″
Up-feed Riser	(4)	80	1¼″
Up-feed Riser	(2) & (3)	50	1¼″
Supply Main	C to R	280	2″
Supply Main	C to S	255	2″
Supply Main	B to C	535	2½″
Boiler Header	A to B	535	2½″
Wet Return Main	R to T	280	1″
Wet Return Main	S to T	235	1″

more than two bathrooms. Consult I=B=R recommendation for additional allowance when required.

TABULATION OF STEAM PIPE SIZES FOR FIG. 2

PIPE SECTION LOCATION	SQUARE FEET EDR		PIPE SIZE (Inches)	REMARKS
	ACTUAL LOAD	PERMITTED LOAD		
D6 to D1	520	936	2	Table 3—Col. 6
C4 to C2	900	936	2	Table 3—Col. 6
C1 to B1	1450	1512	2½	Table 3—Col. 6
Boiler Header	1450	1512	2½	Table 3—Col. 6
Runout to Riser[1]	115	152.4	1¼	Table 4—5″ Pitch
Riser[1]	115	148	1	Table 3—Col. 6
Runout to Riser[3]	150	152.4	1¼	Table 4—5″ Pitch
Riser[3]	150	312	1¼	Table 3—Col. 6

TABULATION OF RETURN PIPE SIZES FOR FIG. 2

PIPE SECTION LOCATION	SQUARE FEET EDR		PIPE SIZE (Inches)	REMARKS
	ACTUAL LOAD	PERMITTED LOAD		
D6 to D2	455	460	1	Table 7—Col. U
D1 to C2	900	964	1¼	
C1 to B1	1450	1512	1½	
B1 to Pump Receiver	1450	1512	1½	
Runout to Riser[1]	115	460	1 (Min.)	
Riser[1]	115	460	1	
Runout to Riser[3]	150	460	1	
Riser[3]	150	460	1	

TABULATION OF RADIATOR CONNECTION FOR FIG. 2

SQUARE FEET EDR LOAD	STEAM SUPPLY*			RETURN PIPE**		
	RUNOUT FROM RISER OR MAIN (Inches)	VERTICAL CONNECTION (Inches)	VALVE SIZE (Inches)	RUNOUT FROM RISER OR MAIN (Inches)	VERTICAL CONNECTION (Inches)	TRAP SIZE (Inches)
50	1	¾	¾	¾	½	½
65	1	¾	¾	¾	½	½
80	1	¾	¾	¾	½	½

*From Table 8. [1]See Fig. 3.
**From Table 9. [3]See Fig. 4.
Note: Fig. 5 shows detailed pipe sizing at End of Main and Riser.

DESIGNING A TWO PIPE SYSTEM

Two pipe steam heating systems can be designed with piping arranged for an up-feed system or for a down-feed system from a steam supply main located in an attic or pipe space. All two pipe systems are similar in all basic respects. Attention must be given to the details of dripping the ends of the steam main, the bottom or heel of up-feed risers through steam drip traps.

The use of a condensate pump for gravity open vented return system or the use of vacuum pumps for a vacuum system is the preferred method of returning condensate to the boiler. Using these mechanical means eliminates the problem for the system.

Capacity tables for sizing steam and return piping are contained in the section of Engineering Data and Technical Information.

Additional information for the piping arrangement for two pipe systems is as follows:

1. Steam and return mains should be pitched not less than 1/4″ in 10 ft.
2. Horizontal run-outs to riser and radiators should be pitched not less than 1/2″ per foot. Where this pitch cannot be obtained for run-outs over 8 feet in length increase the pipe one size larger than called for in the capacity table.
3. It's not desirable to have a steam supply main smaller than 2″
4. When required, the ends of steam mains, bottom of up-feed riser or run-outs to up-feed risers should be dripped through steam traps into the dry return main. The ends of all down-feed steam risers should also be dripped through steam traps to the dry return.
5. Return mains should always be pitched to provide gravity flow of condensate to the pump receiver inlet.

The following eight (8) steps are required for a complete design of a two-pipe steam heating system. The example shown in Fig. 2 will be used to illustrate the step by step procedure. Figs. 3, 4, and 5 are enlarged sections of Fig. 2.

1. Calculate the heat loss for each room or space. Many publications are available which can be used to determine the heat loss, such as ASHRAE GUIDE & DATA BOOK, and I=B=R Heat Loss Calculation Guide. Heat loss calculations are not performed for this example.

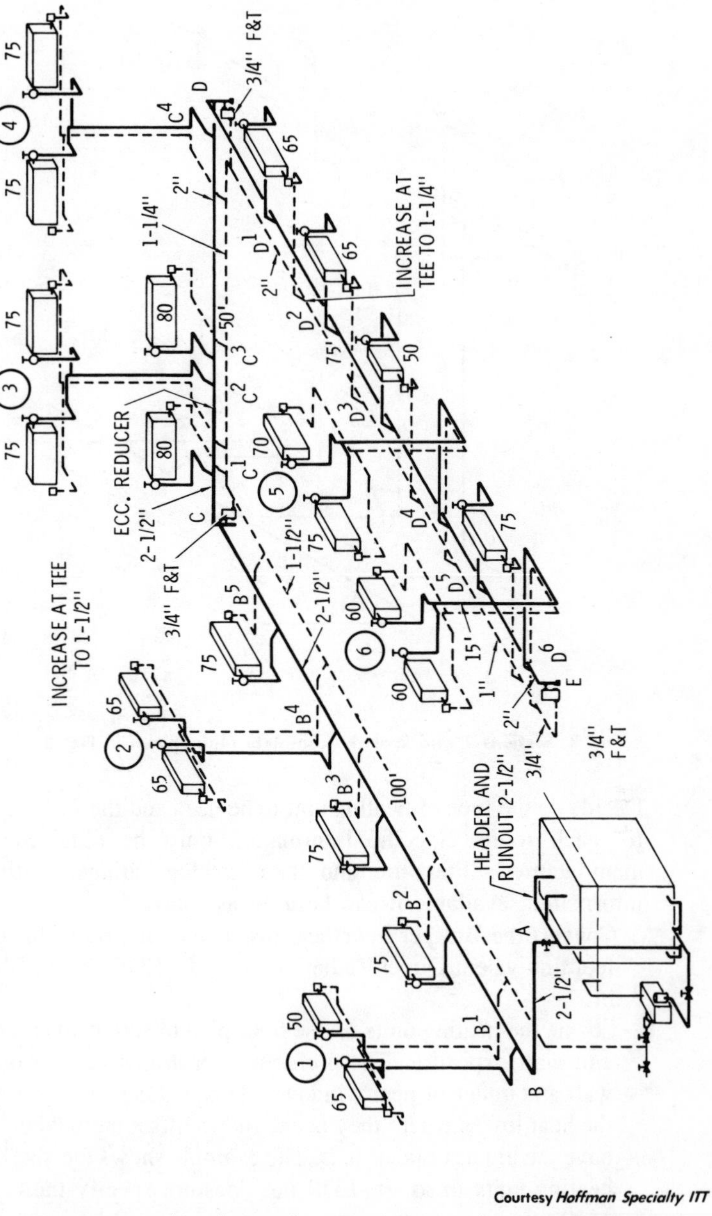
INCREASE AT TEE TO 1-1/2"
3/4" F&T
2-1/2"
ECC. REDUCER
1-1/2"
2-1/2"
1-1/4"
2"
50'
3/4" F&T
INCREASE AT TEE TO 1-1/4"
2"
75'
15'
1"
2"
3/4"
3/4" F&T
100'
HEADER AND RUNOUT 2-1/2"
2-1/2"
A
B
C
D
E

Courtesy *Hoffman Specialty ITT*

Fig. 2. A complete design of a two pipe steam heating system.

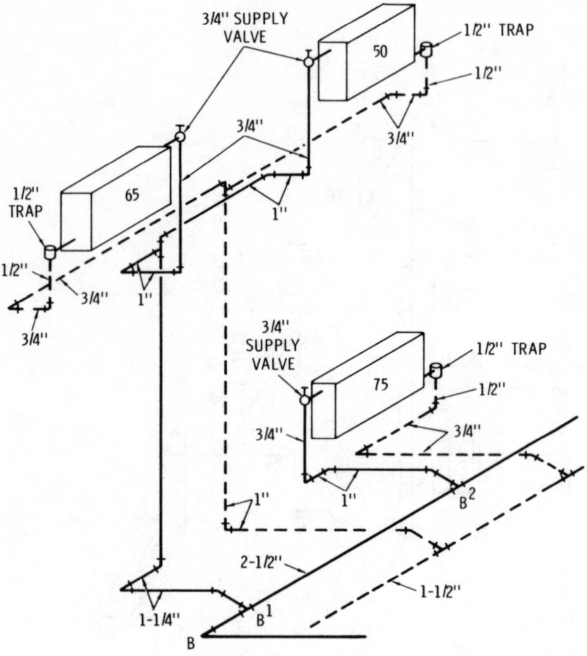
3/4" SUPPLY VALVE
1/2" TRAP
50
1/2"
3/4"
3/4"
65
1/2" TRAP
1"
1/2"
3/4"
1"
3/4"
3/4" SUPPLY VALVE
1/2" TRAP
75
1/2"
3/4"
3/4"
1"
1"
B2
2-1/2"
1-1/2"
1-1/4"
B1
B

Courtesy *Hoffman Specialty ITT*

Fig. 3. Sections 1 and 2 of the complete plan shown in Fig. 2.

2. Decide on the type of heating unit to be used and the size required for each room. Specific information must be obtained from manufacturer's literature and their catalog ratings. With this information available it can be used as follows.
   A. Convert the Btu per hour heat loss from each room to square feet Equivalent Direct Radiation. 1 sq. ft. EDR = 240 Btu per hr.
   B. Locate the heating units on the floor plan to suit available space and wall exposure. The preferable location is on an outside wall and under or near windows. This is done so as to replace the heat losses where they occur and to offset them where they have the highest rate of loss. The example shows the size of the heating units in sq. ft. EDR but does not specify the type of unit.

3. Determine equivalent length of piping.
   A. Make a piping layout to obtain the most direct piping arrangement for supplying steam from the boiler to the most remote radiator and for the return of condensate to the boiler. For this example, the application of a heating system to a long, narrow, two-story building was chosen, as shown by Fig. 2, with the boiler located in one end of the building. The supply and return piping for the radiators are connected to obtain the shortest length of pipe for steam flow from the radiators. The steam supply piping begins at the boiler and ends at the most remote radiator, while the return piping begins at the most remote radiator and ends at the condensate pump adjacent to the boiler. When the piping arrangement has been determined, the procedure is as follows:

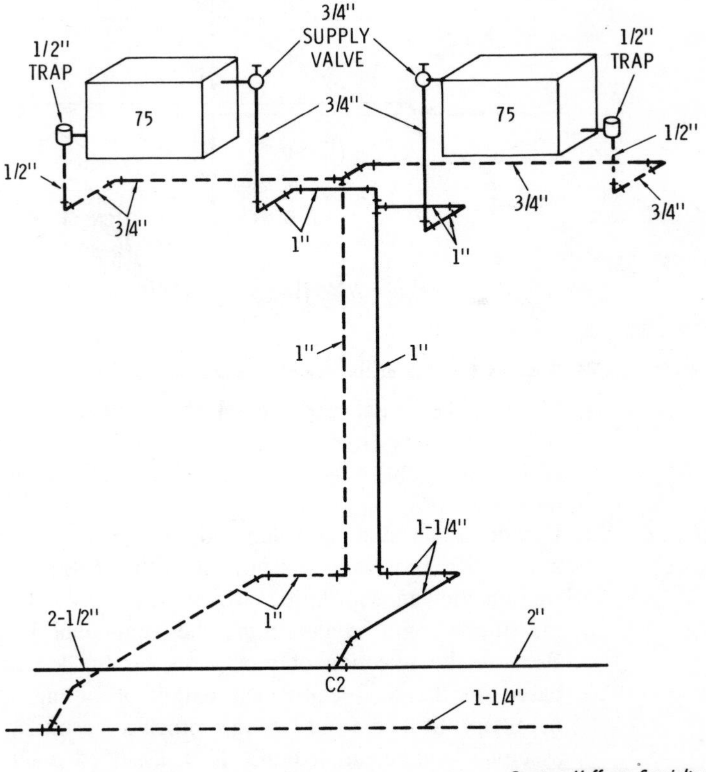
3/4"
SUPPLY
VALVE
1/2"
TRAP
1/2"
TRAP
75
75
3/4"
1/2"
1/2"
3/4"
3/4"
3/4"
1"
1"
1"
1"
1-1/4"
2-1/2"
1"
2"
C2
1-1/4"

Courtesy *Hoffman Specialty ITT*

Fig. 4. Sections 3 and 4 of the complete plan shown in Fig. 2.

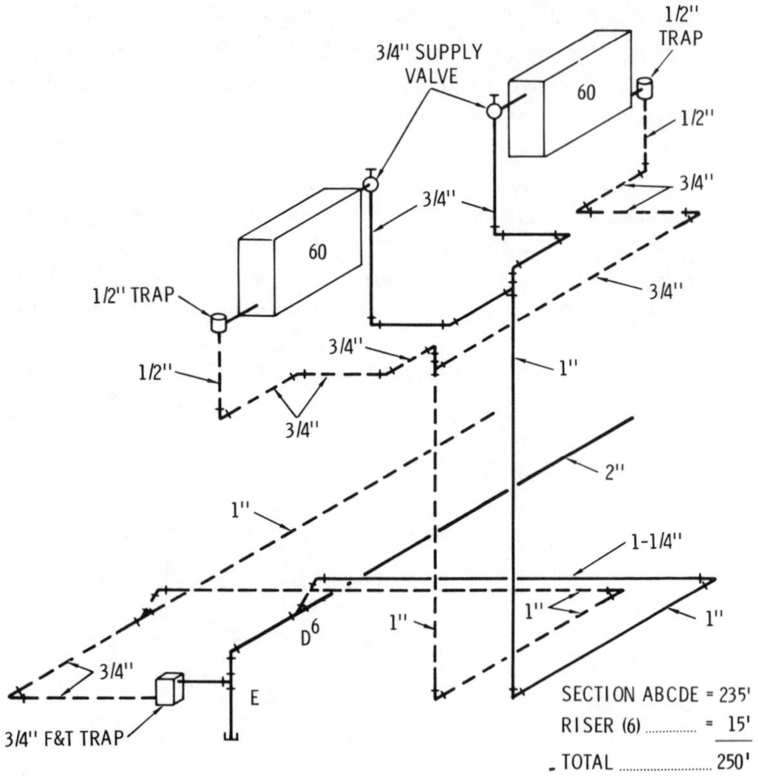

1/2" TRAP
3/4" SUPPLY VALVE
60
1/2"
3/4"
3/4"
60
1/2" TRAP
3/4"
1/2"
3/4"
3/4"
1"
2"
1"
1-1/4"
1"
1"
D6
3/4"
E
1"
3/4" F&T TRAP
SECTION ABCDE = 235'
RISER (6) = 15'
TOTAL 250'

Courtesy *Hoffman Specialty ITT*

Fig. 5. Sections 5 and 6 of the complete plan shown in Fig. 2.

(1). Measure the actual length of steam supply main, run-outs and risers from the boiler to the most remote radiator to establish the maximum pipe length, as shown by Fig.2.

(2). Determine the total equivalent length of piping by making proper allowance for fittings. Table 6 shows the friction allowance for fittings which must be added to the measured length to determine the equivalent length. Because the selection of pipe sizes for the system is based on the total equivalent length of piping, it is necessary to begin the design process based on an assumed equivalent length. This assumed value is usually considered to be double the actual measured length. For this example the assumed equivalent length will be 250 ft. x 2 = 500 ft. The actual equivalent length of pipe can be checked after the piping has been sized, based on the assumed length.

4. Determine system pressure. Before pipe sizes can be selected for the system, several factors must be considered which pertain to the steam pressure at the boiler and system pressure drop under full load conditions. These factors include the following:
   A. The total system pressure drop must not exceed one-half the boiler pressure when steam and condensate flow in the same direction.
   B. The pressure drop must be kept at a value which will not create excessive velocities. For systems operating below 15 psi the velocities should not be more than 5000 ft. per minute for quiet operation for concurrent flow. For the example shown in Fig. 2 the boiler pressure of 3 psi was selected for full load conditions. The system pressure drop, therefore, must not exceed 1.5 psi (24 oz.).

5. Size the supply and return piping.
   A. From step 3 the equivalent length of piping was assumed to be 500 ft. From step 4 the maximum system pressure drop has been established as 1.5 psi (24 oz.). Therefore, for 500 ft. equivalent length, the pressure drop per 100 ft. of length will be approximately 5 oz. However, Table 3, Col. 6 shows the nearest pressure to be 4 oz. per 100 ft. and the capacities shown in this column will be used to size the steam main. When sizing supply mains for low pressure steam heating systems it is not desirable to have a horizontal supply main smaller than 2″.
   B. The horizontal steam run-out to risers and risers will be sized from Table 3, Col. 6 and Table 4, 5″ pitch column.
   C. The return mains, horizontal run-outs to risers, and the return risers will be sized from Table 7, Col. U for a pressure drop of 2 psi (4 oz.) which is the same as that used for sizing the steam piping.
   D. The horizontal steam run-out to radiators from steam main or riser and the supply valve vertical connection and valve size are selected from Table 8. The horizontal return run-outs to radia-

tors from return mains or risers and radiator trap sizes are selected from Table 9.

E. Referring to Fig. 2 and using the pipe capacity tables, the following tabulation shows the pipe sizes selected for the design example. The sizes are also shown in the piping layout of Fig. 2.

F. Pipe sizing up to this point has been made on the basis of an assumed equivalent length. The actual equivalent length can now be determined and the pressure drop per 100 ft. of pipe length can also be determined. The following tabulation shows the result of this check for system pressure drop.

Steam main measured length 250 ft.

Allowance for fittings

Steam Main	
1—2-1/2″ Ell	5.0
2—2-1/2″ S.O. Tee	22.0
1—2-1/2″ G.V.	1.1
3—2-1/2″ Ell	15.0
1—2-1/2″ S.O. Tee	11.0
1—2″ Ell	4.3
2—2″ S.O. Tee	16.0
Runout from Main	
2—1″ Ell	4.4
Runout from Riser	
1—1″ S.O. Tee	5.0
2—1″ Ell	4.4
1—3/4″ Angle Valve	10.0
	348.2 ft.

As can be seen, with an actual total equivalent length of 348.2 ft. and a total system pressure drop of 24 oz. (1.5 psi), the pressure drop per 100 feet is 6.8 oz. Therefore, a study of Table 3 shows that there will be no changes in pipe size if the sizing is made on the basis of an 8 oz. drop (Col. 8) instead of a 4 oz. drop (Col. 6).

G. Determine steam velocity in the system with 3 psi boiler pressure, 24″ main pipe size, 362 lb./hr. (lb./hr. = EDR/4) load in system. 5000 ft. per minute maximum. By calculation using steam velocity formula in Engineering Data Section we get a velocity of 3996 ft. per minute.

6. Select a boiler from manufacturer's catalog having a net rating (I=B=R or SBI) equal to or greater than total connected load to 1450 sq. ft. EDR. Often, domestic water is heated by the same boiler used to supply steam for heating. The size of the heater and storage tank depends on the water requirements. Proper sizing information is available from many sources such as ASHRAE GUIDE, I=B=R, SBI, and several manufacturers of heaters and coils.

7. Select float and thermostatic traps to drip the main.

MAIN A-C = 470 sq. ft.
MAIN C-D = 460 sq. ft.
MAIN D-E = 520 sq. ft.

A Hoffman 3/4" No. 53FT trap will handle 560 sq. ft. EDR (140 lb. of condensate per hour) with a pressure differential of 1 pound per square inch.

8. Select a condensate pump from Hoffman Condensate catalog to meet system requirements.

Heating system load	1450 sq. ft.
Boiler pressure	3 psig

PIPING ILLUSTRATIONS FOR ONE PIPE AND TWO PIPE STEAM SYSTEMS

Figs. 6-21, which follow, are illustrations made to supply additional information required for the design of one pipe and two pipe steam heating systems.

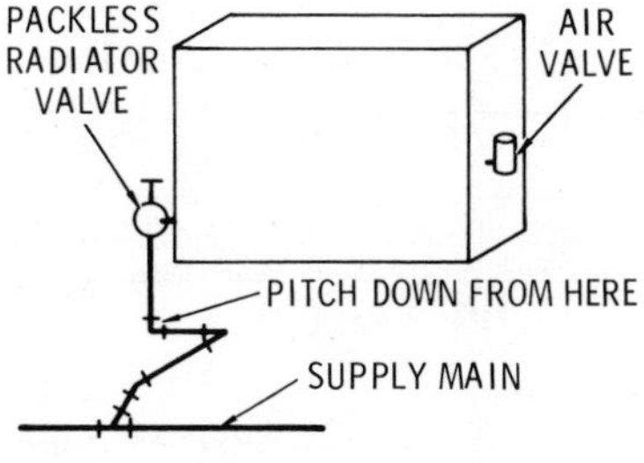
PACKLESS
RADIATOR
VALVE
AIR
VALVE
PITCH DOWN FROM HERE
SUPPLY MAIN

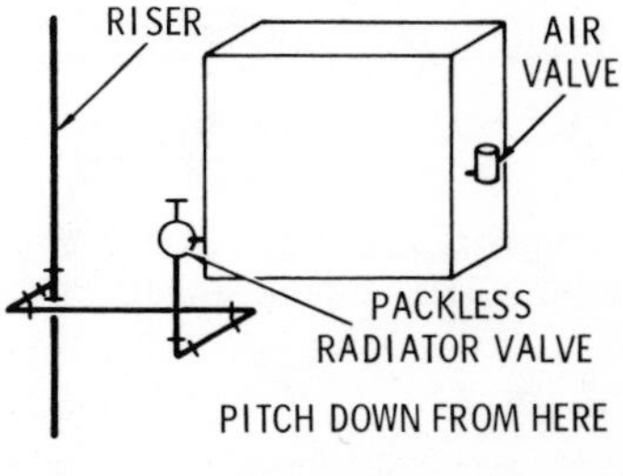
RISER
AIR
VALVE
PACKLESS
RADIATOR VALVE
PITCH DOWN FROM HERE

Courtesy *Hoffman Specialty ITT*

(A) Correct method of making up-feed connection to radiator from one pipe steam system.

(B) Correct method of connecting radiator to up-feed or down-feed riser from one pipe steam system.

Fig. 6. Radiator connections for one pipe steam systems.

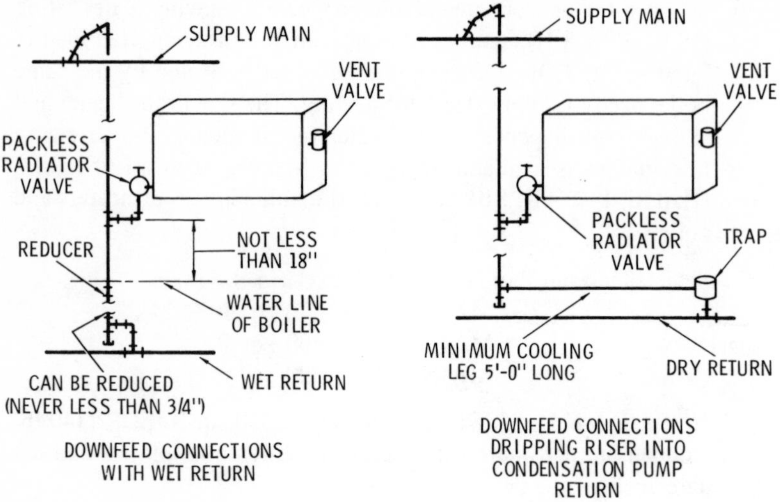
SUPPLY MAIN
VENT VALVE
PACKLESS RADIATOR VALVE
REDUCER
NOT LESS THAN 18"
WATER LINE OF BOILER
CAN BE REDUCED (NEVER LESS THAN 3/4")
WET RETURN
DOWNFEED CONNECTIONS WITH WET RETURN
SUPPLY MAIN
VENT VALVE
PACKLESS RADIATOR VALVE
TRAP
MINIMUM COOLING LEG 5'-0" LONG
DRY RETURN
DOWNFEED CONNECTIONS DRIPPING RISER INTO CONDENSATION PUMP RETURN

Courtesy *Hoffman Specialty ITT*

(A) Correct method of making down-feed connection to radiator and to wet return.

(B) Correct method of making down-feed connection to radiator and to dry return.

Fig. 7. Radiator connections for one pipe systems.

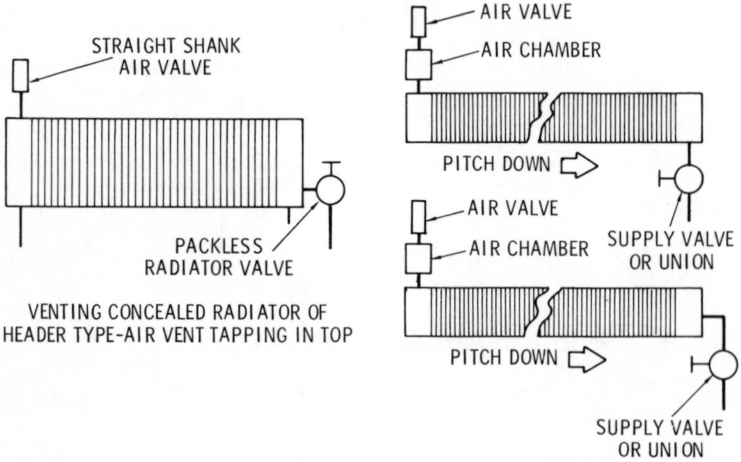
STRAIGHT SHANK AIR VALVE
PACKLESS RADIATOR VALVE
VENTING CONCEALED RADIATOR OF HEADER TYPE-AIR VENT TAPPING IN TOP
AIR VALVE
AIR CHAMBER
PITCH DOWN
SUPPLY VALVE OR UNION
AIR VALVE
AIR CHAMBER
PITCH DOWN
SUPPLY VALVE OR UNION

Courtesy *Hoffman Specialty ITT*

Fig. 8. Three methods of connecting radiation to suit varying job conditions (one pipe system).

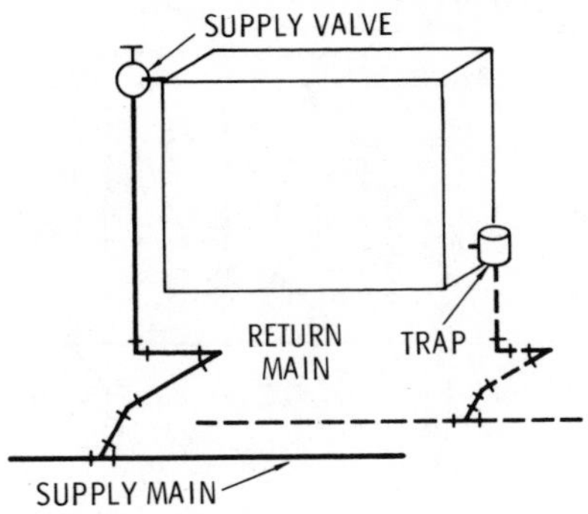
SUPPLY VALVE
RETURN MAIN
TRAP
SUPPLY MAIN

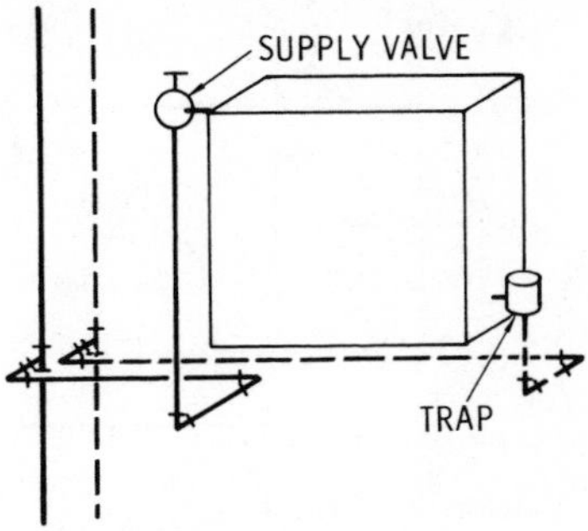
SUPPLY VALVE
TRAP

Courtesy *Hoffman Specialty ITT*

(A) The proper method of making up-feed connections to a radiator.

(B) The proper method of making radiator connections taken from up-feed or down-feed risers.

Fig. 9. Radiator connections for two pipe systems.

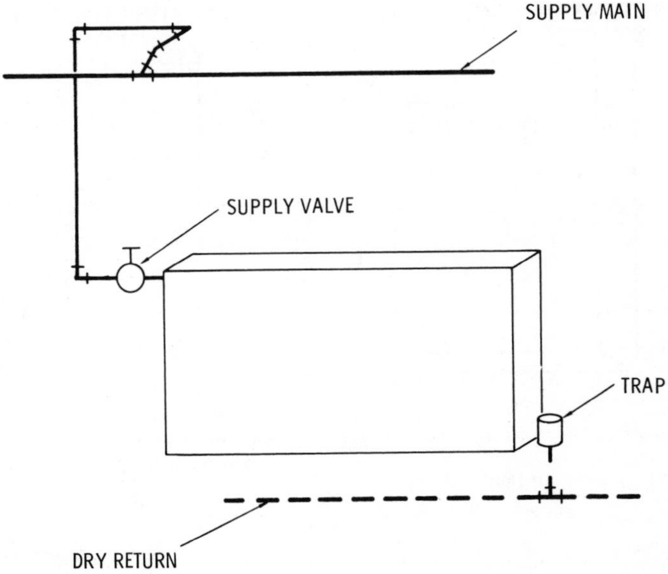
SUPPLY MAIN
SUPPLY VALVE
TRAP
DRY RETURN

Courtesy *Hoffman Specialty ITT*

Fig. 10. The proper method for connecting a radiator to an overhead supply main (two pipe system).

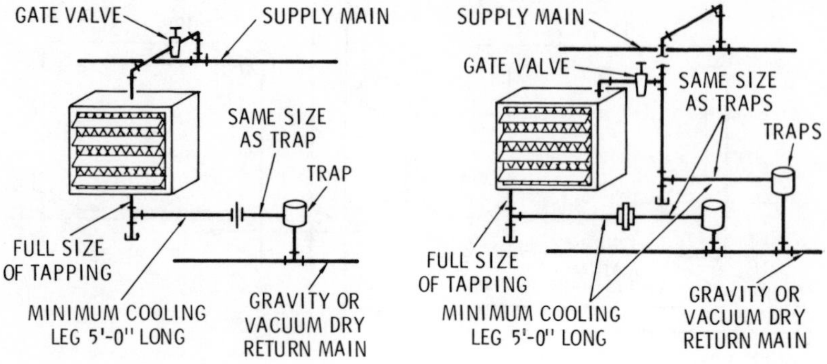
GATE VALVE
SUPPLY MAIN
SAME SIZE AS TRAP
TRAP
FULL SIZE OF TAPPING
MINIMUM COOLING LEG 5'-0" LONG
GRAVITY OR VACUUM DRY RETURN MAIN
SUPPLY MAIN
GATE VALVE
SAME SIZE AS TRAPS
TRAPS
FULL SIZE OF TAPPING
MINIMUM COOLING LEG 5'-0" LONG
GRAVITY OR VACUUM DRY RETURN MAIN

Courtesy *Hoffman Specialty ITT*

(A) Unit heater connections for a two pipe gravity or vacuum system.

(B) Unit heater connections for a two pipe gravity or vacuum system with the supply branch dripped through a trap.

Fig. 11. Unit heater connections for a two pipe system.

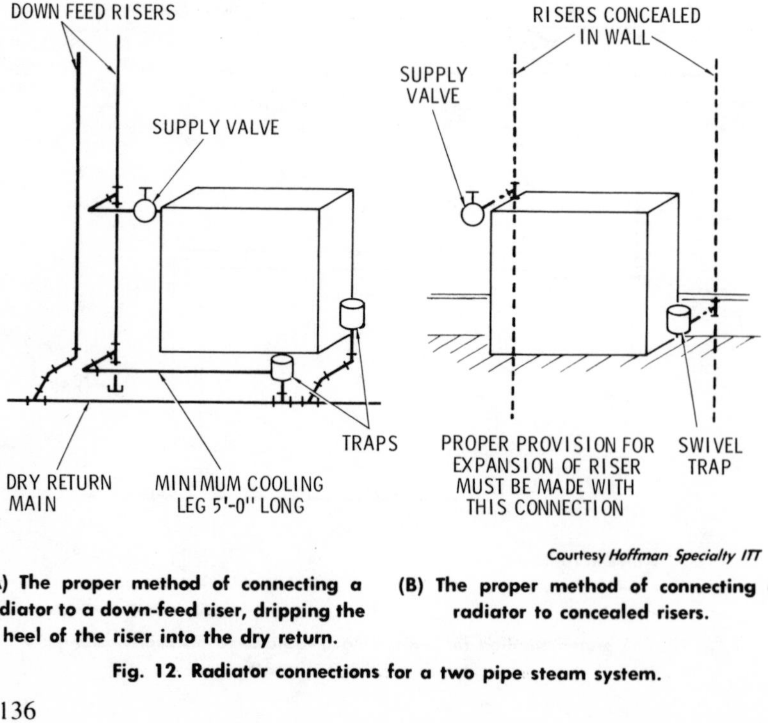
DOWN FEED RISERS
SUPPLY VALVE
TRAPS
DRY RETURN MAIN
MINIMUM COOLING LEG 5'-0" LONG
RISERS CONCEALED IN WALL
SUPPLY VALVE
PROPER PROVISION FOR EXPANSION OF RISER MUST BE MADE WITH THIS CONNECTION
SWIVEL TRAP

Courtesy *Hoffman Specialty ITT*

(A) The proper method of connecting a radiator to a down-feed riser, dripping the heel of the riser into the dry return.

(B) The proper method of connecting a radiator to concealed risers.

Fig. 12. Radiator connections for a two pipe steam system.

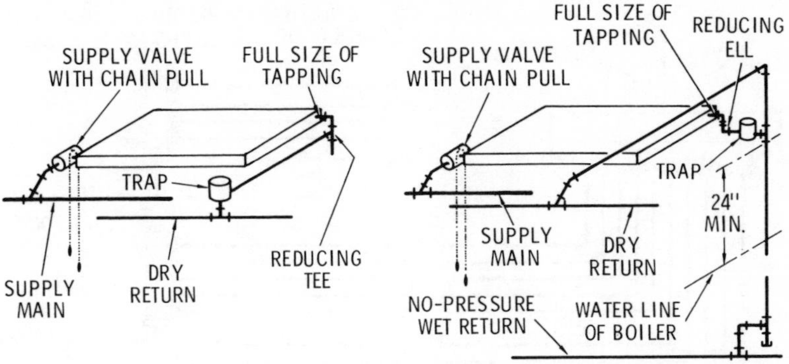
SUPPLY VALVE
WITH CHAIN PULL
FULL SIZE OF
TAPPING
TRAP
SUPPLY
MAIN
DRY
RETURN
REDUCING
TEE
FULL SIZE OF
TAPPING
REDUCING
ELL
SUPPLY VALVE
WITH CHAIN PULL
TRAP
24"
MIN.
SUPPLY
MAIN
DRY
RETURN
NO-PRESSURE
WET RETURN
WATER LINE
OF BOILER

Courtesy *Hoffman Specialty ITT*

(A) The proper method of connecting a ceiling radiator located above the supply and return piping.

(B) The proper method of connecting a ceiling radiator with the return bled into the wet return.

Fig. 13. Radiator connections for a two pipe steam system.

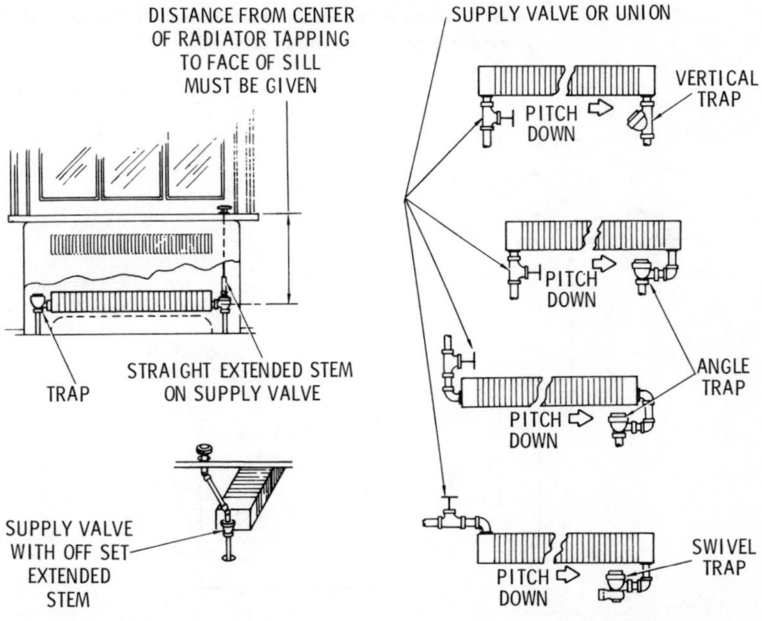
DISTANCE FROM CENTER
OF RADIATOR TAPPING
TO FACE OF SILL
MUST BE GIVEN
TRAP
STRAIGHT EXTENDED STEM
ON SUPPLY VALVE
SUPPLY VALVE
WITH OFF SET
EXTENDED
STEM
SUPPLY VALVE OR UNION
VERTICAL
TRAP
PITCH
DOWN
PITCH
DOWN
ANGLE
TRAP
PITCH
DOWN
PITCH
DOWN
SWIVEL
TRAP

Courtesy *Hoffman Specialty ITT*

Fig. 14. Various methods of connecting convertors to supply and return piping (two pipe systems).

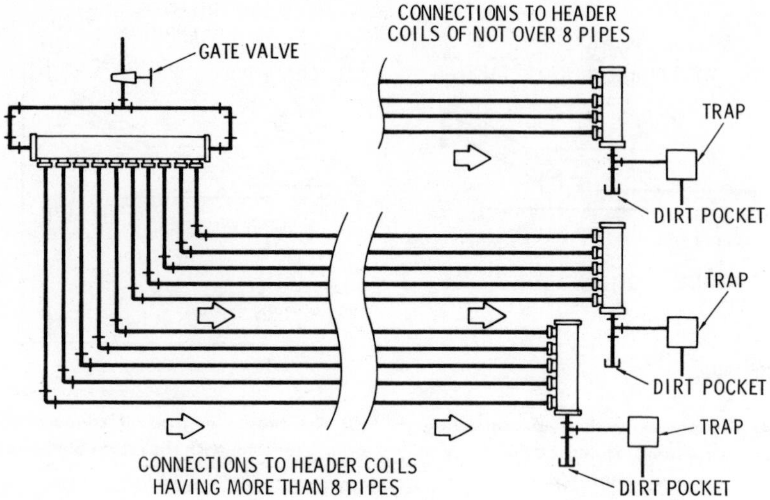
CONNECTIONS TO HEADER
COILS OF NOT OVER 8 PIPES
GATE VALVE
TRAP
DIRT POCKET
TRAP
DIRT POCKET
TRAP
DIRT POCKET
CONNECTIONS TO HEADER COILS
HAVING MORE THAN 8 PIPES

Courtesy *Hoffman Specialty ITT*

Fig. 15. The correct method of connecting exposed steam coils to supply and return piping (two pipe systems).

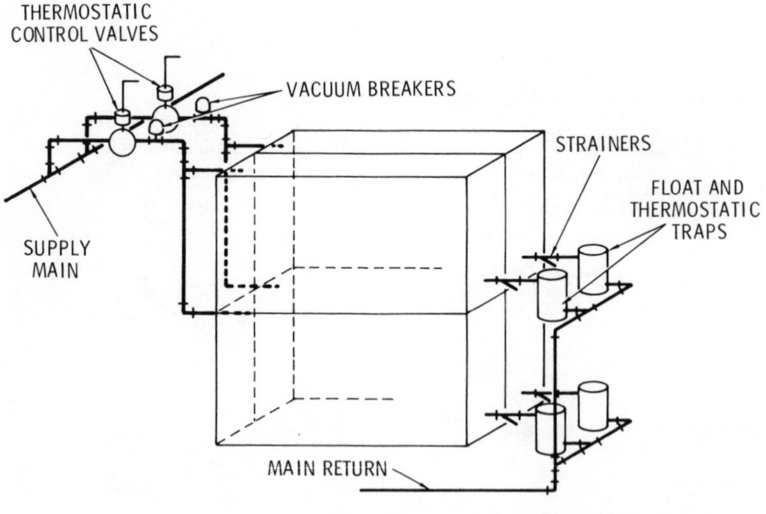
THERMOSTATIC
CONTROL VALVES
VACUUM BREAKERS
STRAINERS
FLOAT AND
THERMOSTATIC
TRAPS
SUPPLY
MAIN
MAIN RETURN

Courtesy *Hoffman Specialty ITT*

Fig. 16. The correct method of connecting blast coils controlled by thermostatic supply valves to supply and return piping.

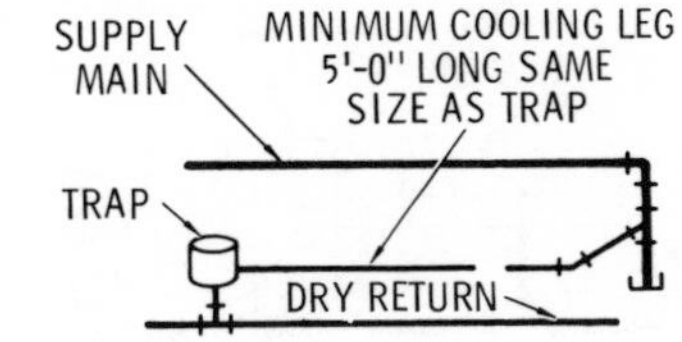
SUPPLY
MAIN
MINIMUM COOLING LEG
5'-0" LONG SAME
SIZE AS TRAP
TRAP
DRY RETURN

(A) Correct method of dripping end of supply main into the dry return main.

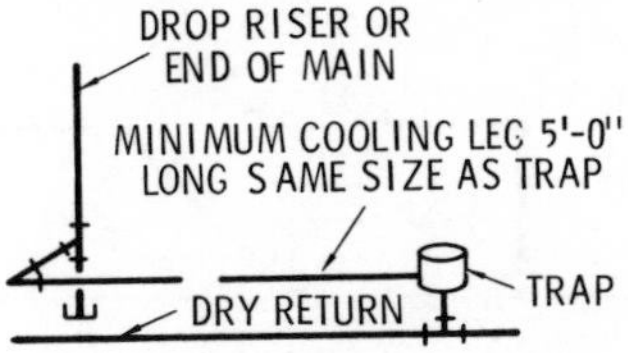
DROP RISER OR
END OF MAIN
MINIMUM COOLING LEG 5'-0"
LONG SAME SIZE AS TRAP
TRAP
DRY RETURN

(B) Correct method of dripping drop riser or end of main into return.

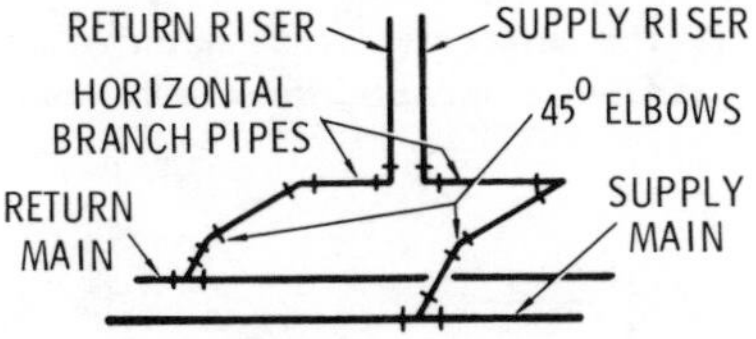
RETURN RISER
SUPPLY RISER
HORIZONTAL
BRANCH PIPES
45° ELBOWS
RETURN
MAIN
SUPPLY
MAIN

(C) Correct method to connect up-feed risers.

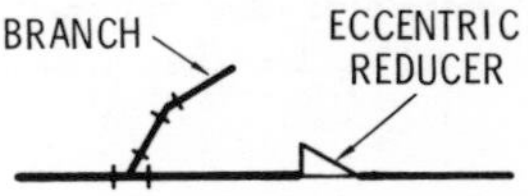
BRANCH
ECCENTRIC
REDUCER

(D) Correct method for reducing size of main.

Courtesy *Hoffman Specialty ITT*

Fig. 17. Trap installations for two pipe systems.

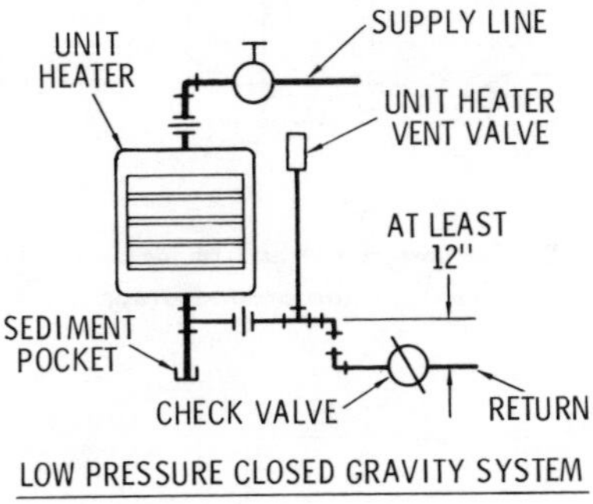
UNIT
HEATER
SUPPLY LINE
UNIT HEATER
VENT VALVE
AT LEAST
12"
SEDIMENT
POCKET
CHECK VALVE
RETURN
LOW PRESSURE CLOSED GRAVITY SYSTEM

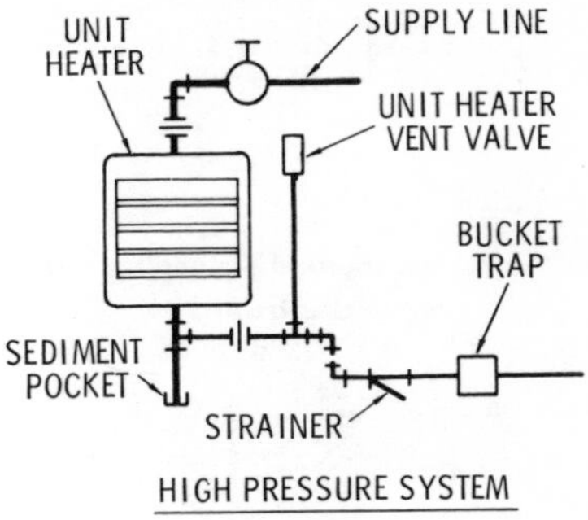
UNIT
HEATER
SUPPLY LINE
UNIT HEATER
VENT VALVE
BUCKET
TRAP
SEDIMENT
POCKET
STRAINER
HIGH PRESSURE SYSTEM

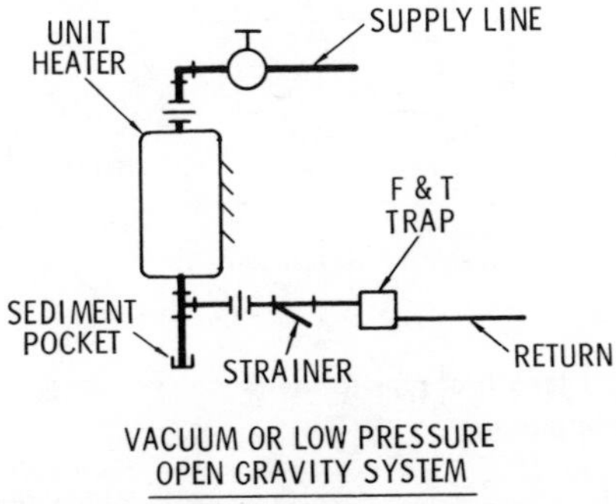
UNIT
HEATER
SUPPLY LINE
F & T
TRAP
SEDIMENT
POCKET
STRAINER
RETURN
VACUUM OR LOW PRESSURE
OPEN GRAVITY SYSTEM

Courtesy *Hoffman Specialty ITT*

Fig. 18. The correct methods of connecting unit heaters to low pressure closed systems, low pressure open systems and high pressure systems.

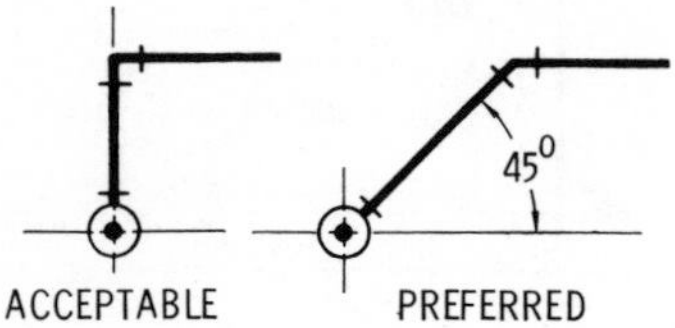
45°
ACCEPTABLE
PREFERRED

(A) The acceptable and the preferred methods of taking branches from mains.

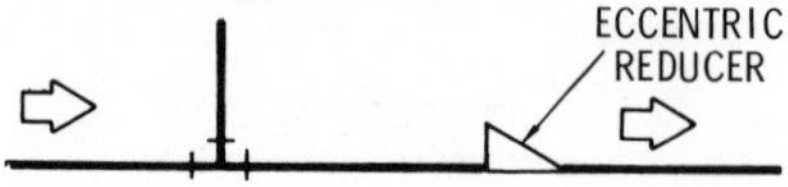
ECCENTRIC
REDUCER

(B) The correct method, using an eccentric reducer, of reducing the size of mains.

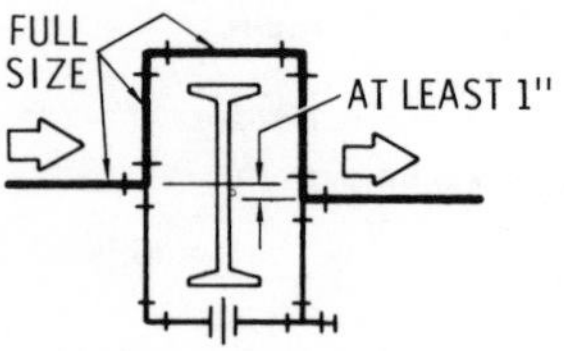
FULL
SIZE
AT LEAST 1"

(C) The correct method of looping a main around a beam.

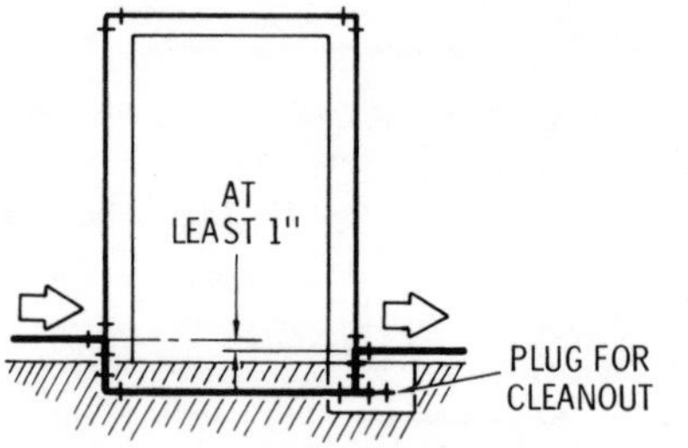
AT
LEAST 1"
PLUG FOR
CLEANOUT

(D) The correct method of looping a dry return main around a door.

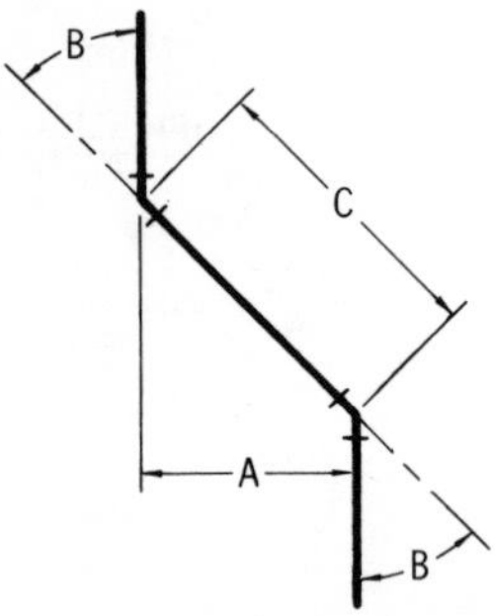
B
C
A
B

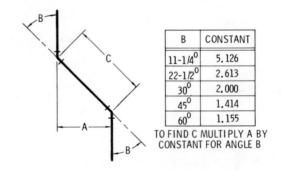

B	CONSTANT
11-1/4°	5.126
22-1/2°	2.613
30°	2.000
45°	1.414
60°	1.155

TO FIND C MULTIPLY A BY CONSTANT FOR ANGLE B

(E) How to use constants to determine length of pipe in break when making offsets in piping.

Courtesy *Hoffman Specialty ITT*

Fig. 19. The correct methods of solving common piping problems.

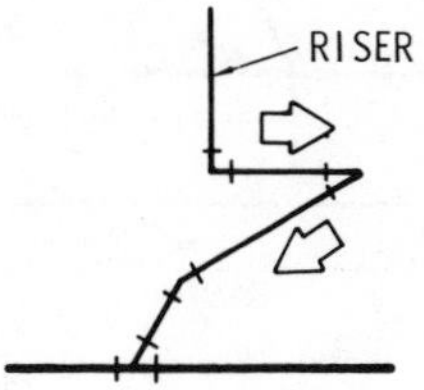
RISER

(A) An up-feed branch connection taken from main at a 45° angle.

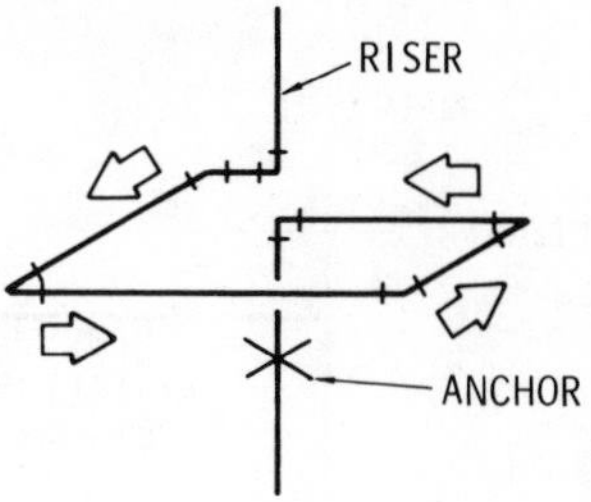
RISER
ANCHOR

(B) An expansion joint in a riser, made up of pipe.

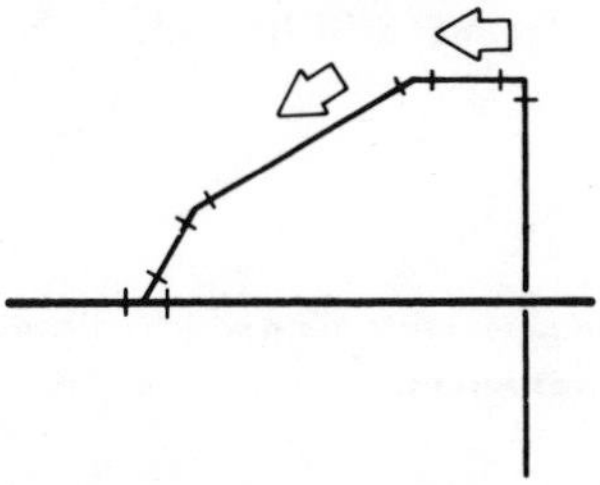

(C) A drop riser branch taken from the top of a main at a 45° angle.

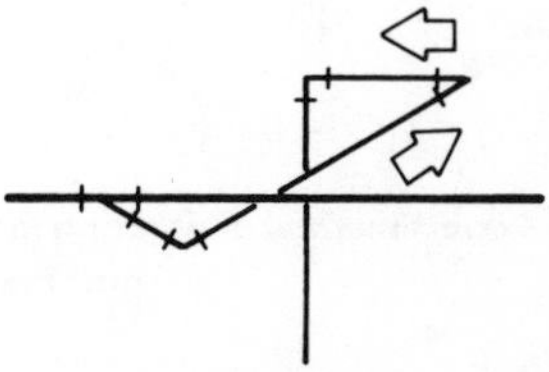

(D) A drop riser branch taken from the bottom of a main.

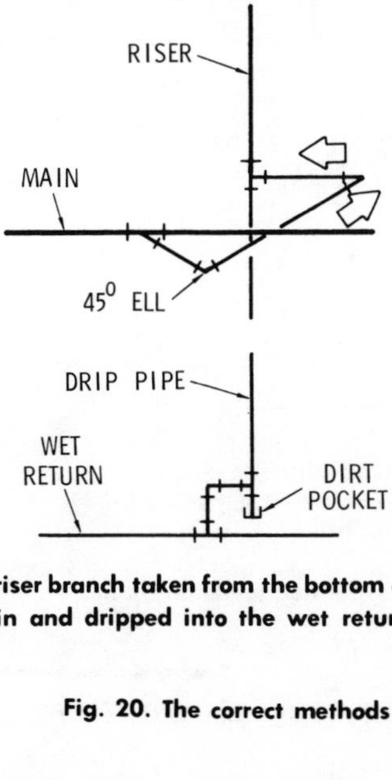
RISER
MAIN
45^0 ELL
DRIP PIPE
WET RETURN
DIRT POCKET

(E) A riser branch taken from the bottom of a main and dripped into the wet return.

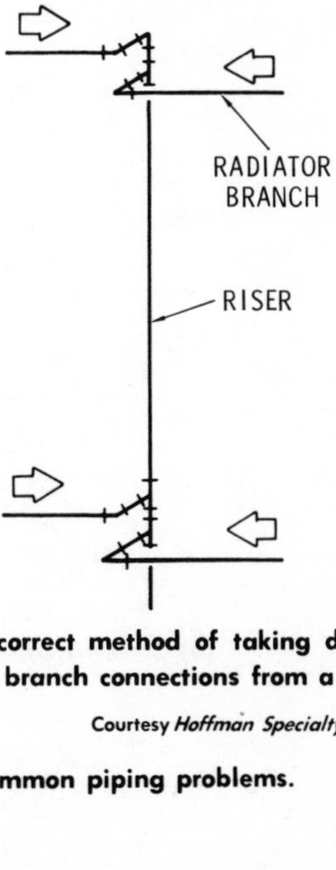
RADIATOR BRANCH
RISER

(F) The correct method of taking double radiator branch connections from a riser.

Courtesy *Hoffman Specialty ITT*

Fig. 20. The correct methods of solving common piping problems.

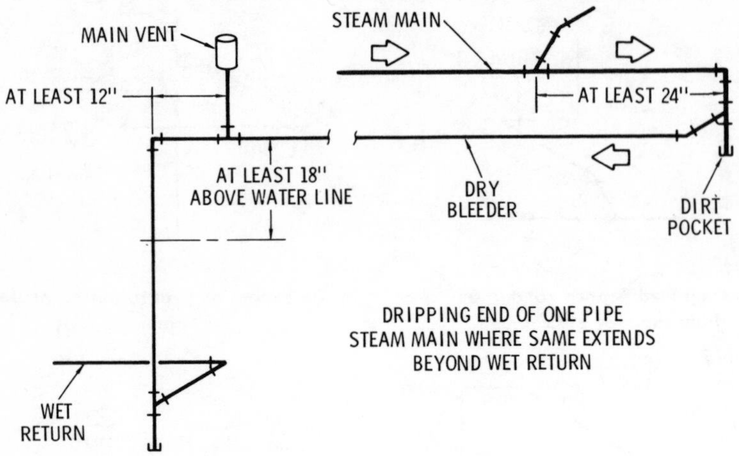
MAIN VENT
STEAM MAIN
AT LEAST 12"
AT LEAST 24"
AT LEAST 18"
ABOVE WATER LINE
DRY
BLEEDER
DIRT
POCKET
DRIPPING END OF ONE PIPE
STEAM MAIN WHERE SAME EXTENDS
BEYOND WET RETURN
WET
RETURN

(A) Correct method of dripping the end of a one pipe steam main which extends past the end of the wet return.

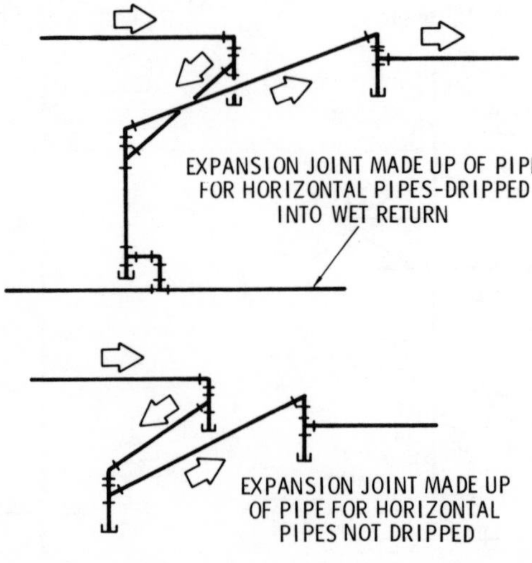
EXPANSION JOINT MADE UP OF PIPE
FOR HORIZONTAL PIPES-DRIPPED
INTO WET RETURN
EXPANSION JOINT MADE UP
OF PIPE FOR HORIZONTAL
PIPES NOT DRIPPED

(B) Correct method of making expansion joints of piping.

Courtesy *Hoffman Specialty ITT*

Fig. 21. The correct methods of solving common piping problems.

ENGINEERING DATA AND TECHNICAL INFORMATION

This section contains necessary engineering technical information used in the design, installation, maintenance or revamping of low pressure steam heating systems. The tables, charts, and examples, including properties of saturated steam, steam capacities of ASTM Schedule 40 (S) pipe and return pipe capacities used in one pipe and two pipe steam heating systems. Also included is data for determining pipe friction allowance, steam velocity, system pressure drops and similar information.

Velocity of Steam

To find the approximate velocity of low pressure steam (feet per second) through a pipe, multiply the condensation in pounds per hour by the volume of steam in cubic feet per pound corresponding to the

Table 1. Properties of Saturated Steam (Approx.)

Absolute Pressure (INCHES OF MERCURY)	Gage Reading at Sea Level (VACUUM—INCHES OF MERCURY)	Temp. °F.	Heat in Water B.T.U. per Lb.	Latent Heat in Steam (Vaporiza-tion) B.T.U. per Lb.	Volume of 1 Lb. Steam Cu. Ft.	Wgt. of Water Lbs. per Cu. Ft.
0.18	29.7	32	0.0	1076	3306	62.4
0.50	29.4	59	27.0	1061	1248	62.3
1.0	28.9	79	47.0	1049	653	62.2
2.0	28	101	69	1037	341	62.0
4.0	26	125	93	1023	179	61.7
6.0	24	141	109	1014	120	61.4
8.0	22	152	120	1007	93	61.1
10.0	20	161	129	1002	75	60.9
12.0	18	169	137	997	63	60.8
14.0	16	176	144	993	55	60.6
16.0	14	182	150	989	48	60.5
18.0	12	187	155	986	43	60.4
20.0	10	192	160	983	39	60.3
22.0	8	197	165	980	36	60.2
24.0	6	201	169	977	33	60.1
26.0	4	205	173	975	31	60.0
28.0	2	209	177	972	29	59.9
29.0	1	210	178	971	28	59.9
30.0	0	212	180	970	27	59.8

Courtesy *Hoffman Specialty ITT*

Table 1. Properties of Saturated Steam (Approx.) (Contd.)

Absolute Pressure POUNDS PER SQ. INCH	Gage Reading at Sea Level PRESSURE—POUNDS PER SQ. INCH	Temp. °F.	Heat in Water B.T.U. per Lb.	Latent Heat in Steam (Vaporization) B.T.U. per Lb.	Volume of 1 Lb. Steam Cu. Ft.	Wgt. of Water Lbs. per Cu. Ft.
14.7	0	212	180	970	27	59.8
15.7	1	216	184	968	25	59.8
16.7	2	219	187	966	24	59.7
17.7	3	222	190	964	22	59.6
18.7	4	225	193	962	21	59.5
19.7	5	227	195	960	20	59.4
20.7	6	230	198	958	19	59.4
21.7	7	232	200	957	19	59.3
22.7	8	235	203	955	18	59.2
23.7	9	237	205	954	17	59.2
25	10	240	208	952	16	59.2
30	15	250	219	945	14	58.8
35	20	259	228	939	12	58.5
40	25	267	236	934	10	58.3
45	30	274	243	929	9	58.1
50	35	281	250	924	8	57.9
55	40	287	256	920	8	57.7
60	45	293	262	915	7	57.5
65	50	298	268	912	7	57.4
70	55	303	273	908	6	57.2
75	60	308	277	905	6	57.0
85	70	316	286	898	5	56.8
95	80	324	294	892	5	56.5
105	90	332	302	886	4	56.3
115	100	338	309	881	4	56.0
140	125	353	325	868	3	55.5

steam pressure. Divide this result by 25 times the internal area of the pipe.

The pipe area is found in the Table "Standard Pipe Dimensions" and the volume per cubic foot is found in the Table "Properties of Saturated Steam."

Example. What is the velocity of steam at 5 lbs. per sq. inch flowing through a 2″ pipe at a rate to produce 175 lbs. of condensate per hour?

1 pound of steam at 5 lbs. pressure = 20 cu. ft.

Internal area of 2″ pipe = 3.36

The velocity of steam is:

$$\frac{175 \times 20}{25 \times 3.36} = \frac{3500}{84} = 41.7 \text{ ft. per sec.}$$

Table 3 gives the capacity of ASTM Schedule 40 (S) pipe expressed in square feet EDR. The values are obtained from charts published by the American Society of Heating, Refrigerating and Air Conditioning Engineers' 1967 GUIDE.

Where condensate must flow counter to steam flow, the governing factor is the velocity which will not interfere with the condensate flow. The capacity limit for horizontal pipe at various pitches is given in Table 4.

Table 2. Relations of Altitude, Pressure, and Boiling Point

Altitude Feet	ATMOSPHERIC PRESSURE ABSOLUTE		BOILING POINT OF WATER °F. (GAGE PRESSURE PSI)				
	Inches of Mercury (Barometer)	Lbs. per Sq. in.	0	1	5	10	15
−500	30.46	14.96	212.8	216.1	227.7	239.9	250.2
−100	30.01	14.74	212.3	215.5	227.2	239.4	249.9
Sea Level	29.90	14.69	212.0	215.3	227.0	239.3	249.7
500	29.35	14.42	211.0	214.4	226.3	238.7	249.2
1000	28.82	14.16	210.1	213.5	225.5	238.1	248.6
1500	28.30	13.90	209.4	212.7	225.0	237.6	248.2
2000	27.78	13.65	208.2	211.7	224.1	236.8	247.7
2500	27.27	13.40	207.3	210.9	223.4	236.3	247.2
3000	26.77	13.15	206.4	210.1	222.7	235.7	246.7
3500	26.29	12.91	205.5	209.2	222.1	235.1	246.2
4000	25.81	12.68	204.7	208.4	221.4	234.6	245.7
4500	25.34	12.45	203.7	207.5	220.7	234.0	245.2
5000	24.88	12.22	202.7	206.8	220.1	233.4	244.7
6000	23.98	11.78	200.9	205.0	218.7	232.4	243.8
7000	23.11	11.35	199.1	203.3	217.3	231.3	242.9
8000	22.28	10.94	197.4	201.6	216.1	230.3	242.0
9000	21.47	10.55	195.7	200.0	214.8	229.3	241.3
10000	20.70	10.17	194.0	198.4	213.5	228.3	240.4
11000	19.95	9.80	192.2	196.8	212.3	227.3	239.6
12000	19.23	9.45	190.6	195.2	211.1	226.3	238.7
13000	18.53	9.10	188.7	193.6	209.9	225.4	237.9
14000	17.86	8.77	187.2	192.3	208.8	224.5	237.2
15000	17.22	8.46	185.4	190.6	207.6	223.6	236.4

Courtesy *Hoffman Specialty ITT*

Table 3. Steam Capacity[a] of ASTM[b] Schedule 4D(S) Pipe at Internal Pressures of 3.5 and 12PS1G[c]

FLOW EXPRESSED IN SQ. FT. EDR

NOMINAL PIPE SIZE INCHES	PRESSURE DROP—PSI PER 100 FT. IN LENGTH													
	1/16 PSI (1 oz.)		1/8 PSI (2 oz.)		1/4 PSI (4 oz.)		1/2 PSI (8 oz.)		3/4 PSI (12 oz.)		1 PSI		2 PSI	
	SAT. PR. 3.5	PSIG 12	SAT. PR. 3.5	PSIG 12	SAT. PR. 3.5	PSIG 12	SAT. PR. 3.5	PSIG 12	SAT. PR. 3.5	PSIG 12	SAT. PR. 3.5	PSIG 12	SAT. PR. 3.5	PSIG 12
1	2	3	4	5	6	7	8	9	10	11	12	13	14	15
3/4"	36	44	56	64	80	96	116	140	144	172	168	200	240	292
1"	68	84	104	124	148	184	216	264	272	328	324	380	456	548
1¼"	144	180	212	264	312	344	444	552	560	680	648	800	928	1120
1"½	224	280	326	400	480	588	696	840	872	1040	984	1216	1440	1720
2"	432	536	648	776	936	1140	1344	1640	1680	2040	1920	2360	2840	3400
2½"	696	860	1032	1240	1512	1840	2160	2640	2720	3280	3120	3800	4600	5480
3"	1272	1520	1860	2200	2640	3240	3840	4640	4760	5720	5520	6680	7800	9600
3½"	1848	2200	2680	3200	3960	4872	5640	6800	6960	8400	8000	9680	11800	13800
4"	2560	3200	3800	4640	5640	6760	7920	9600	9800	12000	11520	13840	16800	19600
5"	4800	5720	6720	8400	9760	12000	14280	17000	17520	21000	20400	24400	30000	34400
6"	7680	9200	11280	13200	15840	19400	22800	28000	28800	34400	33600	40000	47600	56800
8"	15600	19200	22280	28000	32400	40000	45600	57200	58000	70800	66000	82000	96000	118000
10"	28800	35200	40800	50400	60000	72800	84000	104000	104800	128000	120000	148000	170800	208000
12"	45600	54800	66000	78000	93600	113600	13200	160000	164000	198000	192000	230000	271200	324000

a—Based on Moody Friction where flow of condensate does not inhibit the flow of steam.

b—American Society for Testing Materials Schedule. The number 40 refers to the ASTM Schedule. The letter (S) refers to the former designation of standard weight pipe.

c—The flow rates at 3.5 psig can be used to cover saturated pressures from 1 to 6 psig, and the rates at 12 psig can be used to cover saturated pressures from 8 to 16 psig with an error not exceeding 8%.

Table 4. Comparative Capacity of Steam Lines at Various Pitches for Steam and Condensate Flowing in Opposite Directions[a]

PITCH of Pipe	(PITCH OF PIPE IN INCHES PER 10 FT. / VELOCITY IN FT. PER SEC. / CAPACITY IN SQ. FT. EDR)															
	¼ Inch		½ Inch		1 Inch		1½ Inch		2 Inch		3 Inch		4 Inch		5 Inch	
PIPE SIZE INCHES	Capacity	Max. Vel.	Capacity	Max. Vel.	Capacity	Max. Vel.	Capacity	Max. Vel.	Capacity	Max. Vel.	Capacity	Max. Vel.	Capacity	Max. Vel.	Capacity	Max. Vel.
¾	12.8	8	16.4	11	22.8	13	25.6	14	28.4	16	33.2	17	38.6	22	42.0	22
1	27.2	9	36.0	12	46.8	15	51.2	17	59.2	19	69.2	22	76.8	24	82.0	25
1¼	47.2	11	63.6	14	79.6	17	98.4	20	108.0	22	125.2	25	133.6	26	152.4	31
1½	79.2	12	103.6	16	132.0	19	149.6	22	168.0	24	187.2	26	203.2	28	236.8	33
2	171.6	15	216.0	18	275.2	24	333.2	27	371.6	30	398.4	32	409.6	32	460.0	33

a—From research sponsored by ASHRAE

Courtesy *Hoffman Specialty ITT*

The capacity of a steam pipe depends on these factors:

1. The quantity of condensate in the pipe
2. The direction of condensate flow
3. The pressure drop in the pipe. The total pressure drop of a system should not exceed one-half of the supply pressure when steam and condensate are flowing in the same direction.

When steam and condensate flow in the same direction, only the pressure drop need be considered. When steam flows counter to condensate the velocities in Table 4 must not be exceeded.

Table 5 gives steam pipe capacities in square feet EDR when steam flow is counter to condensate flow in either one pipe or two pipe systems.

Table 5. Steam Pipe Capacities for Low Pressure Systems

(For Use on One Pipe Systems or Two Pipe Systems in which Condensate Flows Against the Steam Flow)

NOMINAL	CAPACITY IN SQUARE FEET EDR				
	TWO PIPE SYSTEMS		ONE PIPE SYSTEMS		
	Condensate Flowing Against Steam		Supply Risers Up-feed	Radiator Valves & Vertical Connections	Radiator and Riser Horizontal Runouts
	VERTICAL	HORIZONTAL			
A	B	C[c]	D[b]	E	F[c]
3/4	32	28	24	....	28
1	56	56	44	28	28
1 1/4	124	108	80	64	64
1 1/2	192	168	152	92	64
2	388	362	288	168	92
2 1/2	636	528	464	....	168
3	1128	800	800	....	260
3 1/2	1548	1152	1144	....	476
4	2044	1700	1520	....	744
5	4200	3152	....	....	1112
6	7200	5600	....	....	2180
8	15000	12000	....	....	....
10	28000	22800	....	....	....
12	46000	38000	....	....	....

NOTES:

a—Do not use Col. B for pressure drops of less than 1 oz./100 ft. of equivalent length of run. Use Table 3 Instead.

b—Do not use Col. D for pressure drops less than 2/3 oz./100 ft. of equivalent length of run except for pipe size 3″ and over. Use Table 3 instead.

c—Pitch of horizontal runouts to risers and radiators should not be less than 1/2 inch per ft. Where this pitch cannot be obtained, for runouts 8 ft. in length or over, increase one pipe size larger than shown in this table.

Courtesy *Hoffman Specialty ITT*

In all the pipe sizing tables given in this section, the *length of run* is defined as the *equivalent length of run.* The equivalent length consists of the measured length plus the allowance for fittings and valves which add resistance to the regular friction of the straight run of pipe. Table 6 gives the value in feet to be added to the measured length of pipe.

It is not possible at the beginning of the pipe sizing process to know the equivalent length of pipe so it is necessary to make an assumption. The accepted rule is to measure the actual length and double it to get the total equivalent length. After the pipe sizing is done for the proper pressure drop per 100 ft. of length based on the assumption, then a reexamination can be made to determine if any changes are required which will change the pipe sizing.

Table 6. Frictional Allowance for Fittings in Feet of Pipe (To Be Added to Actual Length of Run) Steam

SIZE OF PIPE INCHES	LENGTH IN FEET TO BE ADDED IN RUN				
	STANDARD ELBOW	SIDE OUTLET TEE	GATE VALVE	GLOBE VALVE	ANGLE VALVE
½″	1.3	3	0.3	14	7
¾″	1.8	4	0.4	18	10
1″	2.2	5	0.5	23	12
1¼″	3.0	6	0.6	29	15
1½″	3.5	7	0.8	34	18
2″	4.3	8	1.0	46	22
2½″	5.0	11	1.1	54	27
3″	6.5	13	1.4	66	34
3½″	8	15	1.6	80	40
4″	9	18	1.9	92	45
5″	11	22	2.2	112	56
6″	13	27	2.8	136	67
8″	17	35	3.7	180	92
10″	21	45	4.6	230	112

EXAMPLE:

MEASURED LENGTH = 132.0 FT.
4″ GATE VALVE = 1.9 FT.
4′4″ ELBOWS = 36.0 FT.
EQUIV. LENGTH 169.9 FT.

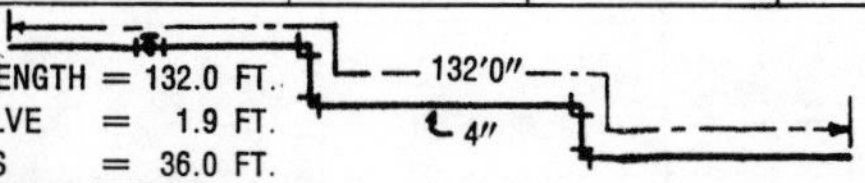
132′0″
℄ 4″

Courtesy *Hoffman Specialty ITT*

Table 7. Return Main and Riser Capacities for Low Pressure Systems—Sq. Ft. EDR

	PIPE SIZE INCHES	1/32 psi or 1/2 oz. Drop per 100 ft.			1/24 psi or 2/3 oz. Drop per 100 ft.			1/16 psi or 1 oz. Drop per 100 ft.			1/8 psi or 2 oz. Drop per 100 ft.			1/4 psi or 4 oz. Drop per 100 ft.			1/2 psi or 8 oz. Drop per 100 ft.		
		WET	DRY	VAC.	WET	DRY	VAC.	WET	DRY	VAC.	WET	DRY	VAC.	WET	DRY	VAC.	WET	DRY	VAC.
	G	H	I	J	K	L	M	N	O	P	Q	R	S	T	U	V	W	X	Y
MAINS	3/4	......	......	......	......	......	168	......	......	400	......	......	568	......	......	800	......	......	1132
	1	500	248	......	580	284	572	700	320	700	1000	412	996	1400	460	1400	......	......	1976
	1 1/4	852	520	......	992	596	976	1200	672	1200	1700	868	1707	2400	964	2400	......	......	3392
	1 1/2	1352	824	......	1572	944	1552	1900	1060	1900	2700	1360	2696	3800	1512	3800	......	......	5360
	2	2800	1800	......	3240	2140	3260	4000	2300	4000	5600	2960	5680	8000	3300	8000	......	......	11320
	2 1/2	4720	3040	......	6320	3472	5440	6720	3800	6720	9400	4920	9520	13400	5440	13400	......	......	18920
	3	7520	5840	......	8520	6240	8720	10720	7000	10720	15000	9000	15200	21400	10000	21400	......	......	30240
	3 1/2	11000	7880	......	13200	8800	13000	16000	10000	16000	22000	12920	22720	32000	14320	32000	......	......	45200
	4	15520	11720	......	18320	13400	18000	22000	15000	22000	31000	19320	31240	44000	21520	44000	......	......	62000
	5	......	......	......	......	......	31520	......	......	38720	......	......	54800	......	......	77600	......	......	109200
	6	......	......	......	......	......	50400	......	......	62000	......	......	88000	......	......	124000	......	......	175200
RISERS	3/4	......	192	......	......	192	572	......	192	700	......	192	996	......	192	1400	......	......	1976
	1	......	452	......	......	452	976	......	452	1200	......	452	1704	......	452	2400	......	......	3392
	1 1/4	......	992	......	......	992	1552	......	992	1900	......	992	2696	......	992	3800	......	......	5360
	1 1/2	......	3000	......	......	1500	3260	......	1500	4000	......	1500	5680	......	1500	8000	......	......	11320
	2	......	1500	......	......	3000	5400	......	3000	6720	......	3000	9520	......	3000	13400	......	......	18920
	2 1/2	......	......	......	......	......	8720	......	......	10720	......	......	15200	......	......	21400	......	......	30240
	3	......	......	......	......	......	13000	......	......	16000	......	......	22720	......	......	32000	......	......	45200
	3 1/2	......	......	......	......	......	17920	......	......	22000	......	......	31240	......	......	44000	......	......	62000
	4	......	......	......	......	......	31520	......	......	38720	......	......	54800	......	......	77600	......	......	109200
	5	......	......	......	......	......	50400	......	......	62000	......	......	88000	......	......	124000	......	......	175200

Courtesy *Hoffman Specialty ITT*

Table 8. Horizontal Steam Supply Run-Outs to Radiator from Main or Riser and Radiator Supply Valve Sizes

	SIZE IN INCHES				
	CAPACITY EDR				
	0-25	26-75	76-150	151-200	201-400
HORIZONTAL RUN-OUT	3/4	1	1 1/4	1 1/2	2
VERTICAL CONNECTION	1/2	3/4	1	1 1/4	1 1/2
SUPPLY VALVE	1/2	3/4	1	1 1/4	1 1/2

Table 9. Horizontal Return Run-Outs to Radiator from Main or Riser and Radiator Trap Size

	SIZE IN INCHES		
	CAPACITY EDR		
	0-200	201-400	401-700
HORIZONTAL RUN-OUT	3/4	3/4	1
VERTICAL CONNECTION	1/2	3/4	1
TRAP	1/2	3/4	1

Courtesy *Hoffman Specialty ITT*

Table 10. Equivalent Capacities of Pipes of Same Length—Steam

Size	NO. OF SMALL PIPES EQUIVALENT TO ONE LARGE PIPE										
	½″	¾″	1″	1¼″	1½″	2″	2½″	3″	3½″	4″	5″	6″
½″	1.00	2.27	4.88	10.0	15.8	31.7	52.9	96.9	140	205	377	620
¾″		1.00	2.05	4.30	6.97	14.0	23.3	42.5	65	90	166	273
1″			1.00	2.25	3.45	6.82	11.4	20.9	30	44	81	133
1¼″				1.00	1.50	3.10	5.25	9.10	12	19	37	68
1½″					1.00	2.00	3.34	6.13	9	13	23	39
2″						1.00	1.67	3.06	4.5	6.5	11.9	19.6
2½″							1.00	1.82	2.70	3.87	7.12	11.7
3″								1.00	1.50	2.12	3.89	6.39
3½″									1.00	1.25	2.50	4.25
4″										1.00	1.84	3.02
5″											1.00	1.65
6″												1.00

This table may be used to find the number of smaller pipes equivalent in steam carrying capacity to one larger pipe. It may also be used to find the size of a larger pipe equivalent to several smaller ones. The pipes in either case must be of the same lengths.

EXAMPLE 1—Find the number of 1″ pipes each 50 ft. long equivalent to one 4″ pipe 50 ft. long.

SOLUTION 1—Follow down column headed 4″ to the point opposite 1″ in vertical column, and it is found that it will take 44 of the 1″ pipes in parallel to equal one 4″ pipe in steam carrying capacity.

EXAMPLE 2—Find the size of one pipe equivalent to four 2″ pipes in steam carrying capacity.

SOLUTION 2—Find 2″ in vertical column headed "Size" and follow across horizontally until closest number to 4 is found. The nearest to 4 is 4.5. Following this column up it is found that the size is 3½″. One 3½″ pipe is, therefore, equivalent in steam carrying capacity to approximately four 2″ pipes.

Courtesy *Hoffman Specialty ITT*

Table 11. Heat Losses from Covered Pipe

85 PERCENT MAGNESIA TYPE

BTU PER LINEAR FOOT PER °F TEMPERATURE DIFFERENCE (SURROUNDING AIR ASSUMED 75°F)

Pipe Size	Insulation Thickness, Inches	MAX. TEMP. OF PIPE SURFACE °F.				
		125	175	225	275	325
½	1	.145	.150	.157	.160	.162
¾	1	.165	.172	.177	.180	.182
1	1	.190	.195	.200	.203	.207
	1½	.160	.165	.167	.170	.175
1¼	1	.220	.225	.232	.237	.245
	1½	.182	.187	.193	.197	.200
1½	1	.240	.247	.255	.260	.265
	1½	.200	.205	.210	.215	.219
2	1	.282	.290	.297	.303	.307
	1½	.230	.235	.240	.243	.247
	2	.197	.200	.205	.210	.217
2½	1	.322	.330	.340	.345	.355
	1½	.260	.265	.270	.275	.280
	2	.220	.225	.230	.237	.242
3	1	.375	.385	.395	.405	.415
	1½	.300	.305	.312	.320	.325
	2	.253	.257	.263	.270	.277
3½	1	.419	.430	.440	.450	.460
	1½	.332	.339	.345	.352	.360
	2	.280	.285	.290	.295	.303
4	1	.460	.470	.480	.492	.503
	1½	.362	.370	.379	.385	.392
	2	.303	.308	.315	.320	.327
5	1	.545	.560	.572	.585	.600
	1½	.423	.435	.442	.450	.460
	2	.355	.360	.367	.375	.382
6	1	.630	.645	.662	.680	.693
	1½	.487	.500	.510	.520	.530
	2	.405	.415	.420	.430	.437
8	1	.790	.812	.835	.850	.870
	1½	.603	.620	.635	.645	.660
	2	.495	.507	.517	.527	.540

Courtesy *Hoffman Specialty ITT*

Table 12. Heat Losses from Horizontal Bare Steel Pipe Sizes 3/4″—6″

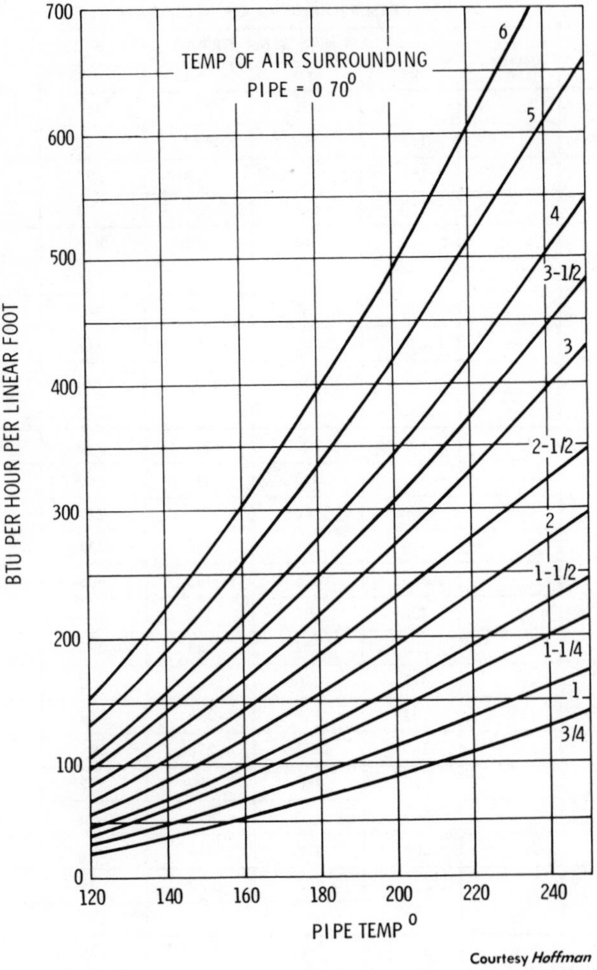
TEMP OF AIR SURROUNDING
PIPE = 0 70⁰
BTU PER HOUR PER LINEAR FOOT
PIPE TEMP ⁰
700
600
500
400
300
200
100
0
120
140
160
180
200
220
240
6
5
4
3-1/2
3
2-1/2
2
1-1/2
1-1/4
1
3/4

Courtesy *Hoffman Specialty ITT*

CHAPTER 5

General and Useful Steam Heating Information

Tables 1-51, which follow, should prove to supply valuable reference data. Although these tables were prepared primarily for steam heating systems, they also contain information related to other types of piping and heating systems.

This chapter also contains definitions of common heating terms and a troubleshooting section which should also be helpful.

DEFINITIONS OF HEATING TERMS

The definitions given in this section are only those applying to heating and particularly as used in this book. It is realized that some do not define the terms for all usages, but in the interest of clearance and space this sacrifice was made.

Absolute Humidity—The weight of water vapor in grains actually contained in one cubic foot of the mixture of air and moisture.

Absolute Pressure—The actual pressure above zero. It is the atmospheric pressure added to the gage pressure. It is expressed as a unit pressure such as lbs. per sq. in. also.

Absolute Temperature—The temperature of a substance measured above absolute zero. To express a temperature as absolute temperature add 460° to the reading of a Fahrenheit thermometer or 273° to the reading of a Centigrade one.

Absolute Zero—The temperature (−460°F. approx.) at which all molecular motion of a substance ceases, and at which the substance contains no heat.

Air—An elastic gas. It is a mechanical mixture of oxygen and nitrogen and slight traces of other gases. It may also contain moisture known as humidity. Dry air weighs 0.075 lbs. per cu. ft.

One Btu will raise the temperature of 55 cu. ft. of air one degree F.

Air expands or contracts approximately 1/490 of its volume for each degree rise or fall in temperature from 32°F.

Table 1. Abbreviations Used in Heating

Absolute	abs	Gallons per Minute	gpm
Alternating-Current	a-c	Gallons per Second	gps
Ampere	amp	Gram	g
Atmosphere	atm	Horsepower	hp
Average	avg	Horsepower-Hour	hp hr
Avoirdupois	avdp	Hour	hr
Barometer	bar.	Inch	in.
Boiling Point	bp	Inch-Pound	in.-lb
Brake Horsepower	bhp	Kilogram	kg
Brake Horsepower-Hour	bhp-hr	Kilowatt	kw
British Thermal Unit	Btu	Melting Point	mp
British Thermal Units per Hour	Btuh	Meter	m
		Miles per Hour	mph
Calorie	cal	Millimeter	mm
Centigram	cg	Minute	mm
Centimeter	cm	Ounce	oz
Cubic	cu	Pound	lb
Cubic Centimeter	cc	Pounds per Square Inch	psi
Cubic Foot	cu ft	Pounds per Square Inch, Gage	psig
Cubic Feet per Minute	cfm		
Cubic Feet per Second	cfs	Pounds per Square Inch, Absolute	psia
Degree	deg or °		
Degree, Centigrade	C	Revolutions per Minute	rpm
Degree, Fahrenheit	F	Revolutions per Second	rps
Diameter	diam	Second	sec
Direct-Current	d-c	Specific Gravity	sp gr
Feet per Minute	fpm	Specific Heat	sp ht
Feet per Second	fps	Square Foot	sq ft
Foot	ft	Square Inch	sq in.
Foot-Pound	ft-lb	Volt	v
Freezing Point	fp	Watt	w
Gallon	gal	Watt Hour	whr

Courtesy *Hoffman Specialty ITT*

Table 2. Inside Temperatures Usually Specified

BUILDING	ROOM	DEG. F.
SCHOOLS	Class Rooms	70-72
	Assembly Rooms	61-72
	Gymnasiums	55-65
	Toilets and Baths	70
	Wardrobe and Locker Rooms	65-68
	Kitchens	66
	Dining and Lunch Rooms	65-70
	Play Rooms	60-65
	Natatoriums	75
THEATRES	Seating Space	68-72
	Lounge Rooms	68-70
	Toilets	68
HOSPITALS	Private Rooms	70-72
	Private Rooms (Surgical)	70-80
	Operating Rooms	70-95
	Wards	68
	Kitchens and Laundries	66
	Toilets	68
	Bath Rooms	70-80
HOTELS	Bed Rooms and Baths	70
	Dining Rooms	70
	Kitchens and Laundries	66
	Ball Rooms	65-68
	Toilets and Service Rooms	68
HOMES		70-72
Stores		65-68
Public Buildings		68-72
Warm Air Baths		120
Steam Baths		110
Factories and Machine Shops		60-65
Foundries and Boiler Shops		50-60
Paint Shops		80

Courtesy *Hoffman Specialty ITT*

Table 3. Heating and Ventilating Symbols

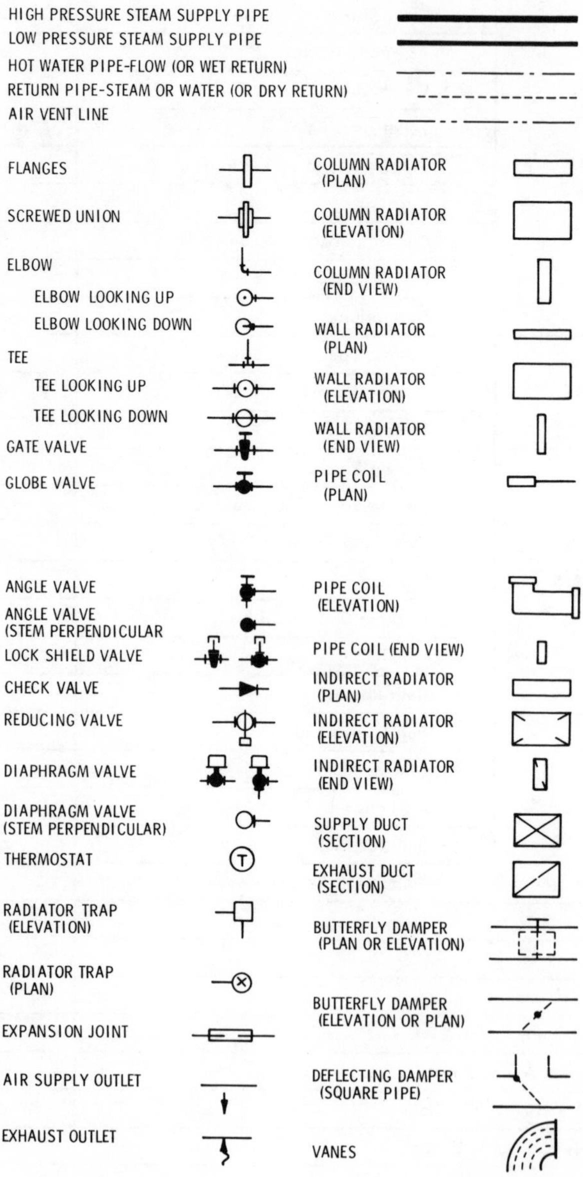
HIGH PRESSURE STEAM SUPPLY PIPE
LOW PRESSURE STEAM SUPPLY PIPE
HOT WATER PIPE-FLOW (OR WET RETURN)
RETURN PIPE-STEAM OR WATER (OR DRY RETURN)
AIR VENT LINE
FLANGES
SCREWED UNION
ELBOW
ELBOW LOOKING UP
ELBOW LOOKING DOWN
TEE
TEE LOOKING UP
TEE LOOKING DOWN
GATE VALVE
GLOBE VALVE
COLUMN RADIATOR (PLAN)
COLUMN RADIATOR (ELEVATION)
COLUMN RADIATOR (END VIEW)
WALL RADIATOR (PLAN)
WALL RADIATOR (ELEVATION)
WALL RADIATOR (END VIEW)
PIPE COIL (PLAN)
ANGLE VALVE
ANGLE VALVE (STEM PERPENDICULAR
LOCK SHIELD VALVE
CHECK VALVE
REDUCING VALVE
DIAPHRAGM VALVE
DIAPHRAGM VALVE (STEM PERPENDICULAR)
THERMOSTAT
T
RADIATOR TRAP (ELEVATION)
RADIATOR TRAP (PLAN)
EXPANSION JOINT
AIR SUPPLY OUTLET
EXHAUST OUTLET
PIPE COIL (ELEVATION)
PIPE COIL (END VIEW)
INDIRECT RADIATOR (PLAN)
INDIRECT RADIATOR (ELEVATION)
INDIRECT RADIATOR (END VIEW)
SUPPLY DUCT (SECTION)
EXHAUST DUCT (SECTION)
BUTTERFLY DAMPER (PLAN OR ELEVATION)
BUTTERFLY DAMPER (ELEVATION OR PLAN)
DEFLECTING DAMPER (SQUARE PIPE)
VANES

Courtesy *Hoffman Specialty ITT*

Table 4. Air Changes to be Provided for Under Average Conditions Exclusive of Air Required for Ventilation

KIND OF ROOM OR BUILDING	Number of Changes Taking Place Per Hour
Rooms, 1—Side Exposed	1
Rooms, 2—Sides Exposed	1½
Rooms, 3—Sides Exposed	2
Rooms, 4—Sides Exposed	2
Rooms with No Windows or Outside Doors	½ to ¾
Entrance Halls	2 to 3
Reception Halls	2
Living Rooms	1 to 2
Dining Rooms	1 to 2
Bath Rooms	2
Drug Stores	2 to 3
Clothing Stores	1
Churches, Factories, Lofts, Etc.	½ to 3

Add 1-air change to above for rooms with a fireplace. While the most accurate way of estimating air infiltration considers leakage through walls and clearance between window frames and sash, the table above is sufficiently accurate for practical purposes. Where windows are equipped with metal weatherstripping, use ½ air change less per hour than shown above. For example:

A room with 2 sides exposed, windows weatherstripped, would be figured at 1-air change instead of 1½.

EXTRA SOURCES OF HEAT

In auditoriums, like a church, a theater, or in factories where a large number of people are gathered a considerable amount of heat is liberated which should be subtracted from the Btu required to heat the room. A person at rest gives off about 400 Btu per hour, while one at work gives off somewhere between 500 and 800.

Electricity used in the room may be quite an item. The total wattage of electricity actually used multiplied by 3.4 should be subtracted from the required heating load (Btu per hour). In factories where gas or oil is used as in process work, the Btu per hour consumed should be subtracted from the required heating load. The heating values of various kinds of gas and oil may be found in the engineering data section.

Courtesy *Hoffman Specialty ITT*

Table 5. Stock Assemblies of Small-Tube Cast Iron Radiators*

Number of Sections	RATING, IN SQUARE FEET							
	3 TUBE	4 TUBE		5 TUBE		6 TUBE		
	25 in.	25 in.	22 in.	25 in.	22 in.	32 in.	25 in.	19 in.
6	9.6	12.0	10.8	14.4	12.6	22.2	18.0	—
10	16.0	20.0	18.0	24.0	21.0	37.0	30.0	23.0
14	22.4	28.0	25.2	33.6	29.4	51.8	42.0	32.2
18	28.8	36.0	32.4	43.2	37.8	66.6	54.0	41.4
22	35.2	44.0	39.6	52.8	46.2	81.4	66.0	50.6
26	41.6	52.0	46.8	62.4	54.6	96.2	78.0	59.8
30	48.0	60.0	54.0	72.0	63.0	—	90.0	69.0
38	60.8	76.0	68.4	91.2	79.8	—	—	87.4

NIPPLES—Sections are assembled with cast- or malleable-iron push nipples.

TAPPINGS—Tappings are standard iron-pipe size. Flow and return tappings are located horizontally opposite the top and bottom nipple ports. Air-vent tappings for water and steam radiators are provided on the end section opposite the supply section.

PIPE FITTINGS—Iron-pipe size plugs or bushings or both, may be furnished with each radiator.

PAINTING—Each section or radiator assembled by the manufacturer is given one priming coat.

NUMBER OF SECTIONS—The stock assemblies are shown in the table. When assemblies of more sections than those listed are required, the maximum number should not exceed 56, to avoid damage in shipping and handling. Consult manufacturers' catalogs.

The square foot of equivalent direct steam radiation is defined as the ability to emit 240 BTU per hour, with steam at 215° F. in air at 70° F. These ratings apply only to radiators installed exposed in a normal manner; not to radiators installed behind enclosures, grilles, etc.

Number of Tubes per Section	Catalog Rating per Section Sq. Ft.	SECTION DIMENSIONS IN INCHES				
		A HEIGHT ①	B WIDTH		C SPACING	D LEG HEIGHT ①
			Min.	Max.		
3	1.6	26	3¼	3½	1¾	2½
4	1.6	19	4 7/16	4 13/16	1¾	2½
4	1.8	22	4 7/16	4 13/16	1¾	2½
4	2.0	25	4 7/16	4 13/16	1¾	2½
5	2.1	22	5⅝	6 5/16	1¾	2½
5	2.4	25	5⅝	6 5/16	1¾	2½
6	2.3	19	6 13/16	8	1¾	2½
6	3.0	25	6 13/16	8	1¾	2½
6	3.7	32	6 13/16	8	1¾	2½

①Over-all height and leg height of radiator as made by some manufacturers is 1 inch greater than shown in Columns A and D. Radiators may be furnished without legs. Where greater than standard leg heights are required, this dimension is to be 4½ in.
*From Simplified Practice Recommendation R174-47, U.S. Dept. of Commerce.

Courtesy *Hoffman Specialty ITT*

Table 6. Output of Direct Cast-Iron Radiators * * *

STEAM PRESSURE (APPROXIMATE)			STEAM OR MEAN WATER TEMP. °F.	BTU PER SQ. FT. E.D.R. PER HOUR ROOM TEMPERATURE °F.						
	GAGE*	ABSOLUTE** Lbs. per Sq. In.		80	75	70	65	60	55	50
VACUUM Inches of MERCURY	22.4	3.7	150	93	102	111	120	129	139	148
	20.3	4.7	160	111	120	129	139	148	158	167
	17.7	6.0	170	129	139	148	158	167	178	188
	14.6	7.5	180	148	158	167	178	188	198	209
	10.9	9.3	190	167	178	188	198	209	218	229
	6.5	11.5	200	188	198	209	218	229	240	250
	3.9	12.8	205	198	209	218	229	240	250	261
Lbs. per Sq. Inch	0.0	14.7	212	211	222	233	242	253	264	276
	1.	15.6	215	218	229	240	250	261	273	282
	2.	17	220	229	240	250	261	273	282	296
	6.	21.	230	250	261	273	282	296	308	316
	10.	25.	240	273	282	296	308	316	329	343
	15.	30.	250	296	308	316	329	343	353	364
	27.	42.	270	343	353	364	375	387	400	414
	52.	67.	300	414	421	436	453	462	471	490

*At sea level only.

**At locations other than sea level use temperature only or convert gage reading to absolute pressure. Add gage reading to atmospheric pressure in lbs. per sq. in. for given altitude. To convert vacuum (inches of mercury) to absolute, multiply inches vacuum by 0.49 and deduct from atmospheric pressure (lbs. per sq. in.) for given altitude.

***These outputs also apply quite closely to the output of the "R" type cast iron radiant baseboards. For exact outputs, the catalogs of the manufacturers of the baseboards should be consulted.

Courtesy *Hoffman Specialty ITT*

Table 7. Standard Column Radiation *

	SINGLE-COLUMN RADIATORS					
		HEATING SURFACE—SQUARE FEET				
No. of Sections	Length 2½"-Sec.	38" 3 Sq. Ft. Sec.	32" 2½ Sq. Ft. Sec.	26" 2 Sq. Ft. Sec.	23" 1⅔ Sq. Ft. Sec.	20" 1½ Sq. Ft. Sec.
3	7½	9	7½	6	5	4½
4	10	12	10	8	6⅔	6
5	12½	15	12½	10	8⅓	7½
6	15	18	15	12	10	9
7	17½	21	17½	14	11⅔	10½
8	20	24	20	16	13⅓	12
9	22½	27	22½	18	15	13½
10	25	30	25	20	16⅔	15
11	27½	33	27½	22	18⅓	16½
12	30	36	30	24	20	18
13	32½	39	32½	26	21⅔	19½
14	35	42	35	28	23⅓	21
15	37½	45	37½	30	25	22½
16	40	48	40	32	26⅔	24
17	42½	51	42½	34	28⅓	25½
18	45	54	45	36	30	27
19	47½	57	47½	38	31⅔	28½
20	50	60	50	40	33⅓	30
21	52½	63	52½	42	35	31½
22	55	66	55	44	36⅔	33
23	57½	69	57½	46	38⅓	34½
24	60	72	60	48	40	36
25	62½	75	62½	50	41⅔	37½
26	65	78	65	52	43⅓	39
27	67½	81	67½	54	45	40½

TWO-COLUMN RADIATORS							
		HEATING SURFACE—SQUARE FEET					
No. of Sections	Length 2½"-Sec.	45" 5 Sq. Ft. Sec.	38" 4 Sq. Ft. Sec.	32" 3⅓ Sqt. Ft. Sec.	26" 2⅔ Sq. Ft. Sec.	23" 2⅓ Sq. Ft. Sec.	20" 2 Sq. Ft. Sec.
3	7½	15	12	10	8	7	6
4	10	20	16	13⅓	10⅔	9⅓	8
5	12½	25	20	16⅔	13⅓	11⅔	10
6	15	30	24	20	16	14	12
7	17½	35	28	23⅓	18⅔	16⅓	14
8	20	40	32	26⅔	21⅓	18⅔	16
9	22½	45	36	30	24	21	18
10	25	50	40	33⅓	26⅔	23⅓	20
11	27½	55	44	36⅔	29⅓	25⅔	22
12	30	60	48	40	33	28	24
13	32½	65	52	43⅓	34⅔	30⅓	26
14	35	70	56	46⅔	37⅓	32⅔	28
15	37½	75	60	50	40	35	30
16	40	80	64	53⅓	42⅔	37⅓	32
17	42½	85	68	56⅔	45⅓	39⅔	34
18	45	90	72	60	48	42	36
19	47½	95	76	63⅓	50⅔	44⅓	38
20	50	100	80	66⅔	53⅓	46⅔	40
21	52½	105	84	70	56	49	42
22	55	110	88	73⅓	58⅔	51⅓	44
23	57½	115	92	76⅔	61⅓	53⅔	46
24	60	120	96	80	64	56	48
25	62½	125	100	83⅓	66⅔	58⅓	50

*This table cover column radiation manufactured prior to 1926.

Courtesy *Hoffman Specialty ITT*

Table 8. Standard Column Radiation *

THREE-COLUMN RADIATORS								FOUR COLUMN RADIATORS							
		HEATING SURFACE SQUARE FEET								HEATING SURFACE SQUARE FEET					
No. of Sections	Length 2½"-Sec.	45" 6 Sq. Ft. Sec.	38" 5 Sq. Ft. Sec.	32" 4½ Sq. Ft. Sec.	26" 3¼ Sq. Ft. Sec.	22" 3 Sq. Ft. Sec.	18" 2¼ Sq. Ft. Sec.	No. of Sections	Length 3"-Sec.	45" 10 Sq. Ft. Sec.	38" 8 Sq. Ft. Sec.	32" 6½ Sq. Ft. Sec.	26" 6 Sq. Ft. Sec.	22" 4 Sq. Ft. Sec.	18" 3 Sq. Ft. Sec.
3	7½	18	15	13½	11¼	9	6¾	3	9	30	24	19½	15	12	9
4	10	24	20	18	15	12	9	4	12	40	32	26	20	16	12
5	12½	30	25	22½	18¾	15	11¼	5	15	50	40	32½	25	20	15
6	15	36	30	27	22½	18	13½	6	18	60	48	39	30	24	18
7	17½	42	35	31½	26¼	21	15¾	7	21	70	56	45½	35	28	21
8	20	48	40	36	30	24	18	8	24	80	64	52	40	32	24
9	22½	54	45	40½	33¾	27	20¼	9	27	90	72	58½	45	36	27
10	25	60	50	45	37½	30	22½	10	30	100	80	65	50	40	30
11	27½	66	55	49½	41¼	33	24¾	11	33	110	88	71½	55	44	33
12	30	72	60	54	45	36	27	12	36	120	96	78	60	48	36
13	32½	78	65	58½	48¾	39	29¼	13	39	130	104	84½	65	52	39
14	35	84	70	63	52½	42	31½	14	42	140	112	91	70	56	42
15	37½	90	75	67½	56¼	45	33¾	15	45	150	120	97½	75	60	45
16	40	96	80	72	60	48	36	16	48	160	128	104	80	64	48
17	42½	102	85	76½	63¾	51	38¼	17	51	170	136	110½	85	68	51
18	45	108	90	81	67½	54	40½	18	54	180	144	117	90	72	54
19	47½	114	95	85½	71¼	57	42¾	19	57	190	152	123½	95	76	57
20	50	120	100	90	75	60	45	20	60	200	160	130	100	80	60
21	52½	126	105	94½	78¾	63	47¼	21	63	210	168	136½	105	84	63
22	55	132	110	99	82½	66	49½	22	66	220	176	143	110	88	66
23	57½	138	115	103½	86¼	69	51¾	23	69	230	184	149½	115	92	69
24	60	144	120	108	90	72	54	24	72	240	192	156	120	96	72
25	62½	150	125	112½	93¾	75	56¼	25	75	250	200	162½	125	100	75
26	65	156	130	117	97½	78	58½								
27	67½	162	135	121½	101¼	81	60¾								

*This table cover column radiation manufactured prior to 1926.

Courtesy *Hoffman Specialty ITT*

Table 9. Heating Surface—Three Tube Radiators (For Steam or Hot Water)

No. of Sections	Length Inches	HEATING SURFACE—SQUARE FEET 240 B.T.U. PER SQ. FT. PER HOUR				
		36″ HEIGHT 3½ Sq. Ft. Per Sec.	30″ HEIGHT 3 Sq. Ft. Per Sec.	26″ HEIGHT 2⅓ Sq. Ft. Per Sec.	23″ HEIGHT 2 Sq. Ft. Per Sec.	20″ HEIGHT 1¾ Sq. Ft. Per Sec.
2	5	7	6	4⅔	4	3½
3	7½	10½	9	7	6	5¼
4	10	14	12	9⅓	8	7
5	12½	17½	15	11⅔	10	8¾
6	15	21	18	14	12	10½
7	17½	24½	21	16⅓	14	12¼
8	20	28	24	18⅔	16	14
9	22½	31½	27	21	18	15¾
10	25	35	30	23⅓	20	17½
11	27½	38½	33	25⅔	22	19¼
12	30	42	36	28	24	21
13	32½	45½	39	30⅓	26	22¾
14	35	49	42	32⅔	28	24½
15	37½	52½	45	35	30	26¼
16	40	56	48	37⅓	32	28
17	42½	59½	51	39⅔	34	29¾
18	45	63	54	42	36	31½
19	47½	66½	57	44⅓	38	33¼
20	50	70	60	46⅔	40	35
21	52½	73½	63	49	42	36¾
22	55	77	66	51⅓	44	38½
23	57½	80½	69	53⅔	46	40¼
24	60	84	72	56	48	42
25	62½	87½	75	58⅓	50	43¾

NOTE—These data apply to obsolete radiators manufactured since 1926 but not manufactured today.

Courtesy *Hoffman Specialty ITT*

Air Change—The number of times in an hour the air in a room is changed either by mechanical means or by the infiltration of outside air leaking into the room through cracks around doors and windows, etc.

Air Cleaner—A device designed for the purpose of removing airborne impurities, such as dust, fumes, and smokes. (Air cleaners include air washers and air filters.)

Air Conditioning—The simultaneous control of the temperature, humidity, air motion, and air distribution within an enclosure. Where human comfort and health are involved, a reasonable air purity with

Table 10. Heating Surface—Four Tube Radiators (For Steam or Hot Water)

No. of Sections	Length Inches	HEATING SURFACE—SQUARE FEET 240 B.T.U. PER SQ. FT. PER HOUR				
		37″ HEIGHT 4¼ Sq. Ft. Per Sec.	32″ HEIGHT 3½ Sq. Ft. Per Sec.	26″ HEIGHT 2¾ Sq. Ft. Per Sec.	23″ HEIGHT 2½ Sq. Ft. Per Sec.	20″ HEIGHT 2¼ Sq. Ft. Per Sec.
2	5	8½	7	5½	5	4½
3	7½	12¾	10½	8¼	7½	6¾
4	10	17	14	11	10	9
5	12½	21¼	17½	13¾	12½	11¼
6	15	25½	21	16½	15	13½
7	17½	29¾	24½	19¼	17½	15¾
8	20	34	28	22	20	18
9	22½	38¼	31½	24¾	22½	20¼
10	25	42½	35	27½	25	22½
11	27½	46¾	38½	30¼	27½	24¾
12	30	51	42	33	30	27
13	32½	55¼	45½	35¾	32½	29¼
14	35	59½	49	38½	35	31½
15	37½	63¾	52½	41¼	37½	33¾
16	40	68	56	44	40	36
17	42½	72¼	59½	46¾	42½	38¾
18	45	76½	63	49½	45	40½
19	47½	80¾	66½	52¼	47½	42¾
20	50	85	70	55	50	45
21	52½	89¼	73½	57¾	52½	47¼
22	55	93½	77	60½	55	49½
23	57½	97¾	80½	63¼	57½	51¾
24	60	102	84	66	60	54
25	62½	106¼	87½	68¾	62½	56¼

NOTE—These data apply to obsolete radiators manufactured since 1926 but not manufactured today.

Courtesy *Hoffman Specialty ITT*

regard to dust, bacteria, and odors is also included. The primary requirement of a good air conditioning system is a good heating system.

Air Infiltration—The leakage of air into a house through cracks and crevices, and through doors, windows, and other openings, caused by wind pressure and/or temperature difference.

Air Valve—See Vent Valve.

Atmospheric Pressure—The weight of a column of air, one square inch in cross section and extending from the earth to the upper level of the blanket of air surrounding the earth. This air exerts a pressure of

Table 11. Heating Surface—Five Tube Radiators (For Steam or Hot Water)

No. of Sections	Length Inches	HEATING SURFACE—SQUARE FEET 240 B.T.U. PER SQ. FT. PER HOUR				
		37" HEIGHT 5 Sq. Ft. Per Sec.	32" HEIGHT 4⅓ Sq. Ft. Per Sec.	26" HEIGHT 3½ Sq. Ft. Per Sec.	23" HEIGHT 3 Sq. Ft. Per Sec.	20" HEIGHT 2⅔ Sq. Ft. Per Sec.
2	5	10	8⅔	7	6	5⅓
3	7½	15	13	10½	9	8
4	10	20	17½	14	12	10⅔
5	12⅓	25	21⅔	17½	15	13½
6	15	30	26	21	18	16
7	17½	35	30⅓	24½	21	18⅔
8	20	40	34⅔	28	24	21⅓
9	22½	45	39	31½	27	24
10	25	50	43⅓	35	30	26⅔
11	27½	55	47⅔	38½	33	29⅓
12	30	60	52	42	36	32
13	32½	65	56½	45½	39	34⅔
14	35	70	60⅔	49	42	37⅓
15	37½	75	65	52½	45	40
16	40	80	69⅓	56	48	42⅔
17	42½	85	73⅔	59½	51	45⅓
18	45	90	78	63	54	48
19	47½	95	82⅓	66½	57	50⅔
20	50	100	86⅔	70	60	53⅓
21	52½	105	91	73½	63	56
22	55	110	05⅓	77	66	58⅔
23	57½	115	99⅔	80½	69	61⅓
24	60	120	104	84	72	64
25	62½	125	108⅓	87½	75	66⅔

NOTE—These data apply to obsolete radiators manufactured since 1926 but not manufactured today.

Courtesy *Hoffman Specialty ITT*

14.7 pounds per square inch at sea level, where water will boil at 212°F. High altitudes have lower atmospheric pressure with correspondingly lower boiling point temperatures.

Boiler—A closed vessel in which steam is generated or in which water is heated by fire.

Boiler Heating Surface—The area of the heat transmitting surfaces in contact with the water (or steam) in the boiler on one side and the fire or hot gases on the other.

Boiler Horse Power—The equivalent evaporation of 34.5 lbs. of water per hour at 212 F. to steam at 212°F. This is equal to a heat output

Table 12. Heating Surface—Six Tube Radiators (For Steam or Hot Water)

No. of Sections	Length Inches	HEATING SURFACE—SQUARE FEET 240 B.T.U. PER SQ. FT. PER HOUR				
		37" HEIGHT 6 Sq. Ft. Per Sec.	32" HEIGHT 5 Sq. Ft. Per Sec.	26" HEIGHT 4 Sq. Ft. Per Sec.	23" HEIGHT 3½ Sq. Ft. Per Sec.	20" HEIGHT 3 Sq. Ft. Per Sec.
2	5	12	10	8	7	6
3	7½	18	15	12	10½	9
4	10	24	20	16	14	12
5	12½	30	25	20	17½	15
6	15	36	30	24	21	18
7	17½	42	35	28	24½	21
8	20	48	40	32	28	24
9	22½	54	45	36	31½	27
10	25	60	50	40	35	30
11	27½	66	55	44	38½	33
12	30	72	60	48	42	36
13	32½	78	65	52	45½	39
14	35	84	70	56	49	42
15	37½	90	75	60	52½	45
16	40	96	80	64	56	48
17	42½	102	85	68	59½	51
18	45	108	90	72	63	54
19	47½	114	95	76	66½	57
20	50	120	100	80	70	60
21	52½	126	105	84	73½	63
22	55	132	110	88	77	66
23	57½	138	115	92	80½	69
24	60	144	120	96	84	72
25	62½	150	125	100	87½	75

NOTE—These data apply to obsolete radiators manufactured since 1926 but not manufactured today.

Courtesy *Hoffman Specialty ITT*

of 33,475 Btu per hour, which is equal to approximately 140 sq. ft. of steam radiation (EDR).

British Thermal Unit (Btu)—The quantity of heat required to raise the temperature of 1 lb. of water 1 F. This is somewhat approximate but sufficiently accurate for any work discussed in this Book.

Bucket Trap (Inverted)—A float trap with an open float. The float or bucket is open at the bottom. When the air or steam in the bucket has been replaced by condensate the bucket loses its buoyancy and when it *sinks* it opens a valve to permit condensate to be pushed into the return.

Table 13. Heating Surface—Seven Tube Radiators (For Steam or Hot Water)

No. of Sections	Length Inches	HEATING SURFACE—SQUARE FEET 240 B.T.U. PER SQ. FT. PER HOUR		
		20" HEIGHT 4¼ Sq. Ft. Per Sec.	16½" HEIGHT 3½ Sq. Ft. Per Sec.	13" HEIGHT 2¾ Sq. Ft. Per Sec.
2	5	8½	7	5½
3	7½	12¾	10½	8¼
4	10	17	14	11
5	12½	21¼	17½	13¾
6	15	25½	21	16½
7	17½	29¾	24½	19¼
8	20	34	28	22
9	22½	38¼	31½	24¾
10	25	42½	35	27½
11	27½	46¾	38½	30¼
12	30	51	42	33
13	32½	55¼	45½	35¾
14	35	59½	49	38½
15	37½	63¾	52½	41¼
16	40	68	56	44
17	42½	72¼	59½	46¾
18	45	76½	63	49½
19	47½	80¾	66½	52¼
20	50	85	70	55
21	52½	89¼	73½	57¾
22	55	93½	77	60½
23	57½	97¾	80½	63¼
24	60	102	84	66
25	62½	106¼	87½	68¾

NOTE—These data apply to obsolete radiators manufactured since 1926 but not manufactured today.

Courtesy *Hoffman Specialty ITT*

Bucket Trap (Open)—The bucket (float) is open at the top. Water surrounding the bucket keeps it floating and the pin is pressed against its seat. Condensate from the system drains into the bucket. When enough has drained into it so that the bucket loses its buoyancy it sinks and pulls the pin off its seat and steam pressure forces the condensate out of the trap.

Calorie (Small)—The quantity of heat required to raise 1 gram of water 1°C (approx.).

Calorie (Large)—The quantity of heat required to raise 1 kilogram of water 1°C (approx.).

Table 14. Heat Emission Standard Column Radiation *

Number of B.T.U. transmitted per hour per sq. ft. of radiation with low pressure steam when heating room to given temperature.

Temperature of Room	TYPE OF RADIATION				
	3 Col. 26″	3 Col. 32″	3 Col. 38″	Wall.	Coil.
Col. A	Col. B	Col. C	Col. D	Col. E	Col. F
40°F.	309	305	293	362	381
45	301	292	281	347	365
50	290	282	271	335	354
55	279	270	261	322	338
60	269	261	250	310	326
65	258	250	240	297	313
70	247	240	231	285	300
75	236	230	220	275	288
80	226	220	211	261	277
85	216	210	200	251	265
90	206	200	190	242	253
95	196	190	180	228	239
100	186	180	170	215	226
105	176	171	162	203	214
110	167	162	155	192	202
115	158	153	147	181	191
120	149	144	139	171	180
125	140	135	130	160	169
130	130	126	121	150	158
135	121	118	113	140	147
140	113	110	106	130	137

If 500 ϕ of wall radiation is heating a room to 50°F, what will its equivalent be in 32″ column radiation setting in 70°F—Refer to Col. E and find one ϕ wall radiation setting in 50°F. = 335 B.T.U. Then 500 × 335 = 167,500 B.T.U. Refer to Col. C at 70°F. for average Col. Rad. and find 240 B.T.U. per sq. ft. Therefore 167,500 ÷ 240 — 698 ϕ direct equivalent.

*This table covers heat emission from radiation manufactured prior to 1926. Courtesy *Hoffman Specialty ITT*

Celsius—A thermometer scale at which the freezing point of water is 0° and its boiling is 100°. In this country it is used only in scientific and laboratory work.

Central Fan System—A mechanical indirect system of heating, ventilating, or air conditioning consisting of a central plant where the air is heated and/or conditioned and then circulated by fans or blowers through a system of distributing ducts.

Chimney Effect—The tendency in a duct or other vertical air passage for air to rise when heated due to its decrease in density.

Table 15. Heat Emission of Direct Pipe Coils (Using Steam at 215° F and Room Temp. 70°)

Wall Coils—Coils Placed Vertical—Pipes Horizontal B.T.U. per linear foot of coil per hour (not lineal feet of pipe)			
Size of Pipe Coil Conductor	**1″**	**1¼″**	**1½″**
Single Row	132	162	185
Two Row	252	312	348
Four Row	440	545	616
Six Row	567	702	793
Eight Row	651	796	907
Ten Row	732	907	1020
Twelve Row	812	1005	1135

Ceiling Coils—Coils placed horizontally—pipes horizontal, emission is equal to that of a single row coil. Allowance must be made, however, if the coil is at the ceiling in a higher temperature. In this case use:

126 B.T.U. per linear foot of pipe for 1″ Coils
156 B.T.U. per linear foot of pipe for 1¼″ Coils
175 B.T.U. per linear foot of pipe for 1½″ Coils

INSTALLATION LIKE STEAM COILS ABOVE								
PIPE SIZE	MEAN WATER °F.	ROWS IN COIL						
		1	2	4	6	8	10	12
1″	170	80	155	270	350	400	450	500
	180	90	175	305	395	450	510	565
	190	105	195	345	445	510	570	635
	200	115	220	385	495	565	635	705
1¼″	170	100	190	335	435	490	560	620
	180	110	215	380	485	555	630	700
	190	125	245	425	550	620	710	785
	200	140	270	475	610	690	790	875
1½″	170	115	215	380	490	560	630	700
	180	130	240	430	550	630	710	790
	190	145	270	480	620	710	795	885
	200	160	305	535	690	790	885	985

Courtesy *Hoffman Specialty ITT*

Circulating Pipe (Hot Water System)—The pipe and orifice in a Hoffman Panelmatic Hot Water Control System through which the return water by-passes the boiler until the temperature of the circulating stream is too low at which time part of it is replaced by the correct quantity of hot boiler water to restore its temperature.

Table 16. Radiator Enclosures

To enclose or partly enclose a radiator reduces its heat output and changes the distribution of heated air in the room. The additional surface usually added to column or tube radiation for various enclosures is indicated below.

*If A is 50% of width of radiator, add 10%; if 150%, add 35%.

‡B = 80% of A. C = 150% of A. D = A.

Example: A room requires 50 sq. ft. radiation radiator recessed flush with wall, $-50\phi + 20\% - 60\ \phi$ radiator required. If radiator for same room is to have grille over entire face only, $-50\ \phi + 30\% - 65\ \phi$ required.

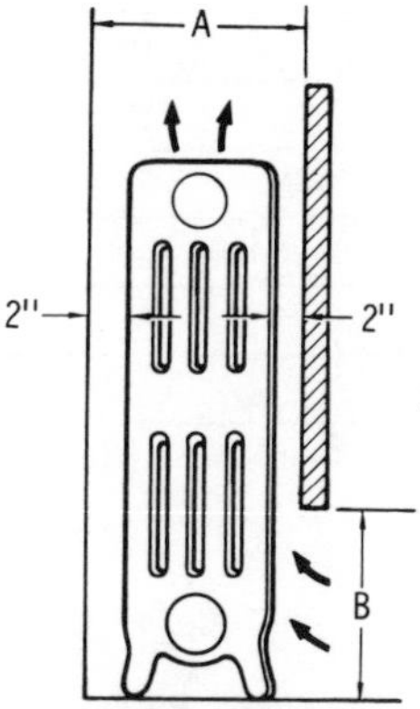
A
2"
2"
B
DEDUCT 10%

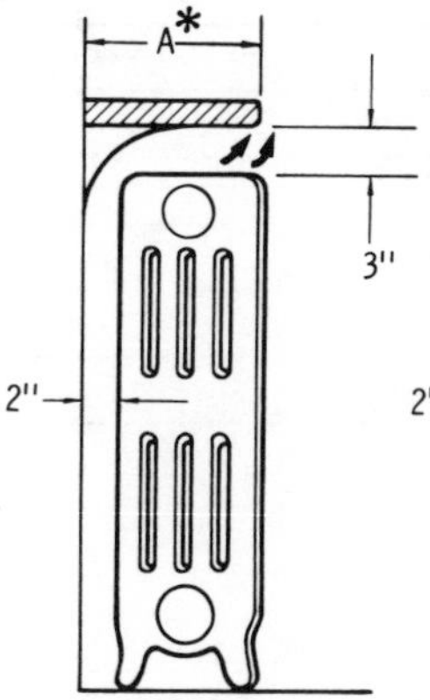
A*
3"
2"
ADD 20%

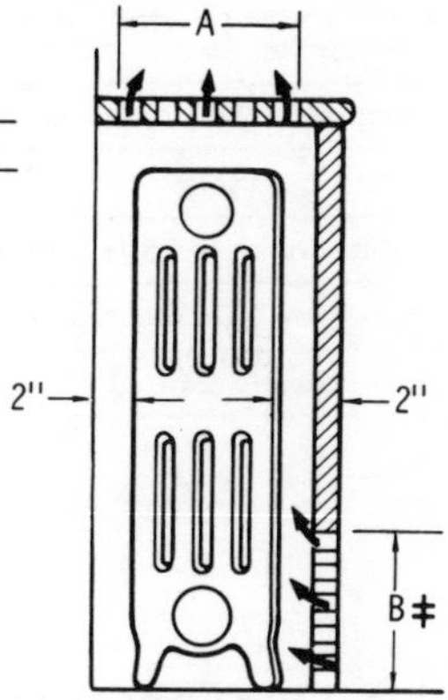
A
2"
2"
B‡
DEDUCT 5%

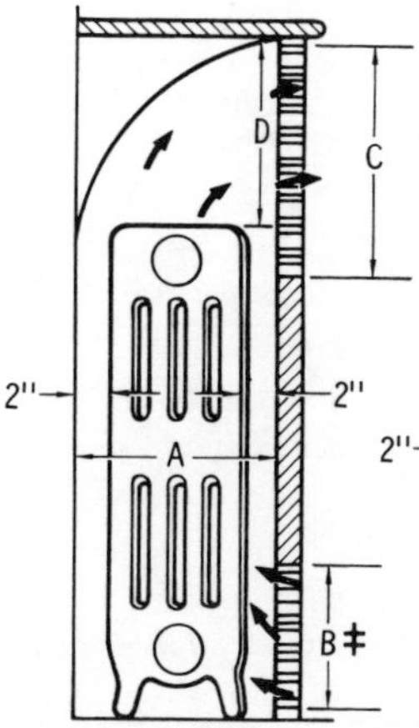
D
C
2"
2"
A
B‡
‡NO CHANGE

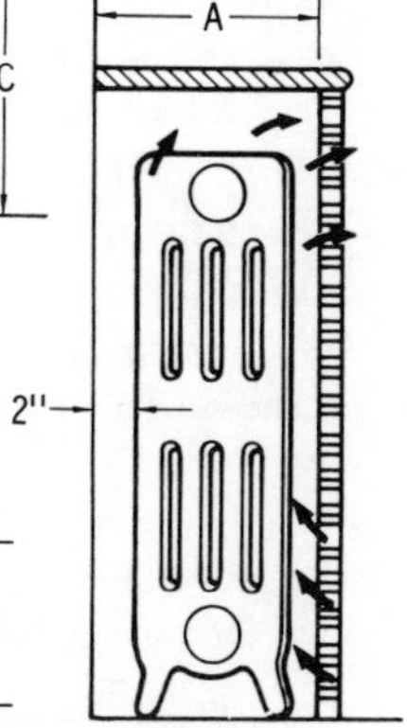
A
2"
ADD 30%

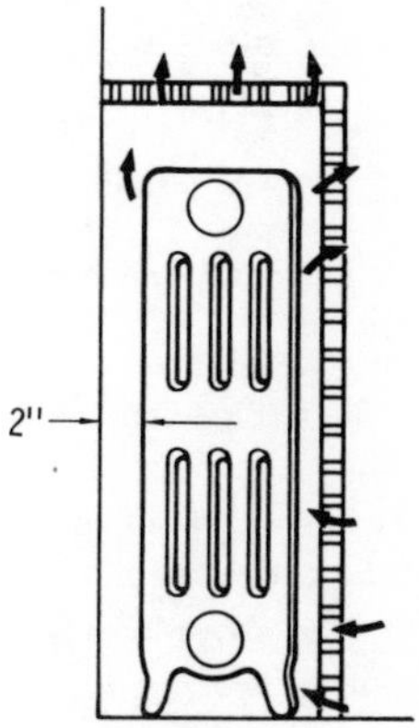
2"
ADD 5%

Table 17. Kitchen Equipment—Pounds of Steam Required Per Hour

	Pounds Per Hour
Bain Marie—Per Foot	12.5
Coffee Urn (3)	50
Egg Boiler—3 Compartment	10
Stock Kettle—40 Gallon	37.5
Steam Table—Each 6 Feet	25
Clam and Lobster Steamer	10
Jet for Pot Sink	10
Vegetable Boiler—Per Section	15
Silver Burnisher	50
Plate Warmer—15 feet long	30
Steamer—3 Compartment	75
Usual Pressure Carried 30-35 lbs.	

Courtesy *Hoffman Specialty ITT*

Table 18. Outdoor Storage Tanks (To Keep from Freezing)

T^1 = Outdoor Temperature.

T^2 = Temperature of Water.

E.S. = Sq. Ft. of Exposed Surface of Tank

Coef. = 2″ Wood Tank—.5 B.T.U. per sq. ft. per hour
Steel Tank—1.5 B.T.U. per sq. ft. per hour

B.T.U. loss from Water per hour = E.S. $\times$ $(T^1 - T^2)$ $\times$ Coef.

$$\text{Pounds Steam Required} = \frac{\text{Total B.T.U. Loss from Water per Hour}}{\text{Latent Heat Steam at Pressure Carried}}$$

Sq. ft. Pipe Coil Required = Lbs. of Steam required $\div$ Lbs. Steam Condensed per Hour per sq. ft. of Pipe (see Table 30)

To obtain sq. ft. of exposed surface for round tanks multiply diameter by 3.1416 $\times$ Height $+$ Area of Bottom and Top.

NOTE—The above is only to be used for water storage tanks, usually located on roof.

Courtesy *Hoffman Specialty ITT*

Coefficient of Heat Transmission (Over-all)-U—The amount of heat (Btu) transmitted *from air to air* in one hour per square foot of the wall, floor, roof, or ceiling for a difference in temperature of one degree Fahrenheit *between the air on the inside and outside of the wall, floor, roof, or ceiling.*

Column Radiator—A type of direct radiator. (This radiator has not been listed by manufacturers since 1926.)

Comfort Line—The effective temperature at which the largest percentage of adults feel comfortable.

Comfort Zone (Average)—The range of effective temperatures over which the majority of adults feel comfortable.

Concealed Radiator—See Convector.

Condensate—In steam heating, the water formed by cooling steam as in a radiator. The capacity of traps, pumps, etc., is sometimes expressed in lbs. of condensate they will handle per hour. One pound of condensate per hour is equal to approximately 4 sq. ft. of steam heating surface (240 Btu per hour per sq. ft.).

Conductance (Thermal)-C—The amount of heat (Btu) transmitted *from surface to surface,* in one hour through one square foot of a material or construction *for the thickness or type under consideration* for a difference in temperature of one degree Fahrenheit between the two surfaces.

Conduction (Thermal)—The transmission of heat through and by means of matter.

Conductivity (Thermal)-k—The amount of heat (Btu) transmitted in one hour through one square foot of a homogenous material one inch thick for a difference in temperature of one degree Fahrenheit between the two surfaces of the material.

Conductor (Thermal)—A material capable of readily transmitting heat by means of conduction.

Convection—The transmission of heat by the circulation (either natural or forced) of a liquid or a gas such as air. If natural, it is caused by the difference in weight of hotter and colder fluid.

Convector—A concealed *radiator.* An enclosed heating unit located (with enclosure) either within, adjacent to, or exterior to the room or space to be heated, but transferring heat to the room or space mainly by the process of convection. A shielded heating unit is also termed a convector. If the heating unit is located exterior to the room or space to be heated, the heat is transferred through one or more ducts or pipes.

Table 19. Heating Power of Low Pressure Steam Pipes in Water for Average Working Conditions

BRASS PIPE					
COL. A	COL. B.	COL. C	COL. A	COL. B	COL. C
Temp. Dif.	B.T.U.	Pounds	Temp. Dif.	B.T.U.	Pounds
6°F.	192	.20	70	13,000	13.55
7	240	.25	75	15,000	15.62
8	300	.31	80	17,000	17.70
9	400	.42	85	19,000	20.00
10	480	.50	90	21,000	21.77
15	800	.83	95	23,000	24.00
16	960	1.00	100	25,000	26.05
20	1,440	1.00	110	30,000	31.25
25	2,300	2.40	120	35,000	36.45
30	3,100	3.23	130	40,000	41.60
35	4,000	4.16	140	45,000	46.90
40	5,000	5.23	150	50,000	52.10
45	6,000	6.25	160	55,000	57.30
50	7,200	7.50	170	61,000	63.54
55	8,500	8.85	180	67,000	70.00
60	10,000	10.50	190	73,500	76.60
65	11,500	12.00	200	80,000	83.23

Col. A = temperature difference between steam in pipe and average temperature of the water in the tank in degrees.

Col. B = B.T.U. transmitted per sq. ft. per hour.

Col. C = lbs. of steam condensed per sq. ft. per hour. Iron pipe will condense ½ as much steam as given in table for brass pipe.

Example: Heat 200 gals. water per hour from 50° to 164°F. with brass pipe and steam at 5 lbs. gauge pressure.

Temperature of steam 5 lbs. pressure	= 227°F.
Average temperature of water (50 + 164) ÷ 2	= 107°F.
Temperature difference	= 120°F.
200 gals. water by weight (200 × 8⅓)	= 1667 lbs.
Temperature rise = 164° − 50°	= 114°F.

B.T.U. required per hr. = 1667 × 114 = 190,038. From Col. A—120°F.—find in Col. B that 1 sq. ft. brass pipe will give up 35,000 B.T.U. and in Col. C. this equals 36.45 lbs. of steam per hour. Then: 190,038 ÷ 35,000 = 5.43 sq. ft. brass pipe and as each sq. ft. will condense 36.45 lbs.—36.45 × 5.43 = 198 lbs. steam required per hour.

Courtesy *Hoffman Specialty ITT*

Convertor—A piece of equipment for heating water with steam without mixing the two. It may be used for supplying hot water for domestic purposes or for a hot water heating system.

Cooling Leg—A length of uninsulated pipe through which the condensate flows to a trap and which has sufficient cooling surface to

Table 20. Water Flow in G.P.M. Thru Pipes for Various Pressure Drops (Not for Heating Systems)

Pressure Drop Lbs. per Sq. In. per 100 Ft. Run	IRON PIPE SIZE IN INCHES								
	3/4"	1"	1 1/4"	1 1/2"	2"	2 1/2"	3"	3 1/2"	4"
5	5.4	11	19	30	62	109	171	252	353
7	6.4	13	23	36	74	129	203	298	418
10	7.6	15	27	43	88	154	242	357	499
20	10.8	22	38	61	125	218	343	504	706
30	13.2	27	47	76	153	267	420	618	864
40	15.0	31	54	86	176	308	485	714	998
50	17.0	35	60	96	197	345	542	800	1115
75	21.0	43	74	117	242	423	665	978	1365
100	24.0	49	85	136	278	485	769	1130	1578
125	27.0	55	96	152	311	544	858	1260	1765
150	30.0	60	105	166	341	598	939	1380	1930

Water Flow in G.P.M. Thru Type L Copper Tubing (for Various Pressure Drops).

Pressure Drop Lbs. per Sq. In. per 100 Ft. Run	TUBING SIZE IN INCHES						
	3/8"	1/2"	3/4"	1"	1 1/4"	1 1/2"	2"
5	1.2	2.4	6.2	14.	23.	37.	78.
7	1.5	2.9	7.6	16.	28.	45.	93.
10	1.8	3.5	9.4	19.	34.	53.	115.
20	2.8	5.2	14.0	29.	50.	80.	170.
30	3.5	6.5	18.0	37.	63.	100.	215.
40	4.1	7.8	21.0	43.	75.	120.	250.
50	4.7	8.9	24.0	49.	85.	135.	290.
75	5.9	11.0	30.0	61.	105.	170.	360.
100	7.0	13.0	36.0	72.	125.	200.	425.

Courtesy *Hoffman Specialty ITT*

permit the condensate to dissipate enough heat to prevent flashing when the trap opens. In the case of a thermostatic trap a cooling leg may be necessary to permit the condensate to drop a sufficient amount in temperature to permit the trap to open.

Degree-Day (Standard)—A unit which is the difference between 65°F. and the daily average temperature when the latter is below 65°F. The "degree days" in any one day is equal to the number of degrees F. that the average temperature for that day is below 65°F.

Dew-Point Temperature—The air temperature corresponding to saturation (100 per cent relative humidity) for a given moisture content. It is the lowest temperature at which air can retain the water vapor it contains.

Table 21. Barometric Pressure in Pounds Per Square Inch

Barometer Inches	Pressure in Lbs. per Sq. In.
28.00	13.75
28.25	13.87
28.50	13.99
28.75	14.12
29.00	14.24
29.25	14.36
29.50	14.48
29.75	14.61
30.00	14.73
30.25	14.85
30.50	14.98
30.75	15.10
31.00	15.22
31.25	15.34

Courtesy *Hoffman Specialty ITT*

Direct-Indirect Heating Unit—A heating unit located in the room or space to be heated and partially enclosed, the enclosed portion being used to heat air which enters from outside the room.

Direct Radiator—Same as *radiator:*

Direct-Return System (Hot Water)—A two-pipe hot water system in which the water after it has passed through a heating unit, is returned to the boiler along a direct path so that the total distance traveled by the water from each radiator is the shortest feasible. There is, therefore, a considerable difference in the lengths of the several circuits composing the system.

Domestic Hot Water—Hot water used for purposes other than for house heating such as for laundering, dishwashing, bathing, etc.

Down-Feed One-Pipe Riser (Steam)—A pipe which carries steam downward to the heating units and into which the condensation from the heating units drains.

Down-Feed System (Steam)—A steam heating system in which the supply mains are above the level of the heating units which they serve.

Dry-Bulb Temperature—The temperature of the air as determined by an ordinary thermometer.

Dry Return (Steam)—A return pipe in a steam heating system which carries both water of condensation and air.

Dry Saturated Steam—Saturated steam containing no water in suspension.

Table 22. Suction Lift of Pumps with Barometric Pressure at Different Altitudes and Equivalent Head of Water in Feet.

Altitude	Barometric Pressure	Equivalent Head of Water in Feet	Practical Suction Lift
Sea Level	14.70 lbs. sq. in.	33.95	22 ft.
¼ Mile Above	14.02 lbs. sq. in.	32.38	21 ft.
½ Mile Above	13.33 lbs. sq. in.	30.79	20 ft.
¾ Mile Above	12.66 lbs. sq. in.	29.24	18 ft.
1 Mile Above	12.02 lbs. sq. in.	27.76	17 ft.
1¼ Miles Above	11.42 lbs. sq. in.	26.38	16 ft.
1½ Miles Above	10.88 lbs. sq. in.	25.13	15 ft.
2 Miles Above	9.88 lbs. sq. in.	22.82	14 ft.

Highest Temp. °F. of Condensate Permissible with Vacuum Pumps.

Vacuum Inches of Mercury in Pump Receiver	Highest Permissible Temperature of Condensate °F.
15	179
12	187
10	192
8	196
6	201
4	205
2	209
1	210

Courtesy *Hoffman Specialty ITT*

Equivalent Direct Radiation (E.D.R.)—See Square Foot of Heating Surface.

Extended Heating Surface—Heating surface consisting of ribs, fins, or ribs which receive heat by conduction from the prime surface.

Extended Surface Heating Unit—A heating unit having a relatively large amount of extended surface which may be integral with the core containing the heating medium or assembled over such a core, making good thermal contact by pressure, or by being soldered to the core or by both pressure and soldering. (An extended surface heating unit is usually placed within an enclosure and therefore functions as a convector.)

Table 23. Length of Expansion Offsets and Bonds for Proper Expansion of Pipe

Total Expansion in Inches*	FEET OF PIPE IN OFFSET OR U-BEND FOR DIFFERENT DIAMETERS OF PIPE									
	2″	3″	4″	5″	6″	8″	10″	12″	14″	16″
1	11	13	15	17	19	21	23	25	27	30
2	15	18	21	23	26	29	32	35	38	42
3	18	22	26	29	32	36	40	43	48	52
4	21	26	30	34	37	42	47	50	56	58
5	24	30	34	38	41	47	53	57	63	65
6	27	33	37	41	45	52	58	63	69	71
7	30	36	40	44	48	56	62	68	74	
8	32	39	43	47	52	60	66	72		

*—This column shows the total expansion the offset will take care of without a cold strain. In general these amounts can be increased 40 percent which increase can be taken up in cold strain of the pipe on being made up.

The length of pipe in the expansion piece should be the same whether in the form of a single right-angle offset or double offset or U-bend.

The lengths of arms figured for 12,000 lb. per square inch tension for wrought iron pipe. If steel pipe is used this is good for 16,000 lb. per inch so that the arm will take care of ⅓ more expansion.

Expansion of Pipes in Inches Per 100 Feet

Temperature Degrees F.	Cast Iron	Wrought Iron	Steel	Brass or Copper
0	0.00	0.00	0.00	0.00
50	0.36	0.40	0.38	0.57
100	0.72	0.79	0.76	1.14
125	0.88	0.97	0.92	1.40
150	1.10	1.21	1.15	1.75
175	1.28	1.41	1.34	2.04
200	1.50	1.65	1.57	2.38
225	1.70	1.87	1.78	2.70
250	1.90	2.09	1.99	3.02
275	2.15	2.36	2.26	3.42
300	2.35	2.58	2.47	3.74
325	2.60	2.86	2.73	4.13
350	2.80	3.08	2.94	4.45

Courtesy *Hoffman Specialty ITT*

Fahrenheit—A thermometer scale at which the freezing point of water is 32° and its boiling point is 212° above zero. It is generally used in this country for expressing temperature.

Flash (Steam)—The rapid passing into steam of water at a high temperature when the pressure it is under is reduced so that its temperature is above that of its boiling point for the reduced pressure. For example: If hot condensate is discharged by a trap into a low

Table 24. Water Required for Humidification

The approximate rule for calculating the amount of water required per hour to maintain any desired humidity in a room is:

Multiply the difference between the number of grains of moisture per cubic foot of air at the required room temperature and humidity and the number of grains per cubic foot of outside air at the given temperature and humidity by the cubic contents of the room by the number of air changes per hour and divide the result by 7000 (this method disregards the expansion of air when heated.)

For most localities, it is customary to assume the average humidity of outside air as 30-40%.

Example:

Grains of moisture at 70° & 40% humidity	=	3.19
Grains of moisture at 0° & 30% humidity	=	.17
Grains of moisture to be added per cu. ft.	=	3.02

Assuming two air changes per hour in a room containing 8000 cu. ft. we have

$$\frac{3.02 \times 8000 \times 2}{7000} = 6.9 \text{ lbs. of water per hour required.}$$

Courtesy *Hoffman Specialty ITT*

pressure return or into the atmosphere, a certain percentage of the water will be immediately transformed into steam. It is also called re-evaporation.

Float & Thermostatic Trap—A float trap with a thermostatic element for permitting the escape of air into the return line.

Float Trap—A steam trap which is operated by a float. When enough condensate has drained (by gravity) into the trap body the float is lifted which in turn lifts the pin off its seat and permits the condensate to flow into the return until the float has been sufficiently lowered to close the port. Temperature does not effect the operation of a float trap.

Furnace—That part of a boiler or warm air heating plant in which combustion takes place. Sometimes also the complete heating unit of a warm air heating system.

Gage Pressure—The pressure above that of the atmosphere. It is the pressure indicated on an ordinary pressure gage. It is expressed as a unit pressure such as lbs. per sq. in. gage.

Grille—A perforated covering for an air inlet or outlet usually made of wire screen, pressed steel, cast-iron or other material.

Head—Unit pressure usually expressed in feet of water or mil-inches of water.

Heat—That form of energy into which all other forms may be changed. Heat always flows from a body of higher temperature to a

Table 25. Properties of Air

DRY AIR				
Temperature Degrees F.	Weight per Cu. Ft. of Dry Air in Lbs.	Ratio to Volume at 70°F.	B.T.U. Absorbed per Cu. Ft. of Air per Deg. F.	Cu. Ft. of Air Raised 1°F. by 1 B.T.U.
0	.08636	.8680	.02080	48.08
10	.08453	.8867	.02039	49.05
20	.08276	.9057	.01998	50.05
30	.08107	.9246	.01957	51.10
40	.07945	.9434	.01919	52.11
50	.07788	.9624	.01881	53.17
60	.07640	.9811	.01846	54.18
70	.07495	1.0000	.01812	55.19
80	.07356	1.0190	.01779	56.21
90	.07222	1.0380	.01747	57.25
100	.07093	1.0570	.01716	58.28
110	.06968	1.0756	.01687	59.28
120	.06848	1.0945	.01659	60.28
130	.06732	1.1133	.01631	61.32
140	.06620	1.1320	.01605	62.31
150	.06510	1.1512	.01578	63.37
160	.06406	1.1700	.01554	64.35
170	.06304	1.1890	.01530	65.36
180	.06205	1.2080	.01506	66.40

Courtesy *Hoffman Specialty ITT*

body of lower temperature. See also: Latent Heat, Sensible Heat, Specific Heat, Total Heat, Heat of the Liquid.

Heat of the Liquid—The heat (Btu) contained in a liquid due to its temperature. The heat of the liquid for water is zero at 32°F. and increases 1 Btu approximately for every degree rise in temperature.

Heat Unit—In the foot-pound-second system, the British thermal unit (Btu) in the centimeter-gram-second system, the calorie (cal.).

Heating Medium—A substance such as water, steam, or air used to convey heat from the boiler, furnace, or other source of heat to the heating units from which the heat is dissipated.

Heating Surface—The exterior surface of a heating unit. See also Extended Heating Surface.

Heating Unit—Radiators, convectors, base boards, finned tubing, coils embedded in floor, wall, or ceiling, or any device which transmits the heat from the heating system to the room and its occupants.

Horsepower—A unit to indicate the time rate of doing work equal to 550 ft.-lb. per second, or 33,000 ft.-lb. per minute. One horsepower equals 2545 Btu per hour or 746 watts.

Table 25. Properties of Air (Contd.)

SATURATED AIR				
Temperature Degrees F.	Vapor Press. Inches of Mercury	Weight of Vapor per Cu. Ft. in Lbs.	B.T.U. Absorbed per Cu. Ft. of Air per Deg. F.	Cu. Ft. of Air Raised 1°F. by 1 B.T.U.
0	0.0383	.000069	.02082	48.04
10	0.0631	.000111	.02039	49.50
20	0.1030	.000177	.01998	50.05
30	0.1640	.000276	.01955	51.15
40	0.2477	.000409	.01921	52.06
50	0.3625	.000587	.01883	53.11
60	0.5220	.000829	.01852	54.00
70	0.7390	.001152	.01811	55.22
80	1.0290	.001576	.01788	55.93
90	1.4170	.002132	.01763	56.72
100	1.9260	.002848	.01737	57.57
110	2.5890	.003763	.01716	58.27
120	3.4380	.004914	.01696	58.96
130	4.5200	.006357	.01681	59.50
140	5.8800	.008140	.01669	59.92
150	7.5700	.010310	.01663	60.14
160	9.6500	.012956	.01664	60.10
170	12.2000	.016140	.01671	59.85
180	15.2900	.019940	.01682	59.45

Courtesy *Hoffman Specialty ITT*

Hot Water Heating System—A heating system in which water is used as the medium by which heat is carried through pipes from the boiler to the heating units.

Humidistat—An instrument which controls the realtive humidity of the air in a room.

Humidity—The water vapor mixed with air.

Insulation (Thermal)—Material with a high resistance to heat flow.

Latent Heat of Evaporation—The heat (Btu per pound) necessary to change 1 pound of liquid into vapor without raising its temperature. In round numbers this is equal to 960 Btu per pound of water.

Latent Heat of Fusion—The heat necessary to melt one pound of a solid without raising the temperature of the resulting liquid. The latent heat of fusion of water (melting 1 pound of ice) is 144 Btu.

Mechanical Equivalent of Heat—The mechanical energy equivalent to 1 Btu which is equal to 778 ft.-lb.

Mil-Inch—One one-thousandth of an inch (0.001″).

One-Pipe Supply Riser (Steam)—A pipe which carries steam to a heating unit and which also carries the condensation from the heating

Table 26. Relative Humidity—Percent

Dry Bulb Thermometer Deg. F.	DIFFERENCE BETWEEN DRY AND WET BULB THERMOMETER																										
	1.	2.	3.	4.	5.	6.	7.	8.	9.	10.	11.	12.	13.	14.	15.	16.	17.	18.	19.	20.	21.	22.	23.	24.	26.	28.	30.
32	89	79	69	59	49	39	30	20	11	2	0																
40	92	83	75	68	60	52	45	37	29	23	15	7	0														
50	93	87	80	74	67	61	55	49	43	38	32	27	21	16	11	5	0										
60	94	89	83	78	73	68	63	58	53	48	43	39	34	30	26	21	17	13	9	5	1	0					
70	95	90	86	81	77	72	68	64	59	55	51	48	44	40	36	33	29	25	22	19	15	12	9	6	0		
80	96	91	87	83	79	75	72	68	64	61	57	54	50	47	44	41	38	35	32	29	26	23	20	18	12	7	
90	96	92	89	85	81	78	74	71	67	65	61	58	55	52	49	47	44	41	39	36	34	31	29	26	22	17	13
100	96	93	89	86	83	80	77	73	70	68	65	62	59	56	54	51	49	46	44	41	39	37	35	33	28	24	21
110	97	93	90	87	84	81	78	75	73	70	67	65	62	60	57	55	52	50	48	46	44	42	40	38	34	30	26
120	97	94	91	88	85	82	80	77	74	72	69	67	65	62	60	58	55	53	51	49	47	45	43	41	38	34	31
140	97	95	92	89	87	84	82	79	77	75	73	70	68	66	64	62	60	58	56	54	53	51	49	47	44	41	38

EXAMPLE:

DRY BULB THERMOMETER = 70°
WET BULB THERMOMETER = 59°
DIFFERENCE = 11°
SATURATION = 100%—BAROMETER 30"

FOLLOW THE HORIZONTAL LINE OPPOSITE 70° DRY BULB TEMP. TI THE VERTICAL LINE UNDER 11 = RELATIVE HUMIDITY—51%

Courtesy *Hoffman Specialty ITT*

Table 27. Grains of Moisture Per Cubic Foot of Air at Various Temperatures and Humidities

TEMPERATURE DEGREES-F.	RELATIVE HUMIDITY, PERCENT									
	100%	90%	80%	70%	60%	50%	40%	30%	20%	10%
75	9.35	8.42	7.49	6.55	5.61	4.68	3.74	2.81	1.87	0.94
72	8.51	7.66	6.81	5.96	5.11	4.25	3.40	2.55	1.70	0.85
70	7.98	7.18	6.38	5.59	4.79	3.99	3.19	2.39	1.60	0.80
67	7.24	6.52	5.79	5.07	4.35	3.62	2.90	2.17	1.45	0.72
65	6.78	6.10	5.43	4.75	4.07	3.39	2.71	2.04	1.36	0.68
60	5.74	5.17	4.60	4.02	3.45	2.87	2.30	1.72	1.15	0.57
50	4.08	3.67	3.26	2.85	2.45	2.04	1.63	1.22	0.82	0.41
40	2.85	2.56	2.28	1.99	1.71	1.42	1.14	0.86	0.57	0.29
30	1.94	1.74	1.55	1.35	1.16	0.97	0.78	0.58	0.39	0.19
20	1.23	1.11	0.99	0.86	0.74	0.62	0.49	0.37	0.25	0.12
10	0.78	0.70	0.62	0.54	0.47	0.39	0.31	0.23	0.16	0.08
0	0.48	0.43	0.39	0.34	0.29	0.24	0.19	0.14	0.10	0.05

7000 Grains of moisture = 1 pound of water.

Table 28. Drying Time and Conditions for Representative Material

KIND AND THICKNESS OF MATERIAL	Temp. Degrees F.	Drying Time
Bedding	150-190	4-6 Hours
Cereals	110-150	24 Hours
Cocoanut	145-155	30 Min.
Coffee	160-180	2½ Hours
Cores—Oil Sand, Molding ½"-1" Thick	300	4½ Hours
Cores—Oil Sand, Molding	480	10 Hours
Black Sand with Goulac Binder about 8"	480	
6/10 of Time for Oil Sand Cores 16"	700	
Feathers	150-180	
Films—Photographic	90	
Fruits and Vegetables	140	2-6 Hours
Furs	110	
Glue	70-90	2-4 Hours
Glue Size on Furniture	130	4 Hours
Gut	150	
Gypsum Wall Board—Start Wet	350	1 Hour
Gypsum Wall Board—Finish	190	
Gypsum Blocks	350-190	8-16 Hours
Hair Goods	150-190	1 Hour
Hats—Felt	140-180	
Hops	120-180	
Hides—Thin Leather	90	2-4 Hours
Ink—Printing	70-300	
Knitted Fabrics	140-180	
Leather—Thick Sole	90	4-6 Hours
Lumber—Green hardwood	100-180	3-180 Days
Lumber—Green softwood	160-220	2-14 Days
Macaroni	90-110	
Matches	140-180	
Milk and Other Liquid Foods (Spray Dried)	250-300	Instantaneous

Courtesy *Hoffman Specialty ITT*

Table 28. Drying Time and Conditions for Representative Material (Continued)

Molds, Green Sand, C.I. Flasks (1 sur. only ex.) 8″ T.	600	6 Hours
Molds, Green Sand, C.I. Flasks (1 sur. only ex.) 13″ T.	700	13 Hours
Nuts	75-140	24 Hours
Paper—Glued	130-300	
Paper—Treated	140-200	
Rubber	80-90	6-12 Hours
Sand, Loose 1″ deep	300	10-15 Min.
Shade Cloth	240	1-2 Hours
Soap	125	12 Hours
Starch	180-200	1-4 Hours
Stock Feed—Mixed	180-220	20-30 Min.
Sugar	150-200	20-30 Min.
Tannin and Other Chemicals (Spray Dried)	250-300	Instantaneous
Terracotta (Air Drying in Conditioning Room)	150-220	12-96 Hours
Wall Board	200-250	12-24 Hours

Courtesy *Hoffman Specialty ITT*

unit. In an up feed riser steam travels upwards and the condensate downward while in a down feed both steam and condensate travel down.

One Pipe System (Hot Water)—A hotwater heating system in which one-pipe serves both as a supply main and also as a return main. The heating units have separate supply and return pipes but both are connected to the same main.

One Pipe System (Steam)—A steam heating system consisting of a main circuit in which the steam and condensate flow in the same pipe. There is but one connection to each heating unit which must serve as both the supply and the return.

Overhead system—Any steam or hot water system in which the supply main is above the heating units. With a steam system the return must be below the heating units; with a water system, the return *may* be above the heating units.

Panel Heating—A method of heating involving the installation of the heating units (pipe coils) within the wall, floor or ceiling of the room.

Panel Radiator—A heating unit placed on, or flush with, a flat wall surface and intended to function essentially as a radiator. Do not confuse with panel heating system.

Table 29. Approximate Fuel Equivalent for House Heating
Approximate Number of Cubic Feet for Ton of Coal

Tons of Coal At 55% Efficiency 12500 BTU per Lb.	Gallons of Oil At 60% Efficiency 140000 BTU per Gal.	Cubic Feet of Gas At 75% Efficiency 550 BTU per Cu. Ft.
1.0	164.	33,300.
6.1	1000.	203,600.
.030	4.9	1000.

TYPE	SIZE	CU. FT./TON
ANTHRACITE	EGG	38
	STOVE	39
	CHESTNUT	39
	PEA & SMALLER	40
BITUMINOUS	EGG	48
	NUT	44
	LUMP	45
	SCREENINGS	44
	MINE RUN	42
COKE	EGG	74
	NUT	71
	PEA	66
	RANGE	74
PETROLEUM COKE	LUMP	74
	SCREENINGS	44
POCAHONTAS	EGG	41
	NUT	44
	LUMP	38
	STOVE	46
	MINE RUN	37

The above table was compiled from information furnished by the following:
Commercial Testing & Engneerng Co., Chcago
Anthracite Industries, Inc., New York

Courtesy *Hoffman Specialty ITT*

Plenum Chamber—An air compartment maintained under pressure and connected to one or more distributing ducts.

Pressure—Force per unit area such as lb. per sq. inch. See Static, Velocity, and Total Gage and Absolute Pressures. Unless otherwise qualified, it refers to unit static gage pressure.

Pressure Reducing Valve—A piece of equipment for changing the pressure of a gas or liquid from a higher to a lower one.

Table 30. Water Evaporated and Heat Required for Drying

M = PERCENTAGE OF MOISTURE IN MATERIAL TO BE DRIED
Q = LBS. WATER EVAPORATED PER TON (2000 LBS.) OF DRY MATERIAL
H = BRITISH THERMAL UNITS REQUIRED FOR DRYING PER TON OF DRY MATERIAL

M.	Q.	H.	M.	Q.	H.	M.	Q.	H.
1	20.2	85,624	14	325.6	424,884	35	1,077	1,269,240
2	40.8	108,696	15	352.9	458,248	40	1,333	1,555,960
3	61.9	130,424	16	381.0	489,720	45	1,636	1,895,320
4	83.3	156,296	17	409.6	521,752	50	2,000	2,303,000
5	105.3	180,936	18	439.0	554,680	55	2,444	2,800,280
6	127.7	206,024	19	469.1	558,392	60	3,000	3,423,000
7	150.5	231,560	20	500.0	623,000	65	3,714	4,222,680
8	173.9	257,768	21	531.6	658,392	70	4,667	5,290,040
9	197.8	284,536	22	564.1	694,792	75	6,000	6,783,000
10	222.2	311,864	23	597.4	732,088	80	8,000	9,023,000
11	247.2	339,864	24	631.6	770,392	85	11,333	12,755,960
12	272.7	368,424	25	666.7	809,704	90	18,000	20,223,000
13	298.9	397,768	30	857.0	1,022,840	95	38,000	42,623,000

FORMULA:

$Q = \frac{2000M}{100 - M}$. $H = 1120\,Q + 63{,}000$.

The value of H is found on the assumption that the moisture is heated from 62° to 212° and evaporated at that temperature, and that the specific heat of the material is 0.21 $(2000 \times (212 - 62) \times 0.21) = 63{,}000$.

Courtesy *Hoffman Specialty ITT*

Table 31. Heating Values of Fuel

All of the values given here are very approximate but sufficiently close for ordinary calculations. If greater accuracy is required the exact value should be obtained from the producer of the fuel in question. The heating value is also referred to as the "calorific value".

COAL	
TYPE	BTU PER POUND
ANTHRACITE	13,000
SEMI ANTHRACITE	13,700
BITUMINOUS	12,500
LIGNITE	7000
OIL	
GRADE	BTU PER GALLON
No. 1	135,000
No. 2	140,000
No. 4	155,000
No. 5	150,000
No. 6	153,000
GAS	
KIND	BTU PER CUBIC FOOT
NATURAL	1000
MANUFACTURED	550
PROPANE	2250
BUTANE	3000

Courtesy *Hoffman Specialty ITT*

Prime Surface—A heating surface having the heating medium on one side and air (or extended surface) on the other.

Radiant Heating—A heating system in which the heating is by radiation only. Sometimes applied to Panel Heating System.

Radiation—The transmission of heat in a straight line through space.

Radiator—A heating unit located within the room to be heated and exposed to view. A radiator transfers heat by radiation to objects "it can see" and by conduction to the surrounding air which in turn is circulated by natural convection.

Recessed Radiator—A heating unit set back into a wall recess but not enclosed.

Reducing Valve—See Pressure Reducing Valve.

Re-Evaporation—See Flash.

Refrigeration, Ton of—See Ton of Refrigeration.

Register—A grille with a built-in damper or shutter.

Table 32. Specific Heat (Average Values 32° to 212° F.)

Air*	.237	Ice	.505
Air†	.169	Iron, Cast	.113
Alcohol	.615	Kerosene	.500
Aluminum	.212	Lead	.030
Ammonia	1.098	Limestone	.217
Ammonia*	.520	Marble	.206
Ammonia†	.391	Mercury	.033
Antimony	.052	Mica	.208
Asbestos	.195	Nickel	.109
Benzole	.340	Nitrogen*	.244
Bismuth	.030	Nitrogen†	.173
Brass	.092	Oak	.570
Brick	.220	Olive Oil	.471
Bronze	.104	Oxygen*	.224
Carbon, Graphite	.126	Oxygen†	.155
Carbon, Dioxide*	.215	Osmium	.031
Carbon, Dioxide†	.168	Paraffin	.589
Carbon, Monoxide*	.243	Petroleum	.504
Carbon, Monoxide†	.173	Platinum	.032
Cement, Portland	.271	Rubber, Hard	.339
Chalk	.220	Sand	.195
Chloroform (Liquid)	.235	Selenium	.068
Chloroform (Gas)	.147	Silicon	.175
Coal	.201	Silver	.056
Cobalt	.103	Steam*	.480
Coke	.0203	Steam†	.350
Concrete	.156	Stones	.200
Copper	.092	Steel	.118
Cork	.485	Sulphuric Acid	.336
Cotton	.362	Tantalum	.033
Ether	.540	Tin	.054
Fuel Oil	.500	Turpentine	.420

Table 32. Specific Heat (Average Values 32° to 212° F.) (Continued)

Gasoline	.500	Tungsten	.034
Glass	.180	Water	1.000
Gold	.032	Wool	.393
Gypsum	.259	Wood	.327
Hydrogen*	3.41	Zinc	.093
Hydrogen†	2.81		

* = Constant Pressure. † = Constant Volume. Courtesy *Hoffman Specialty ITT*

Table 33. Specific Gravity of Metals

METAL	SPECIFIC GRAVITY
Water (Basis for Comparison)	1.00
Aluminum	2.55-2.80
Tin (Cast)	7.2-7.5
Steel	7.84
Cast Iron	7.03-7.13
Wrought Iron	7.6-7.9
Brass	8.4-8.7
Copper	8.8-8.95
Lead (Cast)	11.35
Mercury	13.60
Platinum	21.50

Courtesy *Hoffman Specialty ITT*

Relative Humidity—The amount of moisture in a given quantity of air compared with the maximum amount of moisture the same quantity of air could hold at the same temperature. It is expressed as a percentage.

Return Mains—The pipes which return the heating medium from the heating units to the source of heat supply.

Reverse-Return System (Hot Water)—A two-pipe hot water heating system in which the water from the several heating units is returned along paths arranged so that all radiator circuits of the system are practically of equal length.

Roof Ventilator—A device placed on the roof of a building to permit egress of air.

Sensible Heat—Heat which only increases the temperature of objects as opposed to latent heat.

Table 34. Lineal Expansion of Solids at Ordinary Temperatures.

MATERIAL	For 1°F. Length = 1"
Aluminum (Cast)	0.00001234
Brass Cast	0.00000957
Brass Plate	0.00001052
Brick (Fire)	0.00000300
Bronze (Copper, 96½; Tin, 2½; Zinc, 1)	0.00000986
Copper	0.00000887
Glass, Hard	0.00000397
Gold, Pure	0.00000786
Iron, Wrought	0.00000648
Iron, Cast	0.00000556
Lead	0.00001571
Mercury (Cubic Expansion)	0.00009984
Nickel	0.00000695
Porcelain	0.00000200
Silver, Pure	0.00001079
Slate	0.00000577
Steel, Cast	0.00000636
Steel, Tempered	0.00000689
Stone (Sandstone), Dry	0.00000652
Tin	0.00001163
Wood, Pine	0.00000276
Zinc	0.00001407

Courtesy *Hoffman Specialty ITT*

Specific Heat—In the foot-pound-second system, the amount of heat (Btu) required to raise one pound of a substance one degree Fahrenheit. In the centimeter-gram-second system, the amount of heat (cal.) required to raise one gram of a substance one degree centigrade. The specific heat of water is 1.

Split System—A system in which the heating is accomplished by radiators or convectors and ventilation by separate apparatus.

Square Foot of Heating Surface—Equivalent direct radiation (EDR). By definition, that amount of heating surface which will give off 240

Table 35. Weights of Materials

MATERIAL	Pounds per Cubic Ft.	MATERIAL	Pounds per Cubic Ft.
Aluminum	166.5	Lead	709.7
Ashes	45-50	Lignite	31-47
Barley	37-40	Lime	50-80
Brass—Copper, Zinc		Limestone	156-162
80 20	536.3	Manganese	450
70 30	523.8	Mercury 32°	849.3
60 40	521.3	Mercury 60°	846.8
50 50	511.4	Mercury 212°	834.4
Bronze—Cop. 95 to 80	552.	Nickel	548.7
Tin 5 to 20		Oats	25-30
Cement	90-118	Ore (Iron)	105-215
Charcoal	17-27	Platinum	1333
Clay	95-169	Rye	44-50
Coal (Lump)	50-54	Sand	75-120
Nut Coal & Screenings	53-60	Silver	655.1
Coke	26-30	Slag (Blast Furnace)	37-63
Earth	75-115	Steel	489.6
Gold, Pure	1200.9	Stone	90-120
Copper	552	Wheat	458.3
Gravel	90-135	Wheat	44-50
Iron, Cast	450	Zinc	448
Iron, Wrought	480		

Courtesy *Hoffman Specialty ITT*

Btu per hour when filled with a heating medium at 215°F. and surrounded by air at 70°F. The equivalent square foot of heating surface may have no direct relation to the actual surface area.

Static Pressure—The pressure which tends to burst a pipe. It is used to overcome the frictional resistance to flow through the pipe. It is expressed as a unit pressure and may be either in absolute or gage pressure. It is frequently expressed in feet of water column or (in the case of pipe friction) in mil-inches of water column per foot of pipe.

Steam—Water in the vapor phase. The vapor formed when water has been heated to its boiling point, corresponding to the pressure it is

Table 36. Melting Points of Metals

	Deg. F.		Deg. F.
Aluminum	1220	Iron (Cast) Gray	2460-2550
Antimony	1167	Iron (Cast) White	1920-2010
Bismuth	520	Iron, Wrought	2460-2640
Brass (Red)	1870	Lead	622
Bronze	1900	Silver (Pure)	1751
Copper	1981	Steel	2370-2550
Glass	2377	Tin	449
Gold (Pure)	1945	Zinc	787
Solder (Lead-Tin)	350-570		

Courtesy *Hoffman Specialty ITT*

Table 37. Boiling Points of Various Fluids

	Deg. F.		Deg. F.
Water (Atmospheric Pressure)	212	Turpentine	320
Alcohol	173	Sulphur	832
Sulphuric Acid	620	Linseed Oil	549

Courtesy *Hoffman Specialty ITT*

under. See also Dry Saturated Steam, Wet Saturated Steam, Super Heated Steam.

Steam Heating System—A heating system in which the heating units give up their heat to the room by condensing the steam furnished to them by a boiler or other source.

Steam Trap—A device for allowing the passage of condensate and air but preventing the passage of steam. See Thermostatic, Float, Bucket Trap.

Superheated Steam—Steam heated above the temperature corresponding to its pressure.

Supply Mains—The pipes through which the heating medium flows from the boiler or source of supply to the run-outs and risers leading to the heating units.

Tank Regulator—See Temperature Regulator.

Temperature Regulator—A piece of equipment for controlling the admission of steam to a hot water (or other liquid) heating device in the correct quantities so that the temperature of the liquid will remain constant.

Table 38. Conversion Table (Metric to English Measure)

METRIC	ENGLISH
MEASURES OF LENGTH	
1 Kilometer 1000 Meters	0.621 Mile 3281. Feet
1 Meter 100 Centimeters 1000 Millimeters	1.094 Yards 3.28 Feet 39.37 Inches
1 Centimeter 10 Millimeters	0.0328 Feet 0.394 Inches
1 Millimeter	0.0394 Inches
MEASURES OF SURFACE	
1 Sq. Kilometer 1,000,000 Sq. Meters	0.386 Sq. Mile 247.1 Acres 1,195,985 Sq. Yards
1 Sq. Meter 10,000 Sq. Centimeters	1.196 Sq. Yards 10.76 Sq. Feet 1550. Sq. Inches
1 Sq. Centimeter 100 Sq. Millimeters 1 Sq. Millimeter	0.155 Sq. Inch 0.0011 Sq. Feet 0.00155 Sq. Inch
MEASURES OF VOLUME AND CAPACITY	
1 Cu. Meter 1000 Liters 1,000,000 Cu. Centimeters	1.308 Cu. Yards 35.31 Cu. Feet 61023.4 Cu. Inches
1 Liter 1,000 Cu. Centimeters	0.264 Gallons (U.S.) 0.220 Gallons (Imperial) 1.057 Quarts (U.S.) 0.880 Quarts (Imperial)
1 Cu. Centimeter 1000 Cu. Millimeters	0.061 Cu. Inches
MEASURES OF WEIGHT	
1 Kilogram 1000 Gram	0.0011 Ton (2000 Lbs.) 2.205 Pounds (Av.)
1 Gram	0.0022 Pounds (Av.) 0.035 Ounces (Av.) 15.43 Grains

NOTE—For conversion of pressures from metric to English units see 'Conversion Table—Pressures."

Courtesy *Hoffman Specialty ITT*

Table 38. Conversion Table (Metric to English Measure)

INCHES		FEET Water	MILLIMETER Mercury (Hg.)	METER Water	POUNDS		GRAMS Sq. Centimeter	KILOGRAMS		Atmos-phere
Water	Mercury				Per Sq. Inch	Per Sq. Ft.		Per Sq. Meter	Per Sq. Cm.	
1.0	0.074	0.083	1.88	0.0254	0.036	5.20	2.53	25.37	0.003	0.0024
13.6	1.0	1.13	25.4	0.344	0.490	70.5	34.4	344.4	0.0344	0.0333
12.0	0.884	1.0	22.4	0.305	0.433	62.4	30.4	304.5	0.0304	0.0295
0.54	0.039	0.045	1.0	0.014	0.019	2.78	1.36	13.6	0.00136	0.00131
39.4	2.89	3.28	73.5	1.0	1.422	204.6	100.0	1000.0	0.10	0.0967
27.7	2.04	2.390	51.8	0.703	1.0	144.0	70.3	703.1	0.070	0.0680
0.19	0.014	0.016	0.36	0.005	0.0069	1.0	0.49	4.88	0.00049	0.00047
0.40	0.03	0.033	0.74	0.01	0.014	2.05	1.0	10.0	0.001	0.00097
0.04	0.003	0.0033	0.074	0.001	0.0014	0.205	0.10	1.0	0.0001	0.0001
393.8	28.96	32.8	735.5	10.0	14.2	2048.	1000.0	10000.0	1.0	0.9678
407.	29.92	33.9	760.0	10.3	14.7	2116.	1033.2	10332.	1.03	1.0

To use table go to column headed by unit to be converted. Follow this column down to the "1.0" in heavy print and read horizontally across. Example: Convert five kilograms per sq. meter to lbs. per sq. inch. Select column headed kilograms per sq. meter and follow down to "1.0", then to left column headed lbs. per sq. inch. The number 0.0014 found in this space is the conversion factor by which the number of kilograms per sq. meter must be multiplied to change this quantity to lbs. per sq. inch. Therefore, five kilograms per meter is equal to five times 0.0014, or 0.007 lb. per sq. inch.

Courtesy *Hoffman Specialty ITT*

Table 40. Conversion Table (Inches to Centimeters to Millimeters)

in.	cm.	mm.
1.00	2.54	25.40
2.00	5.08	50.80
3.00	7.62	76.20
4.00	10.16	101.60
5.00	12.70	127.00
6.00	15.24	152.40
7.00	17.78	177.80
8.00	20.32	203.20
9.00	22.86	228.60
10.00	25.40	254.00
20.00	50.80	508.00
30.00	76.20	762.00
36.00	91.40	914.00
40.00	101.60	1016.00
50.00	127.00	1270.00
60.00	152.40	1524.00
70.00	1778.0	1778.00
80.00	203.20	2032.00
90.00	228.60	2286.00
100.00	254.00	2540.00

Courtesy *Hoffman Specialty ITT*

Thermostat—An instrument which responds to changes in temperature and which directly or indirectly controls the room temperature.

Thermostatic Trap—A steam trap which opens by a drop in temperature such as when cold condensate or air reaches it and closes it when steam reaches it. The temperature sensitive element is usually a sealed bellows or series of diaphragm chambers containing a small quantity of volatile liquid.

Ton of Refrigeration—The heat which must be extracted from one ton (2,000 lbs.) of water at 32°F. to change it into ice at 32°F. in 24 hours. It is equal to 288,000 Btu/24 hours, 12,000 Btu/hour, or 200 Btu/minute.

Total Heat—The latent heat of vaporization added to the heat of the liquid with which it is in contact.

Total Pressure—The sum of the static and velocity pressures. It is also used as the total static pressure over an entire area, that is, the unit pressure multiplied by the area on which it acts.

Table 41. Fahrenheit—Celsius Conversion Table

F.	C.	F.	C.	F.	C.	F.	C.
−20	−28.9	62	16.7	144	62.2	226	107.8
−18	−27.8	64	17.8	146	63.3	228	108.9
−16	−26.7	66	18.9	148	64.4	230	110.
−14	−25.6	68	20.	150	65.6	232	111.1
−12	−24.4	70	21.1	152	66.7	234	112.2
−10	−23.3	72	22.2	154	67.8	236	113.3
− 8	−22.2	74	23.3	156	68.9	238	114.4
− 6	−21.1	76	24.4	158	70.	240	115.6
− 4	−20.	78	25.6	160	71.1	242	116.7
− 2	−18.9	80	26.7	162	72.2	244	117.8
0	−17.8	82	27.8	164	73.3	246	118.9
2	−16.7	84	28.9	166	74.4	248	120.
4	−15.6	86	30.	168	75.6	250	121.1
6	−14.4	88	31.1	170	76.7	252	122.2
8	−13.3	90	32.2	172	77.8	254	123.3
10	−12.2	92	33.3	174	78.9	256	124.4
12	−11.1	94	34.4	176	80.	258	125.6
14	−10.	96	35.6	178	81.1	260	126.7
16	− 8.9	98	36.7	180	82.2	262	127.8
18	− 7.8	100	37.8	182	83.3	264	128.9
20	− 6.7	102	38.9	184	84.4	266	130.
22	− 5.6	104	40.	186	85.6	268	131.1
24	− 4.4	106	41.1	188	86.7	270	132.2
26	− 3.3	108	42.2	190	87.8	272	133.3
28	− 2.2	110	43.3	192	88.9	274	134.4
30	− 1.1	112	44.4	194	90.	276	135.6
32	0.	114	45.6	196	91.1	278	136.7
34	1.1	116	46.7	198	92.2	280	137.8
36	2.2	118	47.8	200	93.3	282	138.9
38	3.3	120	48.9	202	94.4	284	140.
40	4.4	122	50.	204	95.6	286	141.1
42	5.6	124	51.1	206	96.7	288	142.2
44	6.7	126	52.2	208	97.8	290	143.3
46	7.8	128	53.3	210	98.9	292	144.4
48	8.9	130	54.4	212	100.	294	145.6
50	10.	132	55.6	214	101.1	296	146.7
52	11.1	134	56.7	216	102.2	298	147.8
54	12.2	136	57.8	218	103.3	300	148.9
56	13.3	138	58.9	220	104.4		
58	14.4	140	60.	222	105.6		
60	15.6	142	61.1	224	106.7		

Courtesy *Hoffman Specialty ITT*

Table 42. Electrical Units

VOLT—The unit of electrical motive force, force required to send one ampere of current through one ohm of resistance.

OHM—Unit of resistance. The resistance offered to the passage of one ampere, when impelled by 1-volt.

AMPERE—Unit of current, the current which one volt can send through a resistance of one ohm.

COULOMB—Unit of quantity. Quantity of current which impelled by one volt, would pass through one ohm in one second.

JOULE—Unit of work. The work done by one watt in one second.

WATT—The unit of electrical energy, and is the product of ampere and volt. That is, one ampere of current flowing under a pressure of one volt gives one watt of energy.

One electrical horsepower is equal to 746 watts.

ONE KILOWATT—Is equal to 1000 watts, or 3415 B.T.U. when used for heating or the equivalent output of 14.2 sq. ft. of steam radiation.

ONE KILOWATT HOUR—(KW. HR.) equals the consumption of 1000 watts in one hour.

To find the watts consumed in a given electrical circuit, multiply the volts by the amperes.

To find the volts—divide the watts by the amperes.

To find the amperes—divide the watts by the volts.

To find the watts consumed in a given electrical circuit, multiply the volts by motor by 746. With A.C. current multiply the wattage by the power factor, then divide by 74.6.

To find the amperes of a given circuit, of which the volts and ohms resistance are known, divide the volts by the ohms.

To find the volts, when the amperes and ohms are known, multiply the amperes by the ohms.

To find the resistance in ohms, when the volts and amperes are known, divide the volts by the amperes.

Courtesy *Hoffman Specialty ITT*

Trap—See Steam Trap, Thermostatic Trap, Float Trap, and Bucket Trap.

Two-Pipe System (Steam or Water)—A heating system in which one pipe is used for the supply main and another for the return main. The essential feature of a two-pipe hot water system is that each heating unit receives a direct supply of the heating medium which cannot have served a preceding heating unit.

Unit Heater—A heating unit consisting of a heat transfer element, a housing, a fan with driving motor, and outlet deflectors or diffusers. It is

Table 43. Capacity of Round and Rectangular Storage Tanks

Depth or Length	NUMBER OF GALLONS									
	18	24	30	36	42	48	54	60	66	72
1 In.	1.10	1.96	3.06	4.41	5.99	7.83	9.91	12.24	14.81	17.62
2 Ft.	26	47	73	105	144	188	238	294	356	423
2½	33	59	91	131	180	235	298	367	445	530
3	40	71	100	158	216	282	357	440	534	635
3½	46	83	129	184	252	329	416	513	623	740
4	53	95	147	210	288	376	475	586	712	846
4½	59	107	165	238	324	423	534	660	800	952
5	66	119	181	264	360	470	596	734	899	1057
5½	73	130	201	290	396	517	655	808	979	1163
6	79	141	219	315	432	564	714	880	1066	1268
6½	88	155	236	340	468	611	770	954	1156	1374
7	92	165	255	368	504	658	832	1028	1244	1480
7½	99	179	278	396	540	705	889	1101	1335	1586
8	106	190	291	423	576	752	949	1175	1424	1691
9	119	212	330	476	648	846	1071	1322	1599	1903
10	132	236	366	529	720	940	1189	1463	1780	2114
12	157	282	440	634	684	1128	1428	1762	2133	2537
14	185	329	514	740	1008	1316	1666	2056	2490	2960
16	211	376	587	846	1152	1504	1904	2350	2844	3383
18	238	423	660	952	1296	1692	2140	2640	3200	3806
20	264	470	734	1057	1440	1880	2380	2932	3556	4230

RECTANGULAR TANKS

To find the capacity in U.S. gallons of rectangular tanks, reduce all dimensions to inches, then multiply the length by the width by the height and divide the product by 231.

Example:

Tank 56″ long × 32″ wide × 20″ deep
Then 56″ × 32″ × 20″ = 35840 cu. in.
35840 ÷ 231 = 155 gallons capacity

Courtesy *Hoffman Specialty ITT*

usually suspended from the ceiling and its heat output is controlled by starting and stopping the fan by a room thermostat. The circulation of the heating medium (steam or hot water) is usually continuous. It is used mostly for industrial heating.

Unit Pressure—Pressure per unit area as lbs. per sq. in.

Table 44. Useful Data

Diameter	×	3.1416	=	Circumference
Circumference	×	.3183	=	Diameter
Diameter^2	×	.7854	=	Area of Circle
Area of Circle	×	1.2732	=	Area of Circumscribed Square
Area of Circle	×	.63662	=	Area of Inscribed Square
Diameter of Circle	×	.88623	=	Side of Equal Square
Diameter of Circle	×	.7071	=	Side of Inscribed Square
Circumerence of Circle	×	1.1284	=	Perimeter of Equal Square
Side of Square	×	1.4142	=	Diameter of Circumscribed Circle
Side of Square	×	1.1284	=	Diameter of Equal Circle
Perimeter of Square	×	.88623	=	Circumference of Equal Circle
Diameter^2	×	3.1416	=	Surface of Sphere
Diameter	×	.5236	=	Volume of Sphere
Diameter of Sphere	×	.806	=	Dimensions of Equal Cube
Diameter of Sphere	×	.667	=	Length of Equal Cylinder
Area of Base	×	⅓ Height	=	Volume of Pyramid or Cone
Base	×	½ Height	=	Area of Triangle
Radius	×	1.1547	=	Side of Inscribed Cube
Square Inches	×	1.2732	=	Circular Inches
Square Inches	×	.00695	=	Square Feet
Square Feet	×	.111	=	Square Yard
Square Yards	×	.0002066	=	Acres
Cubic Inches	×	.00058	=	Cubic Feet
Cubic Feet	×	.03704	=	Cubic Yards
Cubic Inches	×	.004329	=	U.S. Gallons
Cubic Feet	×	7.4805	=	U.S. Gallons
Cubic Inches	×	.000466	=	U.S. Bushels
Cubic Feet	×	.8036	=	U.S. Bushels
U.S. Bushels	×	2150.42	=	Cubic Inches
U.S. Bushels	×	1.242	=	Cubic Feet
U.S. Bushels	×	.046	=	Cubic Yards
U.S. Gallons	×	231.0	=	Cubic Inches
U.S. Gallons	×	.13368	=	Cubic Feet
Cubic Inches Water	×	.036127	=	Pounds (Avoirdupois)
Cubic Feet Water	×	62.4283	=	Pounds (Avoirdupois)
U.S. Gallons Water	÷	268.8	=	Tons
Column of Water 1″ Diameter x 12″ High			=	.34 Pounds (Avoirdupois)
Cubic Inch	×	.263	=	Pound Average Cast Iron
Cubic Inch	×	.281	=	Pound Average Wrought Iron
Cubic Inch	×	.283	=	Pound Average Cast Steel
Cubic Inch	×	.3225	=	Pound Average Copper
Cubic Inch	×	.3037	=	Pound Average Brass
Cubic Inch	×	.26	=	Pound Average Zinc
Cubic Inch	×	.4103	=	Pound Average Lead

Table 44. Useful Data (Continued)

Cubic Inch × .2636	= Pound Average Tin
Cubic Inch × .4908	= Pound Average Mercury
12 × Weight of Pine Pattern	= Iron Casting
13 × Weight of Pine Pattern	= Brass Casting
14 × Weight of Pine Pattern	= Lead Casitng
1—Calorie	= 3.968 B.T.U.
1—Btu	= 0.252 Calorie
1—Pound per Square Inch	= 703.08 Kilograms per M^2
1—Kilogram per M^2	= 0.00142 Pounds per Square Inch
1—Calorie per M^2	= 0.3687 Btu per Square Foot
1—Btu per Square Foot	= 2.712 Calories per M^2
1—Calorie per M^2 per Degree Difference Centigrade	= 0.2048 Btu per Square Foot per Degree Difference, Fahrenheit
1—Btu per Square Foot per Degree Difference Fahrenheit	= 4.882 Calories per M^2 per Degree Difference Fahrenheit
1—Btu per Pound	= 0.556 Calorie per Kilogram
1—Calorie per Kilogram	= 1.8 Btu per Pound
1—Liter of Coke at 26.3 Lb. per Cu. Ft.	= 0.93 Pounds
1—Lb. of Coke at 26.3 per Cu. Ft.	= 1.076 Liter
Water Expands in Bulk from 40° to 212°	= One Twenty-Third

Courtesy *Hoffman Specialty ITT*

Up-Feed System (Hot Water or Steam)—A heating system in which the supply mains are below the level of the heating units which they serve.

Vacuum Heating System (Steam)—A one- or two-pipe heating system equipped with the necessary accessory apparatus to permit the pressure in the system to go below atmospheric.

Vapor—Any substance in the gaseous state.

Vapor Heating System (Steam)—A two-pipe heating system which operates under pressure at or near atmospheric and which returns the condensation to the boiler or receiver by gravity.

Velocity Pressure—The pressure used to create the velocity of flow in a pipe. It is expressed as a unit pressure.

Ventilation—Air circulated through a room for ventilating purposes. It may be mechanically circulated with a blower system or it may be natural circulation through an open window, etc.

Vent Valve (Steam)—A device for permitting air to be forced out of a heating unit or pipe and which closes against water and steam.

Vent Valve (Water)—A device permitting air to be pushed out of a pipe or heating unit but which closes against water.

Warm Air Heating System—A warm air heating plant consists of a heating unit (fuel-burning furnace) enclosed in a casing, from which the heated air is distributed to the various rooms of the building through ducts. If the motive head producing flow depends on the difference in weight between the heated air leaving the casing and the cooler air entering the bottom of the casing, it is termed a gravity system. A booster fan may, however, be used in conjunction with a gravity-designed system. If a fan is used to produce circulation and the system is designed especially for fan circulation, it is termed a fan furnace system or a central fan furnace system. A fan furnace system may include air washer, filters, etc.

Wet Bulb Temperature—The lowest temperature which a water-wetted body will attain when exposed to an air current.

Wet Return (Steam)—That part of a return main of a steam heating system which is completely filled with water of condensation.

Wet Saturated Steam—Saturated steam containing some water particles in suspension.

Table 45. Weights of Steel and Iron Plates and Sheets (U.S. Standard Gauge)

No. of Gage	Thickness Inches (Approx.)	WEIGHT PER SQUARE FOOT	
		Iron	Steel
0000000	.500	20.00	20.40
000000	.469	18.75	19.12
00000	.437	17.50	17.85
0000	.406	16.25	16.57
000	.375	15.00	15.30
00	.344	13.75	14.02
0	.312	12.50	12.75
1	.281	11.25	11.47
2	.266	10.62	10.84
3	.250	10.00	10.20
4	.234	9.37	9.56
5	.219	8.75	8.92
6	.203	8.12	8.29
7	.187	7.50	7.65
8	.172	6.87	7.01
9	.156	6.25	6.37
10	.141	5.62	5.74
11	.125	5.00	5.10
12	.109	4.37	4.46
13	.094	3.75	3.82

Table 45. Weights of Steel and Iron Plates and Sheets (U.S. Standard Gauge) (Continued)

14	.078	3.12	3.19
15	.070	2.81	2.87
16	.062	2.50	2.55
17	.056	2.25	2.29
18	.050	2.00	2.04
19	.044	1.75	1.78
20	.037	1.50	1.53
21	.034	1.37	1.40
22	.031	1.25	1.27
23	.028	1.13	1.15
24	.025	1.00	1.02
25	.022	0.87	0.89
26	.019	0.75	0.76
27	.017	0.69	0.70
28	.016	0.63	0.64
29	.014	0.56	0.57
30	.012	0.50	0.51
31	.011	0.44	0.45
32	.010	0.40	0.41
33	.009	0.37	0.38
34	.009	0.34	0.35
35	.008	0.31	0.32
36	.007	0.28	0.29
37	.007	0.27	0.27
38	.006	0.25	0.25

Courtesy *Hoffman Specialty ITT*

COMMON HEATING TROUBLES IN STEAM HEATING SYSTEMS

The following are a few of the most common difficulties experienced with steam or hot water heating systems, and the probable causes.

Boiler Troubles

The Boiler Fails to Deliver Enough Heat—The cause of this condition may be: (a) poor draft, (b) poor fuel, (c) improper attention or firing, (d) boiler too small, (e) improper piping, (f) improper arrangement of sections, (g) heating surfaces covered with soot, and (h) improper firing rate of oil or gas burner.

The Water Line Is Unsteady—The cause of this condition may be: (a) grease and dirt in boiler, (b) a water column which is connected to a

Table 46. Areas of Circles

Size	Area	Size	Area	Size	Area	Size	Area
1/64	.00019	5½	23.758	25	490.87	64	3216.0
1/32	.00077	6	28.274	26	530.93	65	3318.3
1/16	.00307	6½	33.183	27	572.55	66	3421.2
3/32	.00690	7	38.484	28	615.75	67	3525.6
1/8	.01227	7½	44.178	29	660.52	68	3631.6
5/32	.01917	8	50.265	30	706.86	69	3739.2
3/16	.02761	8½	56.745	31	754.76	70	3848.4
7/32	.03758	9	63.617	32	804.24	71	3959.2
1/4	.04909	9½	70.882	33	855.30	72	4071.5
5/16	.07670	10	78.54	34	907.92	73	4185.3
3/8	.11045	10½	86.59	35	962.11	74	4300.8
7/16	.15033	11	95.03	36	1017.8	75	4417.8
1/2	.19635	11½	103.86	37	1075.2	76	4536.4
9/16	.24580	12	113.09	38	1134.1	77	4656.0
5/8	.30680	12½	122.71	39	1194.5	78	4778.3
11/16	.37122	13	132.73	40	1256.6	79	4901.6
3/4	.44179	13½	143.13	41	1320.2	80	5026.5
13/16	.51849	14	153.93	42	1385.4	81	5153.0
7/8	.60132	14½	165.13	43	1452.2	82	5281.0
15/16	.69029	15	176.71	44	1520.5	83	5410.6
1	.7854	15½	188.69	45	1590.4	84	5541.7
1⅛	.9940	16	201.06	46	1661.9	85	5674.5
1¼	1.227	16½	216.82	47	1734.9	86	5808.8
1⅜	1.484	17	226.98	48	1809.5	87	5944.6
1½	1.767	17½	240.52	49	1885.7	88	6082.1
1⅝	2.073	18	254.46	50	1963.5	89	6221.1
1¾	2.405	18½	268.80	51	2042.8	90	6361.7
1⅞	2.761	19	283.52	52	2123.7	91	6503.8
2	3.141	19½	298.64	53	2206.1	92	6647.6
2¼	3.976	20	314.16	54	2290.2	93	6792.9
2½	4.908	20½	330.06	55	2375.8	94	6939.7
2¾	5.939	21	346.36	56	2463.0	95	7088.2
3	7.068	21½	363.05	57	2551.7	96	7238.2
3¼	8.295	22	380.13	58	2642.0	97	7389.8
3½	9.621	22½	397.60	59	2733.9	98	7542.9
3¾	11.044	23	415.47	60	2827.4	99	7697.7
4	12.566	23½	433.73	61	2922.4	100	7854.0
4½	15.904	24	452.39	62	3019.0		
5	19.635	24½	471.43	63	3117.2		

To find the circumference of a circle when diameter is given, multiply the given diameter by 3.1416. To find the diameter of a circle when circumference is given, multiply the given circumference by .31831.

Courtesy *Hoffman Specialty ITT*

Table 47. ASTM Schedule 40 (S) Pipe Dimensions

DIAMETER			Nominal Thickness In.	CIRCUMFERENCE		TRANSVERSE AREAS		Length of Pipe Per Sq. Ft. of External Surface Ft.	Length of Pipe Containing 1 Cu. Ft.	Nominal Weight Per Ft. in Lbs.	Number of Threads Per Inch of Screw
Nominal Internal In.	Actual External In.	Approx. Internal In.		External In.	Internal In.	External Sq. In.	Internal Sq. In.				
1/8	0.405	0.27	0.068	1.27	0.85	0.13	0.06	9.44	2513.00	0.24	27
1/4	0.540	0.36	0.088	1.70	1.14	0.23	0.10	7.08	1383.30	0.42	18
3/8	0.675	0.49	0.091	2.12	1.55	0.36	0.19	5.66	751.20	0.57	18
1/2	0.840	0.62	0.109	2.63	1.95	0.55	0.30	4.55	472.40	0.85	14
3/4	1.050	0.82	0.113	3.30	2.59	0.87	0.53	3.64	270.00	1.13	14
1	1.315	1.05	0.134	4.13	3.29	1.36	0.86	2.90	166.90	1.68	11 1/2
1 1/4	1.660	1.38	0.140	5.22	4.34	2.16	1.50	2.30	96.25	2.27	11 1/2
1 1/2	1.900	1.61	0.145	5.97	5.06	2.84	2.04	2.01	70.66	2.72	11 1/2
2	2.375	2.07	0.154	7.46	6.49	4.43	3.36	1.61	42.91	3.65	11 1/2
2 1/2	2.875	2.47	0.204	9.03	7.75	6.49	4.78	1.33	30.10	5.79	8
3	3.500	3.07	0.217	11.00	9.63	9.62	7.39	1.09	19.50	7.57	8
3 1/2	4.000	3.55	0.226	12.57	11.15	12.57	9.89	0.96	14.57	9.11	8
4	4.500	4.03	0.237	14.14	12.65	15.90	12.73	0.85	11.31	10.79	8
5	5.563	5.05	0.259	17.48	15.85	24.31	19.99	0.69	7.20	14.62	8
6	6.625	6.07	0.280	20.81	19.05	34.47	28.89	0.58	4.98	18.97	8
8	8.625	8.07	0.276	27.10	25.35	58.43	51.15	0.44	2.82	24.69	8
8	8.625	7.98	0.322	27.10	25.07	58.43	50.02	0.44	2.88	28.55	8
9	9.625	8.94	0.344	30.24	28.08	72.76	62.72	0.40	2.29	33.91	8
10	10.750	10.19	0.278	33.77	32.01	90.76	81.55	0.36	1.76	31.20	8
10	10.750	10.14	0.306	33.77	31.86	90.76	80.75	0.36	1.78	34.24	8
10	10.750	10.02	0.366	33.77	31.47	90.76	78.82	0.36	1.82	40.48	8
12	12.750	12.09	0.328	40.06	37.98	127.68	114.80	0.30	1.25	43.77	8
12	12.750	12.00	0.375	40.06	37.70	127.68	113.10	0.30	1.27	49.56	8

Courtesy *Hoffman Specialty ITT*

Table 48. Surface Areas and Volumes of Spheres

Diameter	Surface Area	Volume	Diameter	Surface Area	Volume
1/8	.04908	.00102	4 1/4	56.745	40.195
1/4	.19636	.00818	4 1/2	63.617	47.712
3/8	.44180	.02761	4 3/4	70.882	56.115
1/2	.78540	.06545	5	78.540	65.450
5/8	1.2272	.12783	5 1/4	86.590	75.766
3/4	1.7672	.22090	5 1/2	95.033	87.113
7/8	2.4053	.35077	5 3/4	103.87	99.542
1	3.1416	.52360	6	113.10	113.10
1 1/8	3.9760	.74550	6 1/4	122.72	127.83
1 1/4	4.9088	1.0227	6 1/2	132.73	143.79
1 3/8	5.9396	1.3611	6 3/4	143.14	161.03
1 1/2	7.0684	1.7671	7	153.94	179.60
1 5/8	8.2956	2.2467	7 1/4	165.13	199.53
1 3/4	9.6212	2.8062	7 1/2	176.71	220.88
1 7/8	11.045	3.4516	7 3/4	188.69	243.72
2	12.566	4.1887	8	201.06	268.08
2 1/4	15.904	5.9640	8 1/4	213.82	294.00
2 1/2	19.635	8.1812	8 1/2	226.98	321.55
2 3/4	23.758	10.889	8 3/4	240.53	350.77
3	28.274	14.137	9	254.47	381.70
3 1/4	33.183	17.974	9 1/4	268.80	414.40
3 1/2	38.484	22.449	9 1/2	283.53	448.92
3 3/4	44.179	27.612	9 3/4	298.65	485.30
4	50.266	35.511	10	314.16	523.60

SPHERE FORMULA

This table can be used for feet, inches or any metric unit. For example, the volume of a 2" diameter sphere is 4.1887 cu. in., and for a 2 ft. diameter, 4.1887 cu. ft. The figures apply to either the exterior or to the interior of a hollow sphere, provided the diameter is measured at the proper place. For example: the capacity of a spherical tank measuring 10 ft. on the inside is 523.60 cu. ft. A float ball having an outside diameter of 6 in. has a volume of 113.10 cu. in.

The area or the volume of a sphere of a diameter not given in the table may be figured from the following simple formula:

$$S = 4A \text{ and } V = 0.524D^3$$

in which:

D = Diameter of the Sphere
A = Area of a Circle of Diameter D
S = Surface Area of Sphere
V = Volume of Sphere

Courtesy *Hoffman Specialty ITT*

very active section and, therefore, may not be showing actual water level in boiler, and (c) boiler operating at excessive output.

Water Disappears from Gauge Glass—This may be caused by: (a) priming due to grease and dirt in boiler, (b) too great a pressure

Table 49. Decimal Equivalents of Parts of an Inch

1/64	.01563	33/64	.51563
1/32	.03125	17/32	.53125
3/64	.04688	35/64	.54688
1/16	.0625	9/16	.5625
5/64	.07813	37/64	.57813
3/32	.09375	19/32	.59375
7/64	.10938	39/64	.60938
1/8	.125	5/8	.625
9/64	.14063	41/64	.64063
5/32	.15625	21/32	.65625
11/64	.17188	43/64	.67188
3/16	.1875	11/16	.6875
13/64	.20313	45/64	.70313
7/32	.21875	23/32	.71875
15/64	.23438	47/64	.73438
1/4	.25	3/4	.75
17/64	.26563	49/64	.76563
9/32	.28125	25/32	.78125
19/64	.29688	51/64	.79688
5/16	.3125	13/16	.8125
21/64	.32813	53/64	.82813
11/32	.34375	27/32	.84375
23/64	.35938	55/64	.85938
3/8	.375	7/8	.875
25/64	.39063	57/64	.89063
13/32	.40625	29/32	.90625
27/64	.42188	59/64	.92188
7/16	.4375	15/16	.9375
29/64	.45313	61/64	.95313
15/32	.46875	31/32	.96875
31/64	.48438	63/64	.98438
1/2	.5	1	1.00000

Courtesy *Hoffman Specialty ITT*

difference between supply and return piping causing water to back into return, (c) valve closed in return line, (d) connection of bottom of water column into a very active section or thin waterway, (e) improper connections between boilers in battery permitting boiler with excess pressure to push water into boiler with lower pressure, and (f) too high a firing rate.

Table 50. Flue Sizes and Chimney Heights

BOILER CAPACITY SQ. FT. RADIATION		RECTAGULAR FLUE		ROUND FLUE		
Steam 240 BTU per Sq. Ft.	Hot Water 150 BTU Per Sq. Ft.	Nominal Outside Dimensions of Fireclay Lining, In.	Effective Area Square Inches	Inside Diameter of Lining, Inches	Actual and Effective Area, Sq. In.	Height in Feet from Grate
590	973	8½ x 13	70			35
690	1140			10	79	
900	1490	13 x 13	99			
900	1490	8½ x 18	100			
1100	1820			12	113	40
1700	2800	13 x 18	156			
1940	3200			15	177	
2130	3520	18 x 18	195			
2480	4090	20 x 20	234			45
3150	5200			18	254	50
4300	7100			20	314	
5000	8250	24 x 24	346			55

This table gives the minimum flue sizes allowable for steam or hot water systems. In connection with this table, the following points should be noted:

1. The ratings given are based on smooth, lined, straight flues. If there is more than one offset or if any single offset is flatter than 60° with the horizontal, the rating should be reduced accordingly.

2. Flue heights and sizes are based upon an altitude not over 500 feet above sea level. For each one thousand feet above sea level, the height of the flue for a given appliance should be increased 10% above the figures in the table and the capacities in the table should be reduced 5%.

3. The effective area of an unlined round flue shall be based upon a diameter two inches less than the actual diameter. The effective area of an unlined rectangular flue shall be computed from the formula:

$$\text{Effective Area in Square Inches} = \frac{\pi(B-2)^2}{4} + (A-B)(B-2)$$

in which (A) and (B) are respectively the long and short inside dimensions in inches.

4. When two appliances are to be connected to one flue, the rated capacities of the flue of various sizes shall be taken as 60% of those given in the table. When three appliances are to be connected, the capacities shall be taken as 40% of those given in the table. When four appliances are to be connected, the capacities shall be taken as 35% of those given in the table.

5. The flue heights and sizes are for ordinary updraft boilers. For special types of boilers, greater heights and area may be required.

Courtesy *Hoffman Specialty ITT*

Water Is Carried over into Steam Main—This may be caused by: (a) grease and dirt in boiler, (b) type of boiler not adapted to job, (c) outlet connections too small, (d) using boiler beyond rated capacity, (e) water level carried higher than required, and (f) firing rate too high.

Boiler Is Slow to Respond—This may be due to: (a) poor draft, (b) inferior fuel, (c) improper attention, (d) accumulation of clinkers on grate, (e) boiler too small for the load, and (f) improper firing rate.

Table 51. Chimney Data

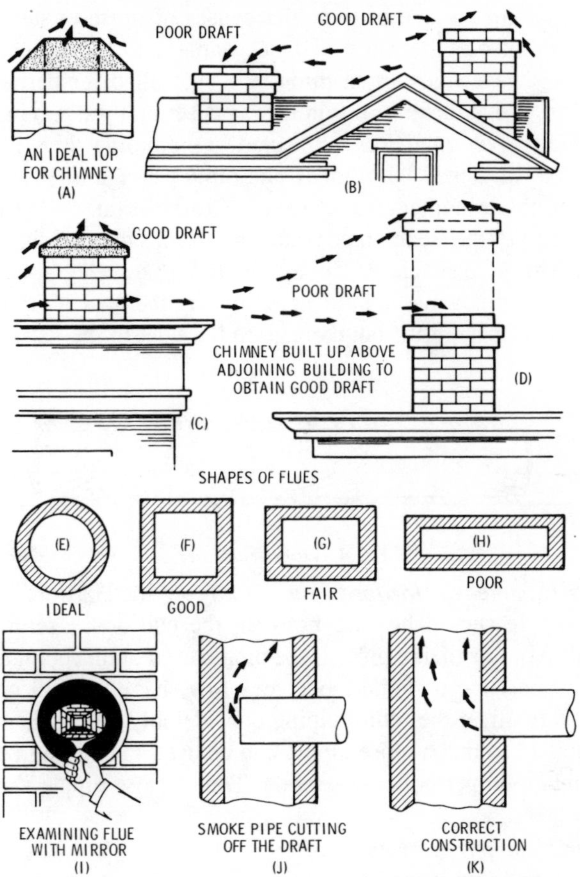
AN IDEAL TOP FOR CHIMNEY
(A)
POOR DRAFT
GOOD DRAFT
(B)
GOOD DRAFT
POOR DRAFT
CHIMNEY BUILT UP ABOVE ADJOINING BUILDING TO OBTAIN GOOD DRAFT
(C)
(D)
SHAPES OF FLUES
(E)
IDEAL
(F)
GOOD
(G)
FAIR
(H)
POOR
EXAMINING FLUE WITH MIRROR
(I)
SMOKE PIPE CUTTING OFF THE DRAFT
(J)
CORRECT CONSTRUCTION
(K)

Courtesy *Hoffman Specialty ITT*

Boiler Flues Collect Soot Quickly—This may be due to: (a) poor draft, (b) smoky combustion, (c) too low a rate of combustion, (d) excess of air in firebox causing chilling of gases, and (f) improper firing rate.

Boiler Smokes through Fire Door—This may be due to: (a) defective draft in chimney or incorrect setting of dampers, (b) air leaks into boiler or breeching, (c) gas outlet from firebox plugged with fuel, (d) dirty or clogged flues, and (e) improper reduction in breeching size.

Pressure Builds up Very Quickly (as indicated by gage) But Steam Does Not Circulate—This is due to grease and dirt in boiler.

Piping Troubles

Water hammer is one of the chief causes of noise in steam heating systems and also is the cause of much damage to traps, vents and the like. It is a wave transmitted through a pipe filled, or partially filled, with water. It may have its origin in waves set up by steam passing at a high velocity over condensate collected in the piping. Fig. 1 illustrates how such a wave may be formed. The rapid passage of steam over the surface of the water causes a wave to form as at "B". The rapid condensation of the steam in pockets "A" brings the two slugs of water together with a considerable force, which may be telegraphed through the piping. If this wave is intercepted by say the float of a vent valve, not only will noise result but damage to the float in the valve is almost certain.

STEAM A
A
WATER B

Fig. 1. Water Hammer Courtesy *Hoffman Specialty ITT*

Water Hammer in Hartford Connection—In the Hartford connection a close nipple should be used between the end of the return and the header drip or equalizing pipe. If the nipple used at this point is too long and the water line of boilers becomes low, hammer will occur. The remedy is to offset the return piping considerably below the water line of the boiler, so that a close nipple can be used in entering header drip and maintain proper boiler water line. Top of close nipple should be 2″ below water line.

Water Hammer in Mains—

1. Water pocket formed by sagging of the main. An easy and reliable method to check this is to stretch a chalk line between fittings and note relation of pipe to line.
2. Improper pitch of mains. Check with level.
3. Water hammer is also caused by too great a pressure drop, due to insufficient pipe sizes, unreamed piping or other restrictions in the line.
4. Insufficient water line difference between the low point of the horizontal main and boiler water line. In one-pipe gravity systems, this distance should normally be 18′ or more and in vapor systems, the ends of dry return mains should be 24″ or more above water line, depending on size of installation and pressure drop.

5. Improper location of air valves for venting steam main. (one pipe gravity or vacuum systems).
6. Excessive quantities of water in main due to priming boiler or improper header construction. All boiler tappings should be used and connected full size to boiler header.

Steam Does Not Circulate to Ends of Mains—

1. Dirty boilers.
2. Insufficient water line difference.
3. Improper venting of mains.

Radiator Troubles

Pounding (One-pipe System)—

1. Radiator supply valve too small or partially closed.
2. Radiator pitched away from supply valve.
3. Vent port of air valve too large, allowing steam to enter radiator too rapidly.

Radiator Does Not Heat—

1. Improper venting of air.
2. Branch supply to radiator too small.
3. Vent port of air valve clogged with dirt (one pipe systems).
4. Drainage tongue of air valve damaged or removed.
5. Branch supply improperly pitched causing water pocket.
6. Steam pressure higher than maximum working pressure of vent valve. This is especially likely to happen where steam is supplied through a reducing valve from high pressure supply.
7. Return branch improperly pitched causing water pocket to form, trapping air (vapor system).
8. In one pipe vacuum system, if gas or oil-fired, some radiators may not heat if on previous firing they were not completely heated. Changing to an open (non-vacuum) system by using open vents on radiators and mains would remedy this.

Radiator Cools Quickly (One-pipe Vacuum System)—

1. Air leakage into system either through leaky joints or through stuffing box of radiator supply valve if ordinary valve is used.
2. Dirt in vacuum valve preventing formation of vacuum.
3. On gas or oil-fired system, radiator cools quickly due to rapid formation of vacuum, and it is better to change to open (non-vacuum) system by changing radiator and main vents to non-vacuum (open) type.

Reaming and Pitching of Piping

Many carefully laid out systems are found to have radiators which cannot be completely heated under test. Improper reaming of pipes can be given as the cause of many such conditions. In mains using unreamed pipe, water pockets are formed which frequently cause trouble.

In many cases troubles caused by burrs can be overcome by repitching the pipe line. If a line having 1/4″ pitch in 10 ft. is increased to 1/2″ in 10 ft., the capacity is increased 20% in 3/4 and 1″ lines and 10-12% in 11/4″ to 2″. See "Comparative Capacity of Steam Lines at Various Pitches" in Section 7. The selection of pipe with smooth interior surface is likewise important in its effect on carrying capacity.

Boiler Cleaning

After a new installation has been in service for a week or so, grease, oil, scale, core sand, etc., will accumulate in the boiler. It should then be thoroughly cleaned with some boiler cleaning compound. There are various types of cleaning compounds on the market and it is important to follow the recommendations of the boiler manufacturers in the selection of the proper one and to follow closely the prescribed method of using it.

Air valves should not be installed in a new system prior to the cleaning of the boiler. If temporary heat is required, the air may be vented with pet cocks or with old vent valves.

In systems using vacuum or condensation pumps, the return to pump should be closed off and all condensation passed to drain until the boiler has been cleaned.

CHAPTER 6

Hot Water Heating Systems

Water has been used as a medium for transmitting heat for many years. The BTU (British thermal unit) is the unit of measurement of heat. Radiation and convectors are designed to furnish a given number of BTU's at specified water temperatures. Hydronics, a coined word, is used for describing hot water heating systems or combination hot water and chilled water heating and cooling systems.

Heat is transmitted in three ways; by conduction, convection, and radiation.

Conduction is the process of transferring heat from one molecule to another. If one end of an iron bar is heated in a flame the heat will travel to the other end by conduction.

Convection is the process of transmitting heat by means of matter. The matter can be either a liquid or a gas (air). Fig. 1 illustrates how air is heated by convection. As the radiation is heated the air surrounding it is also heated. As the air is heated it also becomes lighter and rises, and as it rises it gives off heat. Circulation is established when the heated air cools and drops, flows toward the heating surface, and again rises.

Radiation is the process of transferring heat by means of heat waves. Heat waves are similar to radio waves, the difference being the frequency or length. The earth is heated by the sun, the rays or waves travel through the vacuum of space to warm the earth. The terms *latent* heat and *sensible* heat are often used when working with hydronic systems. Fig. 2 illustrates the meaning of these terms.

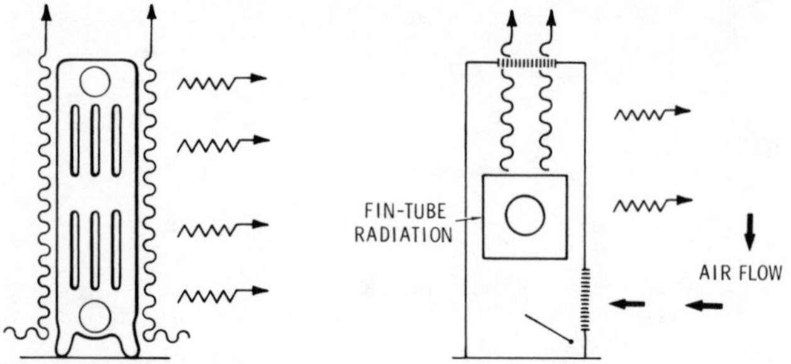
FIN-TUBE
RADIATION
AIR FLOW

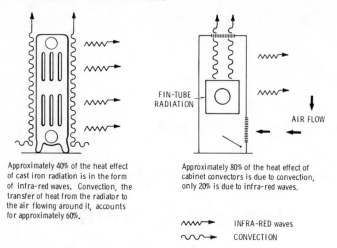

Approximately 40% of the heat effect of cast iron radiation is in the form of infra-red waves. Convection, the transfer of heat from the radiator to the air flowing around it, accounts for approximately 60%.

Approximately 80% of the heat effect of cabinet convectors is due to convection, only 20% is due to infra-red waves.

INFRA-RED waves
CONVECTION

Fig. 1. The heating effect of cabinet convectors and cast-iron radiation.

HOT WATER HEATING BOILERS

Hot water heating boilers are designed to operate at a pressure not to exceed 30 psig. Cast iron is the most widely used material for small hot water boilers; large boilers are usually made of steel. The sections of a cast-iron sectional boiler can be moved up and down stairwells, through doorways, and into and out of areas which would be impossible to negotiate with a steel boiler. Cast-iron boilers can be fired with any of the conventional fuels. Metal cabinets or jackets can be installed to enclose the boiler and its insulation, cutting down on heat loss from the boiler and presenting a pleasing finished appearance. Hot water heating systems can vary from a simple one pipe gravity system to a highly sophisticated zoned reverse-return system. Controls can be installed which will raise or lower the boiler water temperatures as the outside temperature fluctuates.

Water in a gravity heating system circulates because of the difference in density, and therefore a difference in weight, of hot and cold water. One cubic foot of water at a temperature of 200°F. weighs 60.13 lbs. If a container holding one cubic foot of water at 200°F. is cooled to 52°F. it will be found that the water has decreased in volume, or become denser. More water will have to be added to the container to bring the level up to the one cubic foot level and the net weight of the water will then be 62.42 lbs. As water is heated it becomes lighter or less dense, as it cools it becomes heavier or more dense, Below 52°F. no further

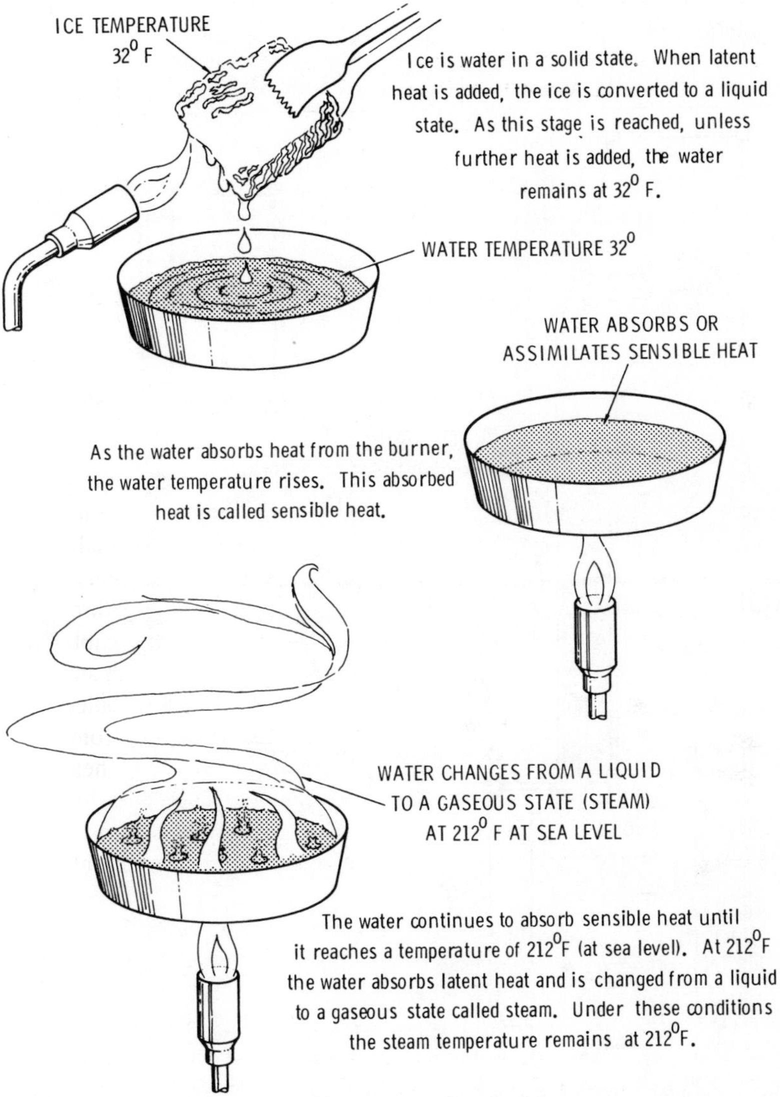
ICE TEMPERATURE
32° F
Ice is water in a solid state. When latent heat is added, the ice is converted to a liquid state. As this stage is reached, unless further heat is added, the water remains at 32° F.
WATER TEMPERATURE 32°
WATER ABSORBS OR ASSIMILATES SENSIBLE HEAT
As the water absorbs heat from the burner, the water temperature rises. This absorbed heat is called sensible heat.
WATER CHANGES FROM A LIQUID TO A GASEOUS STATE (STEAM) AT 212° F AT SEA LEVEL
The water continues to absorb sensible heat until it reaches a temperature of 212°F (at sea level). At 212°F the water absorbs latent heat and is changed from a liquid to a gaseous state called steam. Under these conditions the steam temperature remains at 212°F.

Fig. 2. The effects of latent heat and sensible heat.

shrinkage takes place, at 32°F., just before freezing, water still weighs 62.42 lbs. per cubic foot.

Fig. 3 shows the circulation by gravity of a hot water system. The radiation cools the supply of hot water from the boiler by giving off

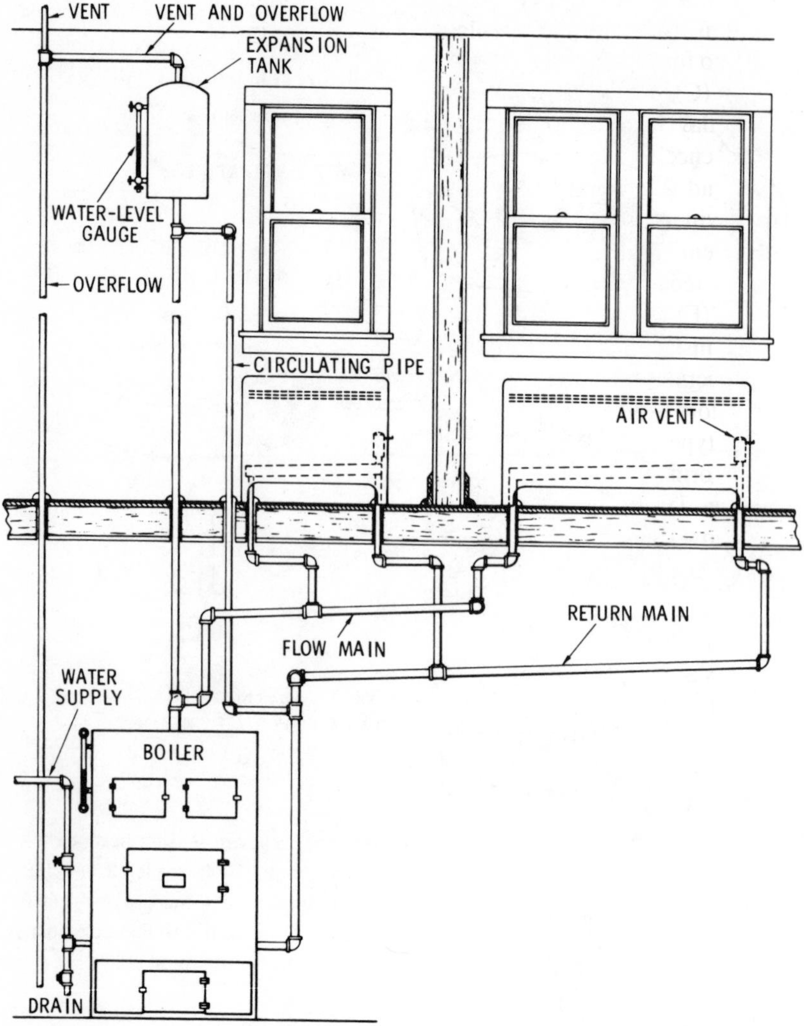
VENT
VENT AND OVERFLOW
EXPANSION TANK
WATER-LEVEL GAUGE
OVERFLOW
CIRCULATING PIPE
AIR VENT
RETURN MAIN
FLOW MAIN
WATER SUPPLY
BOILER
DRAIN

Courtesy *Dunham-Bush, Inc.*

Fig. 3. A typical gravity-pipe hot water heating system. The flow of water takes place because of the difference in weight between the hot water in the supply main and the cooler water in the return main.

heat. As the water from the radiator enters the return main, it being more dense and therefore heavier than the hot water in the boiler, pushes the hot water up and out of the boiler and establishes circulation. Gravity hot water systems are now outmoded and rarely installed.

Fig. 4 shows a typical one pipe forced circulation hot water heating system. Note that the supply main is reduced in size between the tees (B), to force part of the supply to go up and through the first radiator. At point (C), between the radiators, the supply main is again full size. On a cold day with the radiators giving off heat, there could be a marked difference in the temperature of the supply main water between points (A) and (C). Radiator No. 2 has more heating surface because of the two windows but may have to be further increased in size because of the differential in supply water temperature. At point (D) the supply main is again reduced in size to force circulation through radiator No. 2. At point (E) the return main begins and is again full size. Fig. 5 shows this type installation using reducing tees at the supply and return main connections. Balancing cocks may be needed on the returns from the radiators shown in Fig. 5 to insure adequate flow through each radiator. This type system can also be designed using flow fittings (tees) made with orifices and diverters, instead of the cast-iron fittings (tees) as shown. Diverter type fittings take advantage of the stratification of water as shown in Fig. 6.

DIRECT RETURN SYSTEMS

In a two pipe direct return system, shown in Fig. 7, the supply main begins at the first radiator, the one closest to the boiler or supply source, and ends at the last radiator. The return main starts at the last radiator and returns to the boiler or supply source, connecting to the return of each radiator, with the result that each radiator has a different length of supply and return piping. The first radiator off the supply main has the shortest return path back to the boiler or source, while the last radiator on the supply piping has the longest path back. This creates balancing problems and the system must be provided with balancing cocks or valves on the return side of each radiator. The balancing cocks may need to be adjusted very accurately to insure flow, and heat, to the farthest radiator in the system.

REVERSE RETURN SYSTEMS

A two pipe reverse return system is shown in Fig. 8. The reverse return system has many advantages over the direct return. In the reverse

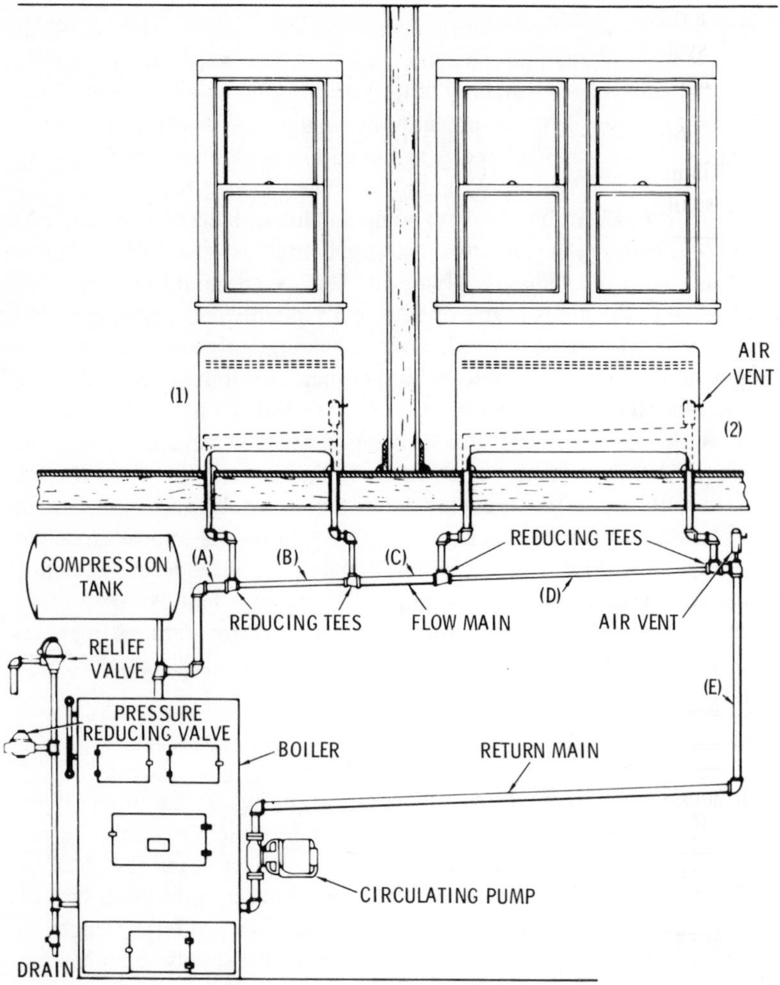
AIR VENT
(1)
(2)
REDUCING TEES
COMPRESSION TANK
(A)
(B)
(C)
(D)
RELIEF VALVE
REDUCING TEES
FLOW MAIN
AIR VENT
(E)
PRESSURE REDUCING VALVE
BOILER
RETURN MAIN
CIRCULATING PUMP
DRAIN

Courtesy *Dunham-Bush, Inc.*

Fig. 4. A typical one pipe forced-circulation hot-water heating system.

return system the first radiator supplied from the supply main has the shortest supply main but has the *longest* return main. The supply main to each radiator in the system is progressively longer but as the supply main is longer, the return main is progressively shorter. The result is that the actual developed length of supply and return main to each radiator is basically the same, or balanced. The initial cost of installation of a reverse return system may be somewhat higher than for a direct return system, due to the extra labor and material involved, but the advantages of a reverse return system are well worth the extra cost.

AIR CONTROL IN HOT WATER HEATING SYSTEMS

Air is necessary in the proper operation of a hot water heating system because of the fact that water is virtually incompressible. In a closed hot

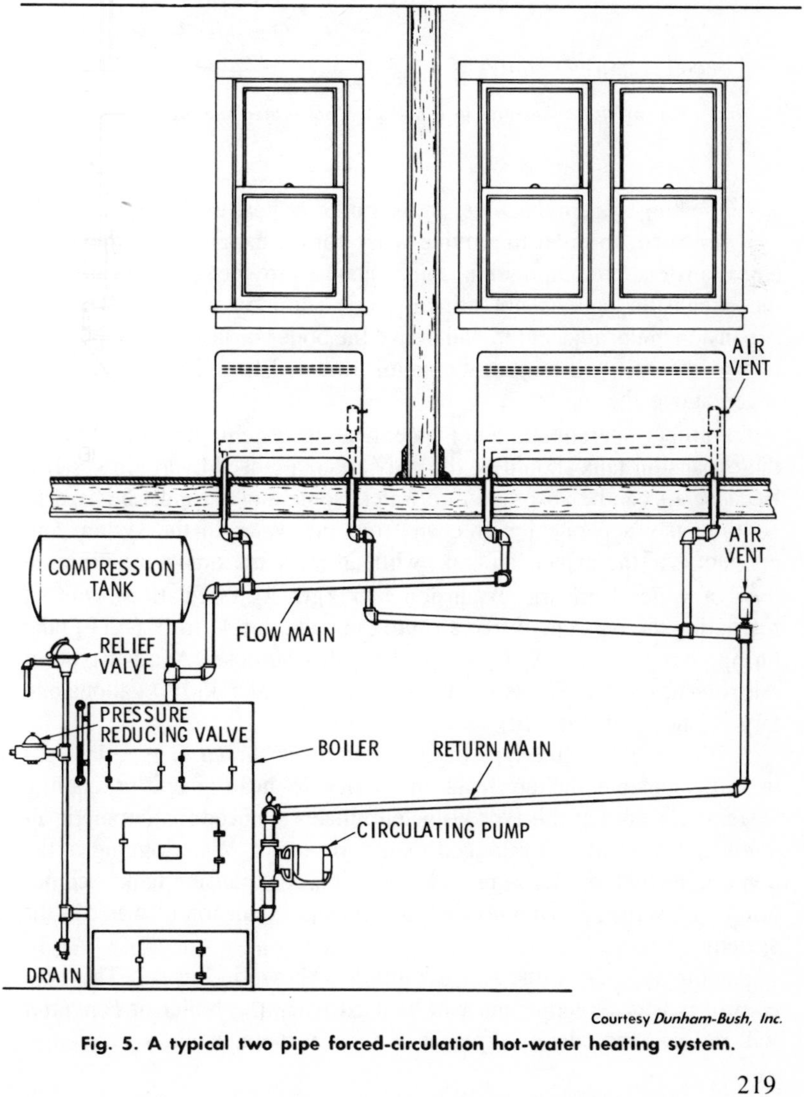
AIR
VENT
AIR
VENT
COMPRESSION
TANK
FLOW MAIN
RELIEF
VALVE
PRESSURE
REDUCING VALVE
BOILER
RETURN MAIN
CIRCULATING PUMP
DRAIN

Courtesy *Dunham-Bush, Inc.*

Fig. 5. A typical two pipe forced-circulation hot-water heating system.

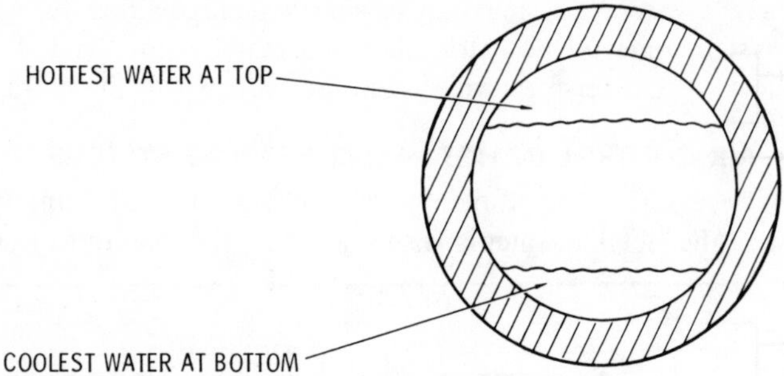
HOTTEST WATER AT TOP
COOLEST WATER AT BOTTOM

Fig. 6. Water stratifies in a one pipe hot water heating system.

water heating system the water in the boiler or heat exchanger expands as it is heated. In order to provide space for the expansion of the water an expansion, or compression, tank must be provided. The location of the expansion tank is not critical, it is common practice to install expansion tanks adjacent to and above the boiler or heat exchanger. An expansion tank will normally contain from 1/3 to 1/2 water, with air space above the water.

For proper operation of a hot water heating system, the air cushion in the expansion tank should be the only air in a closed hydronic system. Because air can be absorbed into water, some method must be used to continuously separate the free air from the water in the system and conduct it to the expansion tank, while at the same time restricting the flow of water from the expansion tank (gravity circulation) without restricting the passage of free air into the tank. An Airtrol® (ATF) tank fitting, shown in Fig. 9, is designed for this purpose. Another type of Airtrol® fitting (ATFL) is used on expansion tanks of 100 gallons and larger (shown in Fig. 10).

A Rolairtrol® in-line type of air separator is shown in Fig. 11. It is used to separate the air from the water in heating and/or cooling systems. The use of this type fitting eliminates the need for automatic air vents and prevents waterlogged expansion tanks. Waterlogging is the condition which results when a compression or expansion tank becomes filled with water, leaving no air space for the expansion of water in the system.

Another type of in-line Airtrol® fitting is shown in Fig. 12. This type control is very effective and can be used when the boiler or converter will not accommodate a dip tube for the purpose of air separation.

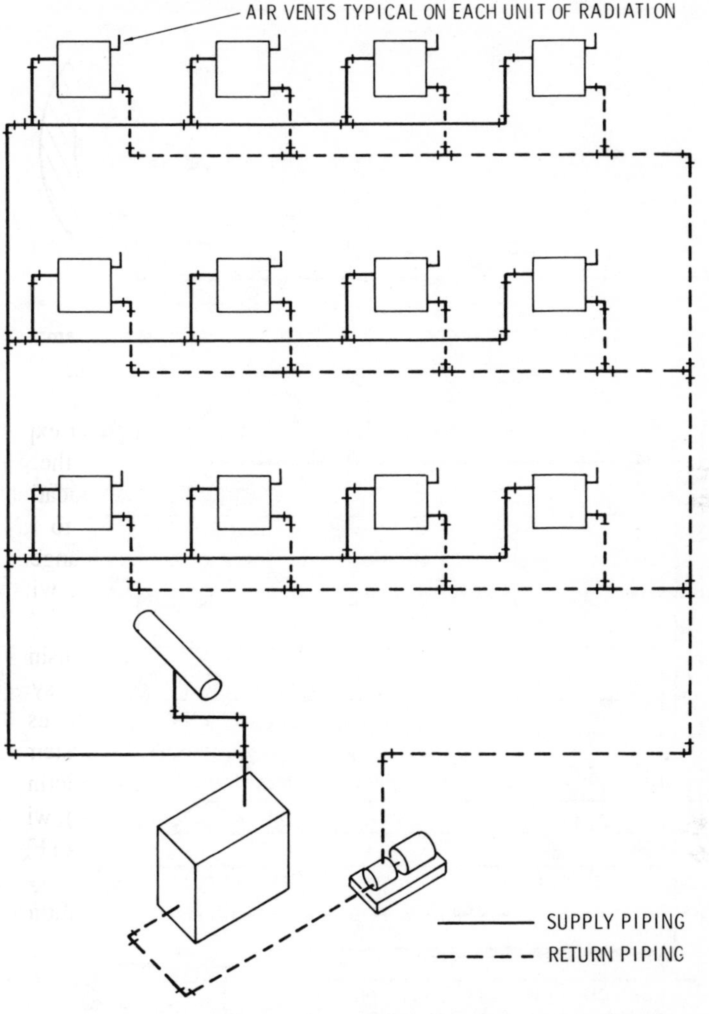
AIR VENTS TYPICAL ON EACH UNIT OF RADIATION
SUPPLY PIPING
RETURN PIPING

Fig. 7. Two pipe direct return system using base mounted circulating pump.

Side or end outlet boilers are generally tapped for forced circulation pipe sizes. An Airtrol® ABF -SO boiler fitting can be used on this type boiler to aid in removing free air from the boiler water. This control is shown in Fig. 13.

The Airtrol® boiler fitting shown in Fig. 14 is designed for use on top outlet boilers.

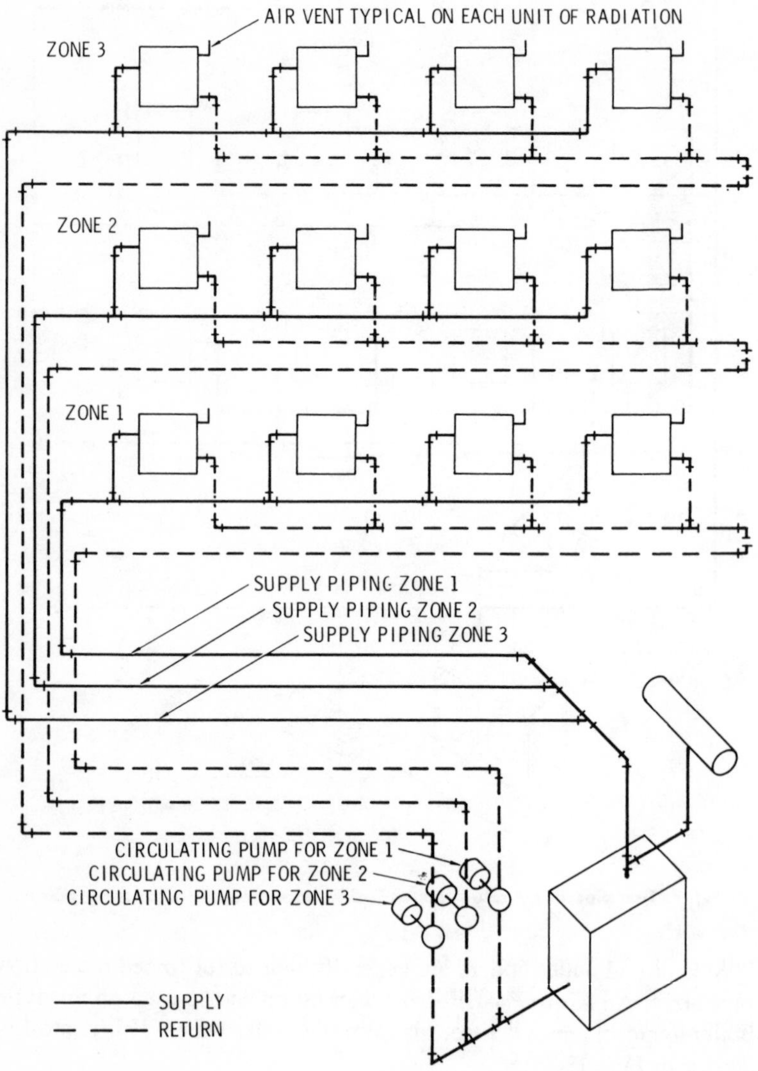
AIR VENT TYPICAL ON EACH UNIT OF RADIATION
ZONE 3
ZONE 2
ZONE 1
SUPPLY PIPING ZONE 1
SUPPLY PIPING ZONE 2
SUPPLY PIPING ZONE 3
CIRCULATING PUMP FOR ZONE 1
CIRCULATING PUMP FOR ZONE 2
CIRCULATING PUMP FOR ZONE 3
SUPPLY
RETURN

Fig. 8. Reverse return system with booster pumps for each zone.

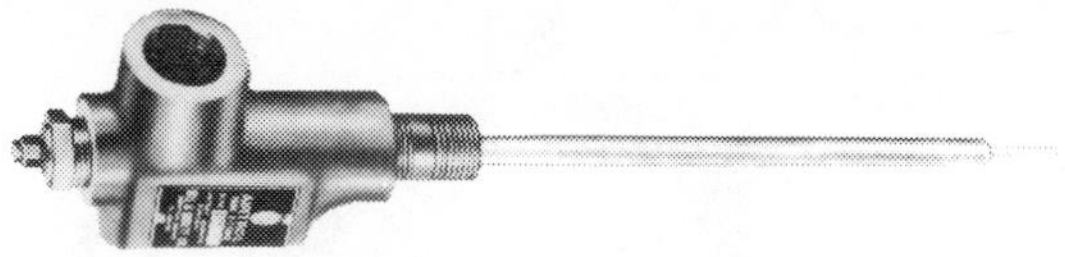

Courtesy *Bell & Gossett ITT*

Fig. 9. Airtrol® tank fitting (ATF).

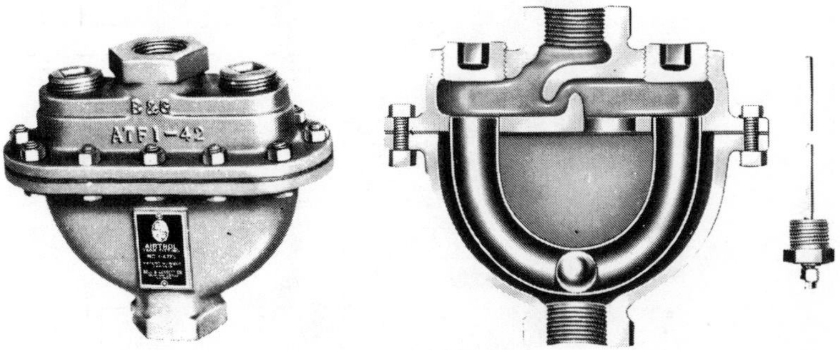
B&G
ATF1-42

Courtesy *Bell & Gossett ITT*

Fig. 10. The ATFL Airtrol® tank fitting is designed for use on expansion tanks of 100 gallons and larger.

In summation, the elimination of air from the boiler or converter and from the piping system is vitally important in order for the system to function as designed and remain trouble free.

AIR VENTS

Free air can cause noise in the piping system and can also interfere with water circulation. On the initial start-up of a heating system, either because it is new and dry or because it has been drained, the air which will be forced to the high points in the system must be removed. Manually operated air vents should be installed at high points in the piping, if these points are readily accessible, otherwise automatic air vents should be installed at these points. A drain line (1/4″ O.D. copper tubing) should be installed from the automatic vents to a suitable receptor. Some of the common types of manual and automatic air vents are shown in Fig. 15. When the manually operated air vent is opened, it should be left open until a steady stream of water is obtained. It may be

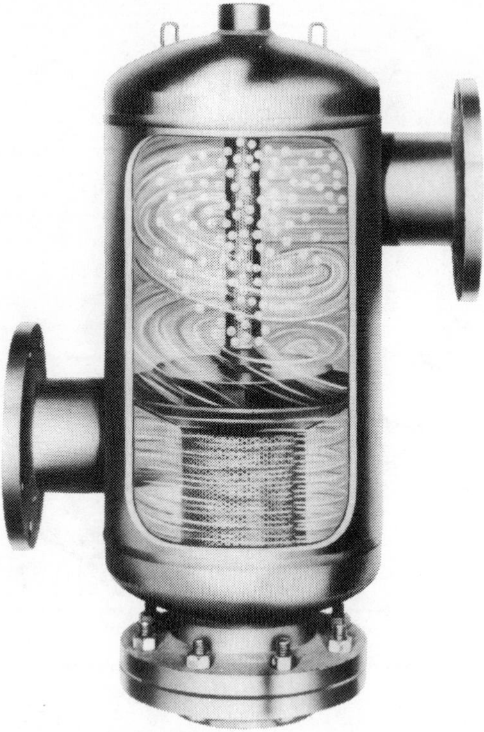

Courtesy *Bell & Gossett ITT*

Fig. 11. Rolairtrol® air separator.

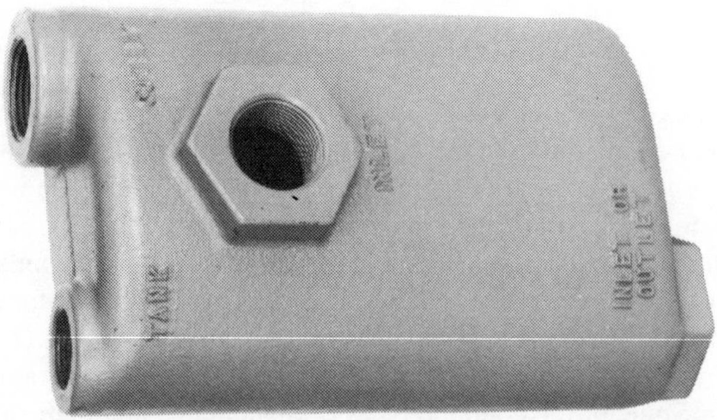

Courtesy *Bell & Gossett ITT*

Fig. 12. In-line type Airtrol® fitting.

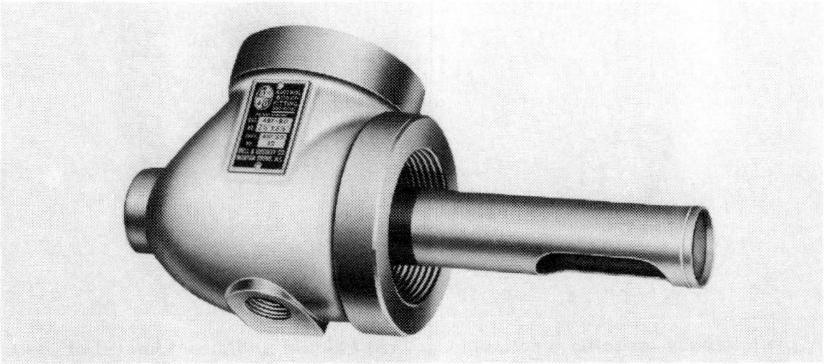

Courtesy *Bell & Gossett ITT*

Fig. 13. ABF-SO boiler fitting.

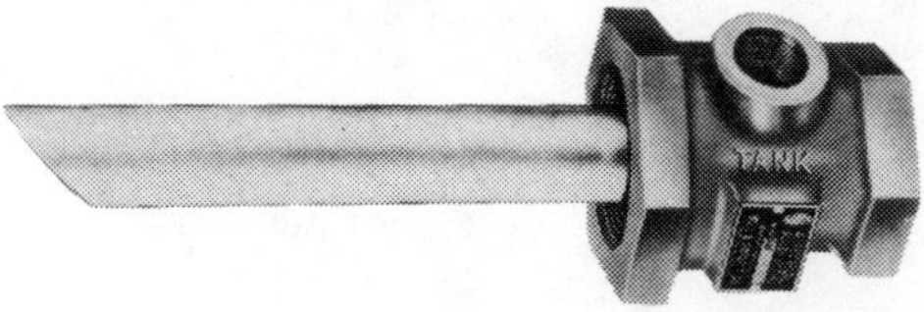
TANK

Courtesy *Bell & Gossett ITT*

Fig. 14. Airtrol® boiler fitting.

necessary to check all the air vents several times to purge all the air from the system. When a hot water heating system is placed in operation, hot water is circulating through the system and the system has been properly balanced; if one or more units of radiation do not heat it is generally because air is trapped in the unit. Venting the air will relieve this problem.

HOT WATER BOILER WATER FEEDERS

Water feeders of the type shown in Fig. 16 are used to supply make-up water to hot water heating boilers. The operating range of this type valve is adjustable to suit varying building heights. The valve shown is designed for fast feeding of a boiler and piping system.

EXPANSION TANKS

When a heating system has been filled with water and the expansion (compression) tank is in normal operating condition, the tank will be from 1/3 to 1/2 full of water, the balance of the tank will contain air. The air in an expansion tank serves as a cushion for thermal expansion of water in the system. The tank must be absolutely air tight; if there is

(A) Manually operated air vent.

(B) Manual and/or automatic air vent.

(C) Automatic air vent.

Courtesy *Bell & Gossett ITT*

Fig. 15. Some common types of manual and automatic air vents.

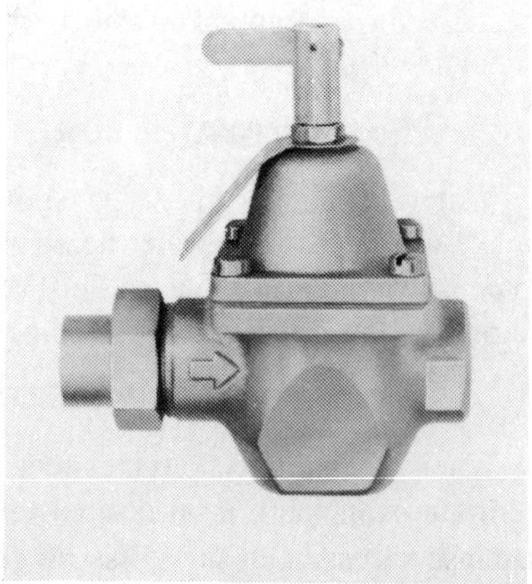

Fig. 16. Watts feed water pressure regulator.

Courtesy *Watts Regulator Co.*

a leak in the tank, the air will escape and be replaced with water. An expansion tank which is filled with water is said to be waterlogged. An expansion tank should be carefully checked at the time it is installed and periodically thereafter, with particular attention given to the gauge cocks and the packing nuts securing the gauge glass.

When a boiler and heating system is placed in operation, with cold water in the system, the water volume will expand when the water is heated. From 40° to 200° water will expand to 1.04 its original volume. Assuming that the expansion tank has been sized correctly there will be sufficient air space in the expansion tank to accommodate this increased volume. As the water in the boiler is heated it expands, and as it expands it builds up pressure. The air cushion in the tank will accommodate the increased pressure, up to 30 lbs. at which point the relief valve on the boiler will open and relieve the built up pressure. A combination temperature, pressure and altitude gauge should be installed on the boiler in order to check operating pressures and temperature. When filling a system, or in normal operation, pressure should not be excessive; 15 to 18 lbs. should be satisfactory and will allow for water volume expansion and pressure build up. When fast filling a system through a manual fill valve, excess pressure can be relieved by opening the relief valve momentarily. In a closed system if the expansion tank becomes waterlogged, with no space for the expanded water volume, the pressure will build up very quickly as the water is heated and the relief valve will open and relieve the excess pressure. When this condition is encountered the reason for the expansion tank becoming waterlogged must be found and corrected. A typical expansion tank is shown in Fig. 17.

The expansion (compression) tank shown in Fig. 17 (A) is typical of expansion tanks used in hydronic systems. The bottom side of the tank is shown, in use the openings shown would be turned down. An expansion tank should also have openings on one end of the tank for gauge cocks and a gauge glass. To help prevent air leakage leading to waterlogging of the tank, the tank should have no openings on the top side.

HOT WATER BOILER RELIEF VALVES

Boiler water safety relief valves are installed on boilers for protection against excessive water pressure caused by the thermal expansion of water. The type shown in Fig. 17 (B) has an emergency Btu steam discharge capacity if run-away firing conditions should occur.

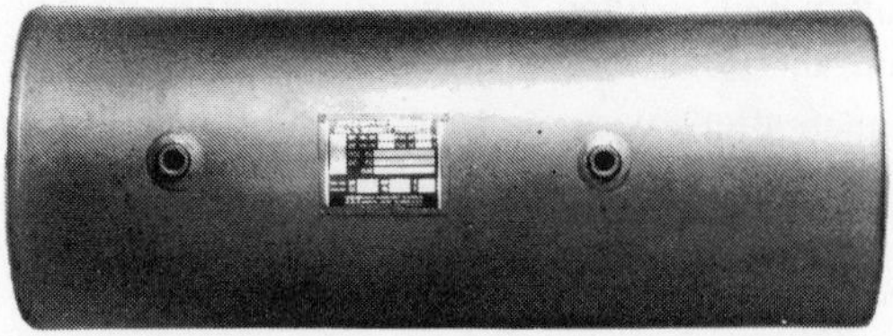

Courtesy *Bell & Gossett ITT*

(A) A typical compression (expansion tank used with hot water heating systems).

INLET

Courtesy *B. Watts Regulator Co.*

(B) A boiler water safety relief valve.

Fig. 17. A typical compression (expansion) tank used with hot water heating systems.

HOT WATER BOILER OPERATING CONTROLS

Two aquastats should be installed in a hot water boiler, one to serve as the operating control, the other to serve as a high limit control. The operating control will maintain a set minimum water temperature and the high limit control will inactivate the heat source if the boiler water temperature exceeds the set point of the high limit control. Both controls are actually thermally operated electrical switches. The operating control is a normally open switch which makes contact or closes a circuit on temperature fall. When the water temperature in the boiler is below the set point of the aquastat the switch in the control will be in closed position, calling for heat. The high limit control is a normally closed switch which opens and breaks contact, breaking the electrical circuit, when the boiler water temperature rises above the set point of this control. Both aquastats are wired in series with the gas valve or other heat source. Other types of safety controls may also be installed on a boiler, in which case they would also be wired in series with the operating control and the high limit control. Opening a switch, or breaking the electrical circuit, of any single unit in a series wired circuit will interrupt the flow of current in that circuit and inactivate the heat source. Operating and high limit aquastats are shown in Fig. 18.

CIRCULATING PUMPS FOR HOT WATER HEATING SYSTEMS

Two Types of pumps are in general use for circulating the water in hydronic heating and cooling systems. Booster pumps of the type shown in Fig. 19 are generally used on smaller size systems for individual zone circulation while the larger base mounted pumps, similar to the type shown in Fig. 20 are used on larger piping systems.

It is considered good piping practice to install valves on the inlet and outlet sides of the circulating pumps in hot water heating systems. Gate valves can be used with booster type pumps to isolate the pumps for repair or replacement, and to hold the water in the system and thus avoid the air problems which would arise if a zone or a system required refilling.

The discharge side of a centrifugal pump used in a heating and/or cooling system should be provided with a shut/off valve, a balancing valve to regulate the flow, and a check valve. The valve shown in Fig. 21 is designed to perform all three functions.

It is a check valve, a shut-off valve, and it is made with a calibrated stem for returning the valve to the "set" position after the valve is used

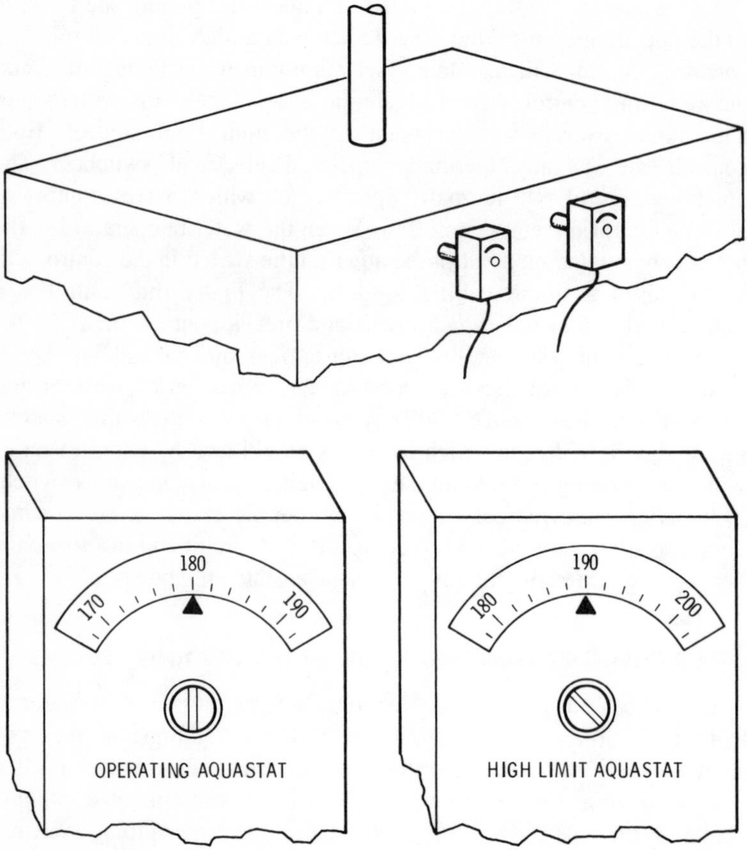
180
170
190
OPERATING AQUASTAT
190
180
200
HIGH LIMIT AQUASTAT

Fig. 18. The high limit aquastat is normally set 10° above the operating aquastat.

as a shut-off. When the valve is in the "set" position it serves as a balancing valve.

The piping shown in Fig. 22 is typical of most baseboard, fin tube convector and cabinet heater piping. Some systems include an automatic temperature control valve. This, if used, would be located between the supply valve and the radiation. The square head balancing cock is used to pinch down the flow through the radiation in order to balance the system.

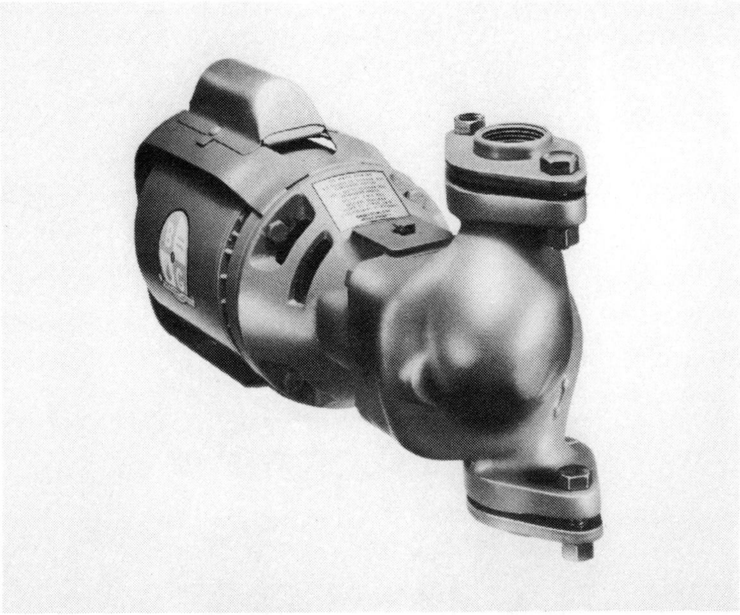

Courtesy *Bell & Gossett ITT*

Fig. 19. Booster pump.

Courtesy *Bell & Gossett ITT*

Fig. 20. Base-mounted circulating pump.

Courtesy *Bell & Gossett ITT*

Fig. 21. Triple-duty valve®.

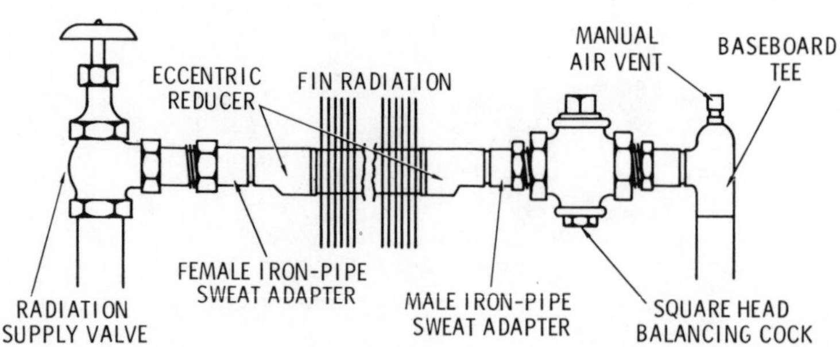
MANUAL
AIR VENT
BASEBOARD
TEE
ECCENTRIC
REDUCER
FIN RADIATION
FEMALE IRON-PIPE
SWEAT ADAPTER
MALE IRON-PIPE
SWEAT ADAPTER
RADIATION
SUPPLY VALVE
SQUARE HEAD
BALANCING COCK

Courtesy *Chase Brass & Copper Co.*

Fig. 22. Typical baseboard radiation connections.

RADIANT-PANEL HEATING

Radiant-panel heating is a method of heating a room by raising the temperature of one or more of its interior surfaces (floor, walls, or ceiling), as in Fig. 23, instead of heating the air, as in Fig. 24. Conventional heating systems heat the air within the room by convection, giving off very little radiant heat. The name *radiant-panel heating* is derived from the fact that tubes or pipes are placed in the floor, walls, or ceiling to form a heated "panel".

There are various methods used to achieve radiant heating. One of the most common, and the one which relates to the plumbing trade, is where hot-water pipes or tubes are embedded in the structure of the building. Other methods of radiant heating are:

1. By circulating warm air through shallow ducts beneath the floor or in spaces formed in the walls.
2. By the use of electrically heated wires, metal plates, or glazed fire-clay panels.
3. By the use of electrically heated tapestry formed into portable screens or hung on the walls of a room.

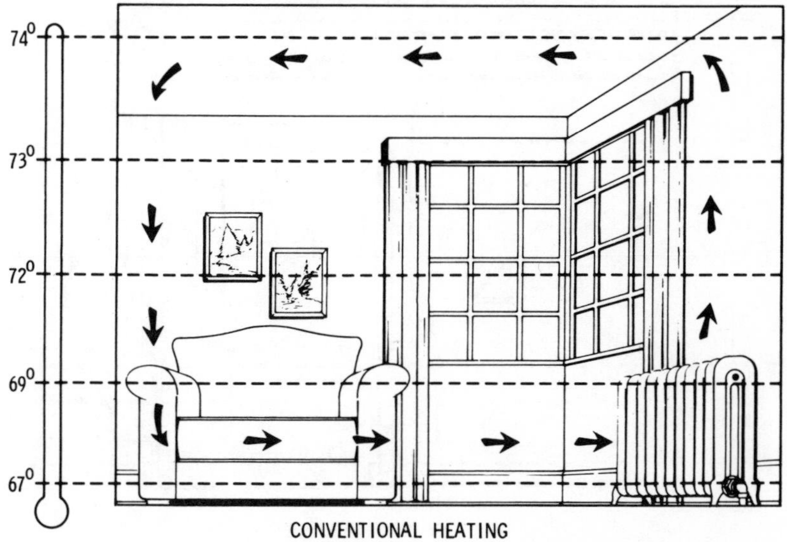
74°
73°
72°
69°
67°
CONVENTIONAL HEATING

Courtesy *Chase Brass & Copper Co.*

Fig. 23. Temperature levels in a room heated by a conventional radiator (or warm-air register). Notice the wide variation in temperature. The air currents often create drafts, spread dust, and cause hot air to collect at the ceiling and cool air at the floor.

4. By attaching fabricated metal plate to walls and ceiling, through which hot water or steam is circulated.

The installation procedure generally accepted for radiant-panel heating is to embed specially constructed pipe coils or lengths of tubing in the floor, walls, or ceiling. These coils generally consist of small-bore wrought-iron, steel, brass, or copper pipe, usually with an inside diameter of 3/8 to 1 inch.

Insulation Requirements

Extra insulation is unnecessary when radiant-panel heat is employed. In fact, a poorly insulated house that is often uncomfortable because of the prevailing low temperature of the walls will be much more comfortable when radiant heat is installed. This is because the inner surfaces of the outside walls will be at a higher temperature than before. Two main factors should be considered, however, before a decision is made to convert to radiant-panel heating.

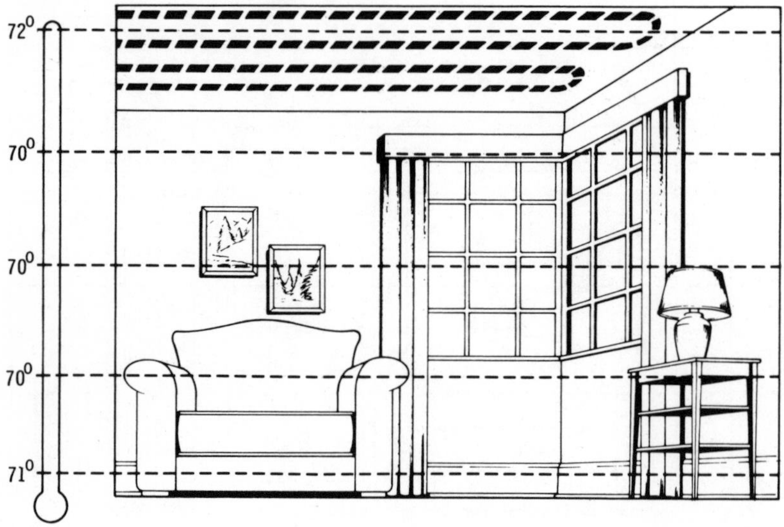
72°
70°
70°
70°
71°

RADIANT PANEL HEATING

Courtesy *Chase Brass & Copper Co.*

Fig. 24. Radiant heating warms the interior surfaces (ceilings, walls, or floor) of a room instead of the air. The result of this is the elimination of cold, drafty floors and a more even temperature from floor to ceiling.

1. Install ceiling systems only in houses having fair insulation. Floor systems will have an excess output under these conditions and will be uncomfortably warm.
2. Consider that insulation will always pay for itself in a short time by the savings in fuel.

Air Venting

Air venting is an essential feature in the control of any panel-heating system. Any small amount of air collecting in either the circuit pipes or pipe coils will cause continued trouble and invariably result in a shortage of heat. An arrangement similar to the one in Fig. 25 will permit uniform venting. Because of the continuous slope to the coil connections, it may prove sufficient to install automatic vents at the top of the return riser only, omitting such vents on the supply riser.

The fill and drain lines in Fig. 25 are connected at the boiler. The difference in elevation *H* should be at least three inches, or sufficient to

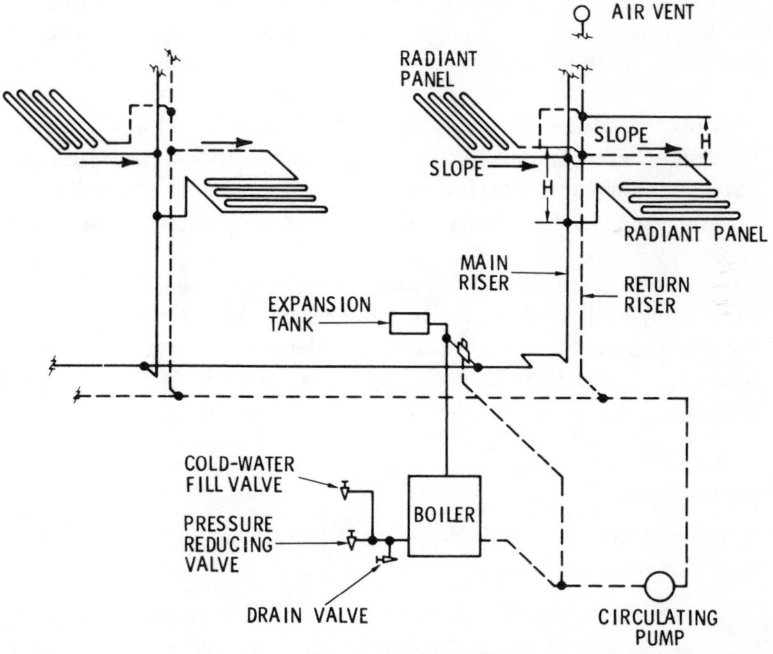
AIR VENT
RADIANT PANEL
SLOPE
H
SLOPE
H
RADIANT PANEL
MAIN RISER
RETURN RISER
EXPANSION TANK
COLD-WATER FILL VALVE
BOILER
PRESSURE REDUCING VALVE
DRAIN VALVE
CIRCULATING PUMP

Courtesy *Chase Brass & Copper Co.*

Fig. 25. Diagram of a radiant-panel heating system showing the draining and venting.

provide for a continuous slope to the coil connection. This slope should not be excessive, however, in order to fully utilize the gravitational forces acting on the system when heated.

System Leakage Test

No troubles should be experienced with leaks in a radiant-heating system if the proper procedure is followed and reasonable care taken in installing. After the installation is completed, a hydraulic pressure test, as recommended by the tube or pipe manufacturer, is applied to the system.

The material immediately surrounding the coil protects it from corrosion. An additional safety measure is to connect the coils to the water-supply source during the installation period. Thus, it will immediately be noticed if the piping becomes damaged.

Panel Temperatures

The temperature conditions that are satisfactory for comfort by the occupant determine the highest allowable panel surface temperature. For rooms occupied continuously, the floor temperatures should never be above 85°F. For ceilings, the maximum allowable temperature depends on the height. Generally, 100°F is considered maximum for seven-foot ceilings, with 10°F added for each additional foot. Thus, the maximum allowable temperature of a nine-foot ceiling is 120°F. With the coils embedded in plaster, the maximum temperature is considered to be 130°F.

The temperature drop of the water in the panels should be limited to maintain the efficiency of the heating system. For practical purposes, the temperature difference between the inlet and outlet water should be between 15° and 30°F for floor panels, and between 20° and 35°F for ceiling panels.

Advantages

The advantages of radiant-panel heating are numerous. Among the most common are:

1. More floor space. Radiant-panel heating eliminates radiators and grills, as well as the need for window recesses to accommodate radiators. Elimination of radiators actually provides more floor space.

2. Interior of rooms can be better decorated. With radiant-panel heating, no restrictions exist as to furniture arrangement and wall decorations.
3. Lower velocities of air currents and therefore less streaking and dust on walls and ceilings.
4. Warm floors in homes without basements. Radiant-panel heating is the only system that eliminates cold floors when there is no basement.
5. Simplifies troublesome architectural and engineering design problems.
6. Lower operating costs. A well-designed radiant-panel heating system should result in the lowest overall operating and general maintenance expense for the heating plant.

Installation

Radiant-heating coils can be installed in the floor, ceiling, or walls. Ceiling installation provides a greater percentage of radiant-heat transmission than either floors or walls. For example, only about 50% of the heat emitted from floor panels is in the form of radiant rays, while about 65% of the heat emitted by ceiling panels is radiant. The percentage for wall panels is even less than for floor panels.

The shape and size of the coils used in hot-water radiant-heating systems depends on such factors as pressure drop, the shape of the space to be heated, the material of which the pipes are made, etc. For reasons of economy, it is important to lay out the lines in long straight runs. In addition, it is necessary to select panel arrangements which result in the best comfort conditions for the rooms to be heated. In most cases, it is advisable to locate the panel close to outside walls and entrances. In rooms having the greatest heat loss, it may be necessary to spread the heating coil over the entire ceiling or floor area. In other rooms, possibly only one-half to three-quarters of the ceiling or floor area will need to be covered by the heating coil.

Copper tubing is the material most often used in the installation of hot-water radiant heating. A minimum of solder joints should be used, so short sections of tubing should be avoided if possible. Instead, continuous coils should be used, as shown in Figs. 26 and 27.

The shape of the radiant-heating coil will usually be determined by the size and shape of the area to be heated. Fig. 28 shows a conventional sinusoidal coil shape. Here, the area immediately inside the door is not

Courtesy *Chase Brass & Copper Co.*

Fig. 26. Workman forming copper tubing for a radiant floor panel.

Courtesy *Chase Brass & Copper Co.*

Fig. 27. Continuous lengths of tubing should be used where possible and the spacing should be uniform.

covered by the coil since the occupants will not be sitting or standing in this part of the room.

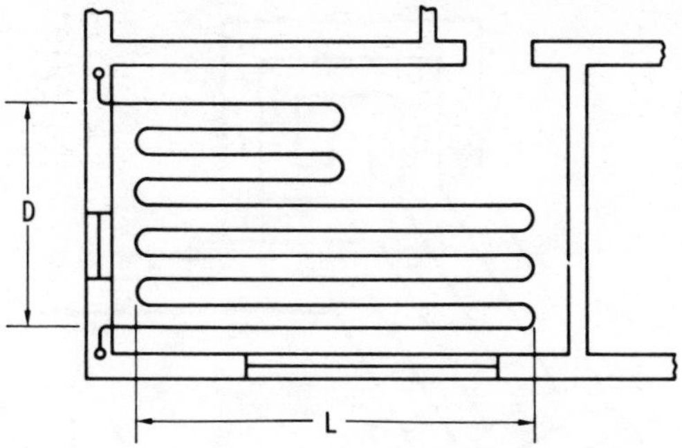
D
L

Fig. 28. A conventional sinusoidal radiant-heating coil.

The tube or pipe spacing should be restricted to 12 inches or less for all radiant-heating panels. Spacings greater than this will result in noticeable temperature variations between various points on the surface of the panel.

Fig. 29 illustrates a grid-type coil for use where a low pressure drop is desired or where a constant temperature of the entire panel surface is required. Irregularly shaped rooms often require odd-shaped coils, such as the one in Fig. 30. In large rooms where several coils must be used, a combination grid and continuous coil system (Fig. 31) is often used. In this type, common supply and return headers are connected to the mains.

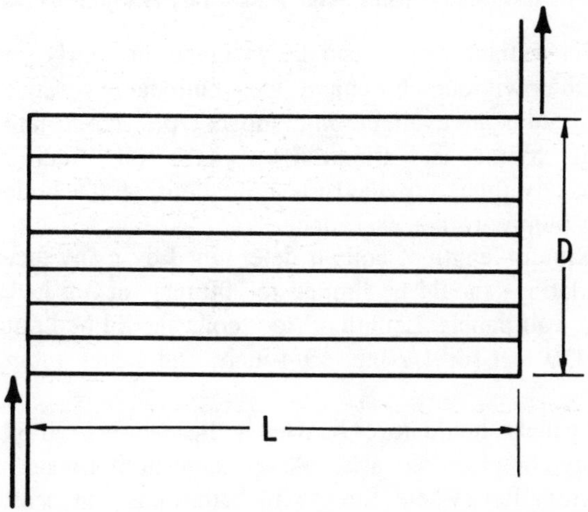
D
L

Fig. 29. A grid-type heating coil used where a low pressure drop is desired or a constant panel surface temperature is required.

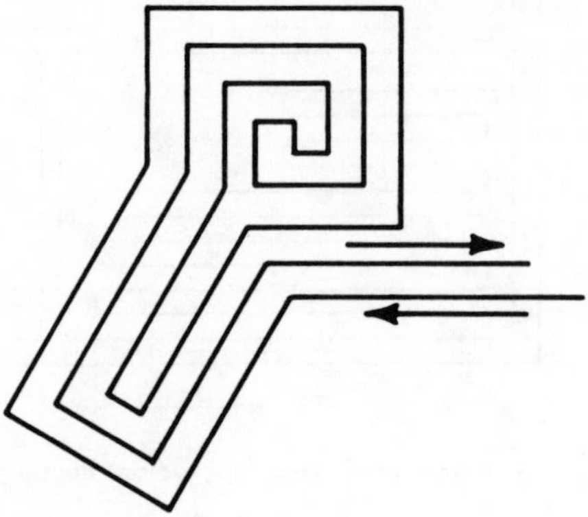

Fig. 30. A coil for an irregularly-shaped room.

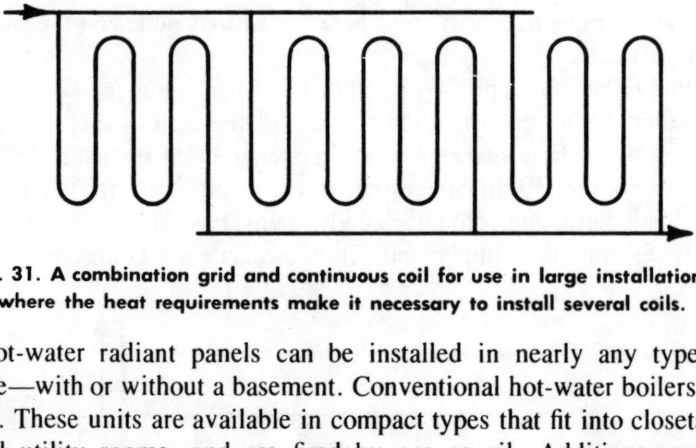

Fig. 31. A combination grid and continuous coil for use in large installations where the heat requirements make it necessary to install several coils.

Hot-water radiant panels can be installed in nearly any type of home—with or without a basement. Conventional hot-water boilers are used. These units are available in compact types that fit into closets or small utility rooms, and are fired by gas or oil. Additions can be installed at any time, provided the limitations of the boiler unit and circulating pump are not exceeded.

The maximum length of coils is determined by many factors, but in most installations should be limited to 200 feet for 3/8-inch tubing in ceiling and wall panels. Length of floor coils should be limited to 180, 280, and 450 feet for 1/2-inch, 3/4-inch, and 1-inch tubing, respectively.

Radiant panels should never be used with steam instead of hot water, as too many complications arise. Also, domestic hot water should not be taken from the system for use in bathrooms, kitchens, etc. The addition of fresh water should be kept to a minimum to avoid accumulation of impurities present in the water.

Provision should be made for draining the system if the need should arise. Care therefore must be taken in the design and installation to ensure that the system can be completely drained, with no water pockets existing that will hold water and cause damage in the event of a freeze. The system should be installed with a vent at the high point and a drain at the low point, and without traps so that it may be easily drained.

Cold water should never be circulated in the system during hot weather in an attempt to provide cooling. Sweating of the panels will almost certainly take place, resulting in damage or generally unsatisfactory conditions.

Ceiling Panels—The most satisfactory location for radiant-heating panels is in the ceiling. These are usually fastened to the underside of the lath and then covered and embedded in the plaster as the room is finished. If properly installed, this type of installation will not cause the plaster to crack, provided the plaster is allowed to cure and dry thoroughly before hot water is introduced into the coil. Insulation should be provided above the ceiling to prevent excessive heat transfer into the attic or rooms above.

Wallpaper or paint can be applied to the plaster in which the heating coils are embedded. The wallpaper or paint should be allowed to dry thoroughly before heating is attempted.

A typical piping diagram for a ceiling panel is shown in Fig. 32. In this arrangement, the hot water enters the panel near the outside walls and leaves from the inside. The control valves may be located in either the supply or return line to suit the convenience of the particular installation. The surface of the lath should be as level as possible. The coils may be preformed and lifted into place, or may be formed directly on the ceiling.

Forming the coils directly on the ceiling is usually the simplest and easiest method of installation. Fig. 33 shows copper tubing being formed on a ceiling that is covered with expanded metal lath. Plaster will later be applied. Fig. 34 shows copper tubing applied to a gypsum lath ceiling. However, copper tube radiant panels can be applied to any type of lathing. As shown in Fig. 35, the ceiling need not be level *if the tubing is installed in such a way as to eliminate pockets which would trap the water.*

Ceiling panels should not be installed above plywood, composition board, or other insulating types of ceiling material. Such surfaces have an undesirable insulating effect that diminishes the full heat output of the panel.

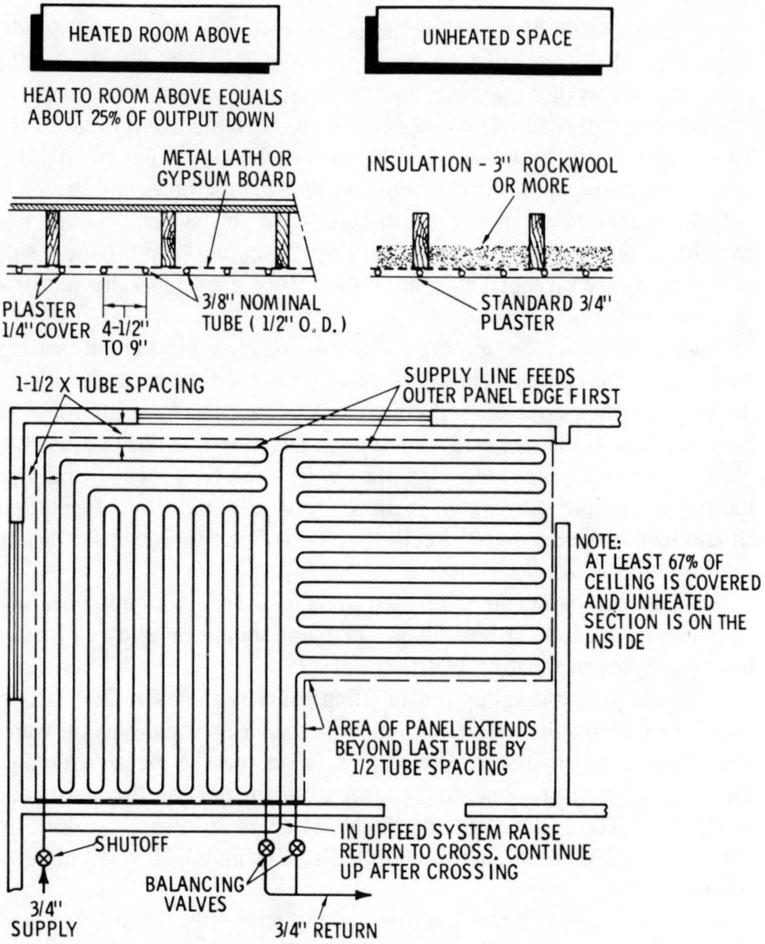
HEATED ROOM ABOVE
UNHEATED SPACE
HEAT TO ROOM ABOVE EQUALS
ABOUT 25% OF OUTPUT DOWN
METAL LATH OR
GYPSUM BOARD
INSULATION - 3" ROCKWOOL
OR MORE
PLASTER
1/4" COVER
4-1/2"
TO 9"
3/8" NOMINAL
TUBE (1/2" O. D.)
STANDARD 3/4"
PLASTER
1-1/2 X TUBE SPACING
SUPPLY LINE FEEDS
OUTER PANEL EDGE FIRST
NOTE:
AT LEAST 67% OF
CEILING IS COVERED
AND UNHEATED
SECTION IS ON THE
INSIDE
AREA OF PANEL EXTENDS
BEYOND LAST TUBE BY
1/2 TUBE SPACING
SHUTOFF
IN UPFEED SYSTEM RAISE
RETURN TO CROSS. CONTINUE
UP AFTER CROSSING
BALANCING
VALVES
3/4"
SUPPLY
3/4" RETURN

Fig. 32. Diagram for a typical ceiling panel.

Floor Panels—Radiant panels are installed in the floor in many instances. When this is done, the most satisfactory arrangement is to embed the coils in the concrete floor slab. The best results are obtained when the tubing or pipe is buried at least two inches below the floor surface, or deeper if a heavy traffic load is anticipated. Allow at least two weeks for the concrete to set before applying heat, and then only gradually. Floor covering of tile, terrazzo, linoleum, carpeting, or any commercial tile can then be installed. No damage to rugs, varnish,

Courtesy *Chase Brass & Copper Co.*

Fig. 33. Workmen installing a ceiling coil for a radiant-panel heating system. Metal lath is shown here, but gypsum or wood lath would be satisfactory from a heating standpoint.

Courtesy *Chase Brass & Copper Co.*

Fig. 34. A radiant panel installed on a gypsum lath ceiling.

Courtesy *Chase Brass & Copper Co.*

Fig. 35. The ceiling panel need not be level if the tubing is installed in such a manner as to eliminate pockets that might trap the water.

polish, or any other material will result if the water temperature is kept below the prescribed maximum of 85°F.

Fig. 36 shows a typical piping diagram for a radiant floor panel. The same care to avoid low places in the coil should be taken as with ceiling-panel coils.

Some loss of heat to the ground from a floor panel laid directly on the ground will be experienced. This will vary with type of construction, fill, etc., but is estimated to be 10% to 20% of the heat given to the room. Heat loss to the outside through the slab edges will usually be greater than the amount lost straight down to the ground. Waterproof insulation that is 1/2 to 2 inches thick should be installed along the edges of the slab.

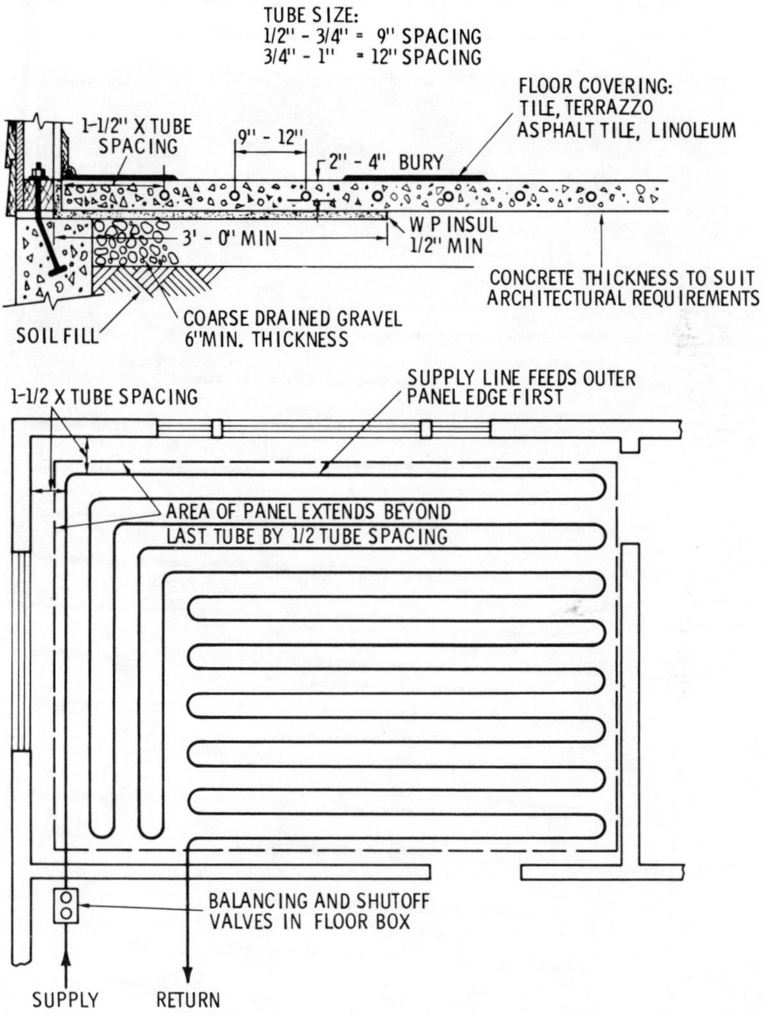
TUBE SIZE:
1/2" - 3/4" = 9" SPACING
3/4" - 1" = 12" SPACING
FLOOR COVERING:
TILE, TERRAZZO
ASPHALT TILE, LINOLEUM
1-1/2" X TUBE
SPACING
9" - 12"
2" - 4" BURY
3' - 0" MIN
W P INSUL
1/2" MIN
CONCRETE THICKNESS TO SUIT
ARCHITECTURAL REQUIREMENTS
COARSE DRAINED GRAVEL
6"MIN. THICKNESS
SOIL FILL
1-1/2 X TUBE SPACING
SUPPLY LINE FEEDS OUTER
PANEL EDGE FIRST
AREA OF PANEL EXTENDS BEYOND
LAST TUBE BY 1/2 TUBE SPACING
BALANCING AND SHUTOFF
VALVES IN FLOOR BOX
SUPPLY
RETURN

Fig. 36. Diagram for a typical radiant floor panel.

Wall Panels—Radiant wall panels are not usually employed except to provide supplementary heat where ceiling and/or floor panels do not give the required degree of comfort. Wall panels are also installed in bathrooms in some homes where the presence of more than normal radiant heat is desirable. Fig. 37 shows a coil installed on a bathroom wall prior to the plastering operation. Low spots that will trap the water must be avoided to permit complete drainage when necessary.

Courtesy *Chase Brass & Copper Co.*

Fig. 37. Bathroom wall panels may be extended to the floor for concentrated heating.

Wall panels installed in areas other than bathrooms need not extend to the floor. They are usually confined to the upper half of the wall to prevent the screening effect on the heat output by furniture placed against or near the wall. Fig. 38 shows the installation details for a typical radiant wall panel.

Panel Testing—After the coils have been fastened in place and the system has been completed in every respect, a pressure test is made shortly before the concrete is poured or before the plaster is applied. *It is important to again check the installation to make certain that no sags or water traps exist in the coil assembly.*

There are two testing methods that are usually used to check radiant heating coils—an *air test* and a *hydraulic test.* For the air test, compressed air at a pressure of at least 100 psi is used. An ordinary dial-type pressure gauge is installed in the line to be tested, along with a shutoff valve on the inlet side of the gauge. Pressure is applied to the system, after which the shutoff valve is closed and the source of air pressure is disconnected. Any leaks in the system will cause the pressure reading to continually drop. The location of the leak can usually be found by listening for the leaking air or by bubbles formed when a soapy solution is applied to the suspected area.

After a satisfactory air test, each coil should undergo a hydraulic pressure test before being covered. The coils should be filled with water, with care being taken to make sure all air is expelled from the

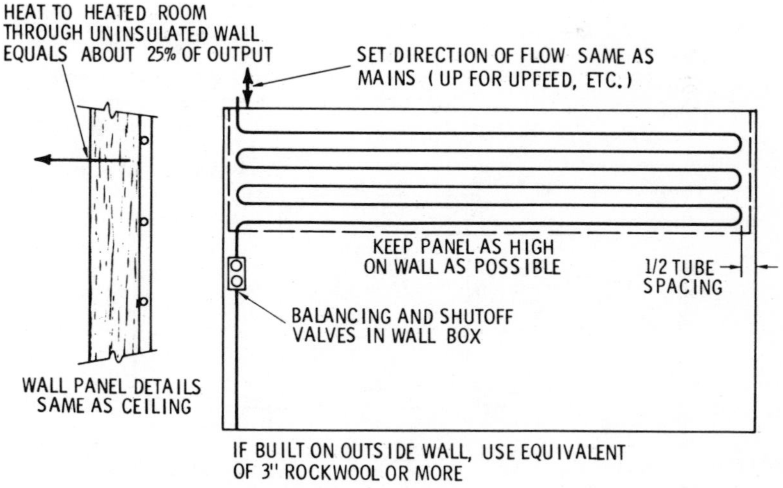
HEAT TO HEATED ROOM
THROUGH UNINSULATED WALL
EQUALS ABOUT 25% OF OUTPUT
SET DIRECTION OF FLOW SAME AS
MAINS (UP FOR UPFEED, ETC.)
KEEP PANEL AS HIGH
ON WALL AS POSSIBLE
1/2 TUBE
SPACING
BALANCING AND SHUTOFF
VALVES IN WALL BOX
WALL PANEL DETAILS
SAME AS CEILING
IF BUILT ON OUTSIDE WALL, USE EQUIVALENT
OF 3" ROCKWOOL OR MORE

Fig. 38. Installation of a typical radiant wall panel.

system. The water pressure should be increased to 300 psi and the system checked for leaks. Maintain the pressure for at least two hours and again check for a pressure drop. In high buildings, the test water pressure should exceed the combined static and pump pressure by 100 psi.

Boilers—The size of the boiler necessary for a particular installation may be determined in the same manner as that employed for conventional convection-type hot-water systems. Add the sum of all panel-heating requirements and the piping losses.

Circulating pumps for use with radiant heating should have a somewhat higher head rating than for convector systems of the same capacity. This is because the coil pressure drop is considerably higher than the drop in a radiator or convector.

It is customary to use a conventional boiler layout, such as the one in Fig. 39, for small buildings. In this type of installation, the return water is mixed with hot water from the boiler by means of the flow control or bypass valve. This mixing valve is regulated by the outdoor and indoor thermostats, and if the boiler circulation is fully stopped, this valve will turn the circulating pump off. The temperature of the boiler water is kept constant.

The rooms may be grouped according to exposure, with all panels having the same exposure connected to a common distributing line, the circulating water in which is regulated by individual mixing valves.

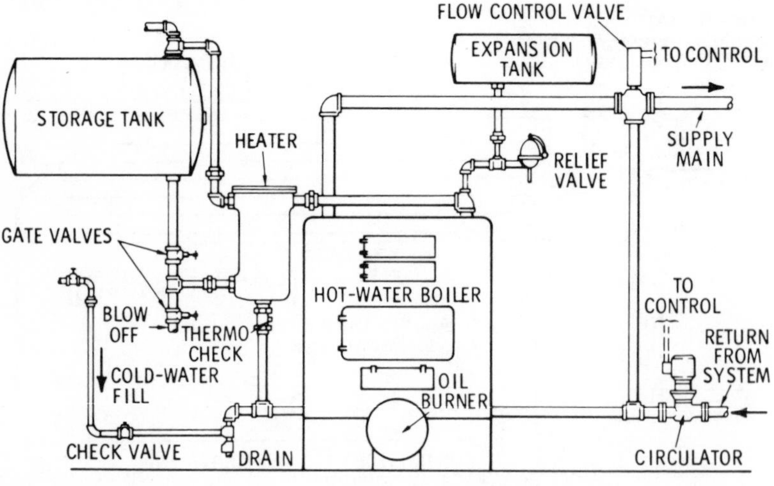
FLOW CONTROL VALVE
EXPANSION TANK
TO CONTROL
STORAGE TANK
HEATER
SUPPLY MAIN
RELIEF VALVE
GATE VALVES
HOT-WATER BOILER
TO CONTROL
BLOW OFF
THERMO CHECK
RETURN FROM SYSTEM
COLD-WATER FILL
OIL BURNER
CHECK VALVE
DRAIN
CIRCULATOR

Fig. 39. Diagram of the piping for a typical small radiant-heating system.

In cases where panels of small heat capacity (lag or inertia) are used, such as ceiling panels with small-diameter tubing and close spacing, the bypass valve at the boiler may be regulated by an aquastat, keeping the hot water temperature constant. The circulation is governed by the room thermostat turning the pump on and off.

With panels of greater inertia, this system is not recommended. Instead, the water temperature should vary inversely with the outside temperature. With such panels, the time lag between the start of the panel heating period and the start of the room heating period is too great. In addition, if the hot water is kept at a constant temperature, the panel stores a great amount of heat on warm days before the room thermostat is affected, causing an excessive rise or "overshoot" in room temperature. For this reason, regulation of the water temperature by both outside and inside thermostats is preferable.

Controls—In the design of any heating system, the primary consideration must be the comfort of the occupants regardless of the weather conditions. To preserve and properly utilize the advantages of radiant heating, it is necessary to maintain a moderate coil temperature that is modulated gradually without violent fluctuations. Thus, it is obvious that intermittent operation cannot fulfill this requirement.

The method of controlling heat by starting and stopping the circulation of the hot water through the coils is simple, but has many serious disadvantages. Chief among these are:

1. Starting and stopping the circulating pump necessarily means a cycling operation—in other words, the system has to reverse itself many times as called for by changing room conditions. This is poor control at best, even in conventional hot water heating. However, it becomes much worse when applied to radiant heating. In the heavy mass of concrete, or even the lighter mass of plaster, it will not be able to reverse itself quickly, and a considerable lag will follow which may result in periodic cycles of discomfort.
2. In a radiant-heating system with several circuits, it is difficult to obtain a perfect balance of circuit resistance and heat reserve for all rooms; therefore, when circulation stops, some rooms cool much more rapidly than others. For the same reason, when circulation is restored, some rooms heat up more rapidly than others, with a resultant lack of overall comfort conditions.
3. The most fundamental objection to the on-and-off principle of control with radiant heat is that it allows the surface of the panel to

cool and thereby defeats, intermittently, the whole purpose of radiant heat. As the panel surfaces are allowed to become cold at intervals, the system ceases to function as radiant heating, and the human body then becomes exposed to cold surfaces with a falling air temperature.

Fig. 40 shows the actual recorded air temperatures and surface temperatures of a concrete floor in which pipes were embedded and controlled by an on-and-off switch operating the pump according to room temperature. The control was set to start the pump when the air temperature dropped to 68°F. In this case, the bottoms of the pipes were two inches below the top surface of the concrete floor. The lag and the amount of variation will, however, vary according to the depth of the pipes below the surface and the type of floor covering used.

It has been proved by many years' experience that a well-controlled continuous-flow system is much more economical than an on-and-off system. Obviously then, the proper way of controlling radiant panels is to allow continuous circulation of water in the coils at all times. The water temperature should be modulated according to outside temperature changes and an indoor thermostat sensitive to radiant heat used as a limit control to prevent overheating without stopping circulation.

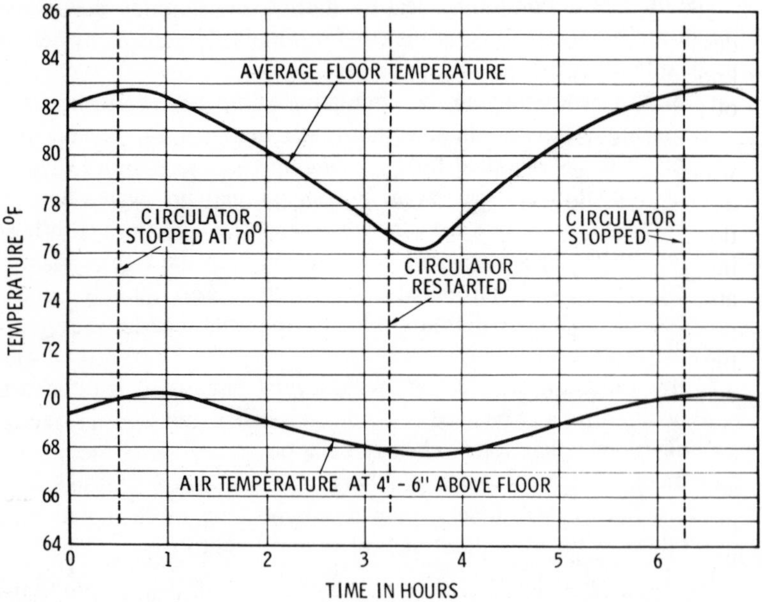

86
84
82
80
78
76
74
72
70
68
66
64
TEMPERATURE °F
AVERAGE FLOOR TEMPERATURE
CIRCULATOR STOPPED AT 70°
CIRCULATOR STOPPED
CIRCULATOR RESTARTED
AIR TEMPERATURE AT 4' - 6" ABOVE FLOOR
0
1
2
3
4
5
6
TIME IN HOURS

Fig. 40. Variation of air and surface temperature with on-and-off control.

A typical automatic water blender controlled by the outside air temperature is shown in Fig. 41. A typical boiler installation using this type of blender is illustrated in Fig. 42.

The water blender consists principally of a thermostatic three-way valve adapted to recirculate a variable portion of the return water, automatically adding the proper amount of hot water from the boiler to maintain the correct coil temperature. A simple liquid expansion thermostat is mounted on an outside wall of the building, preferably with a northwest exposure and away from the direct rays of the sun. Pressure built up in this thermostat as it absorbs heat is transmitted through a flexible tube to element *A* in the valve, pushing piston *B* downward. Hot water enters the valve at *1*, and return water from the system through inlet *2*. The hot and cooler return water pass through the mixing valve *S*, then over element *A*, and to the system through outlet *3*.

Element *A* forms a part of the thermostatic system, so that the position of mixing valve *S* is determined by the combined influence of the temperature at the outside thermostat and temperature of the mixed water flowing over *A*. The two elements (*A* and *S*) of the thermostatic system are so proportioned as to supply the heating coils with water at exactly the temperature needed for the existing weather conditions.

Should it be necessary to raise or lower the water temperature, the desired adjustment can be made by loosening locknut *J* and turning knob *K*. All controls are calibrated at the factory for the particular needs of the system.

Heat loss from buildings is subject to several variables, not all of which can be adequately controlled in accordance with the fluctuations of the outside temperature alone. Exposure, wind velocity and direction, as well as the insulating properties of the walls, all exert their influence. Because there is a certain time lag between changes in atmospheric conditions and the temperature in the rooms, control by outside temperature alone may result in rooms becoming too warm at times. An automatic check against this overheating can be provided by a compensating indoor control.

This control system is similar to that shown in Figs. 41 and 42, except that an additional temperature—sensitive bulb *Q* (Fig. 43) is added. This bulb is fitted with a small electric heater coil (*H*) and a manually adjustable resistance (*R*) mounted in a metal box for convenient attachment to the wall in the boiler room. A thermostat is placed in a key room from which overheating by the system can be controlled. The thermostat has a normally open switch which acts to close the

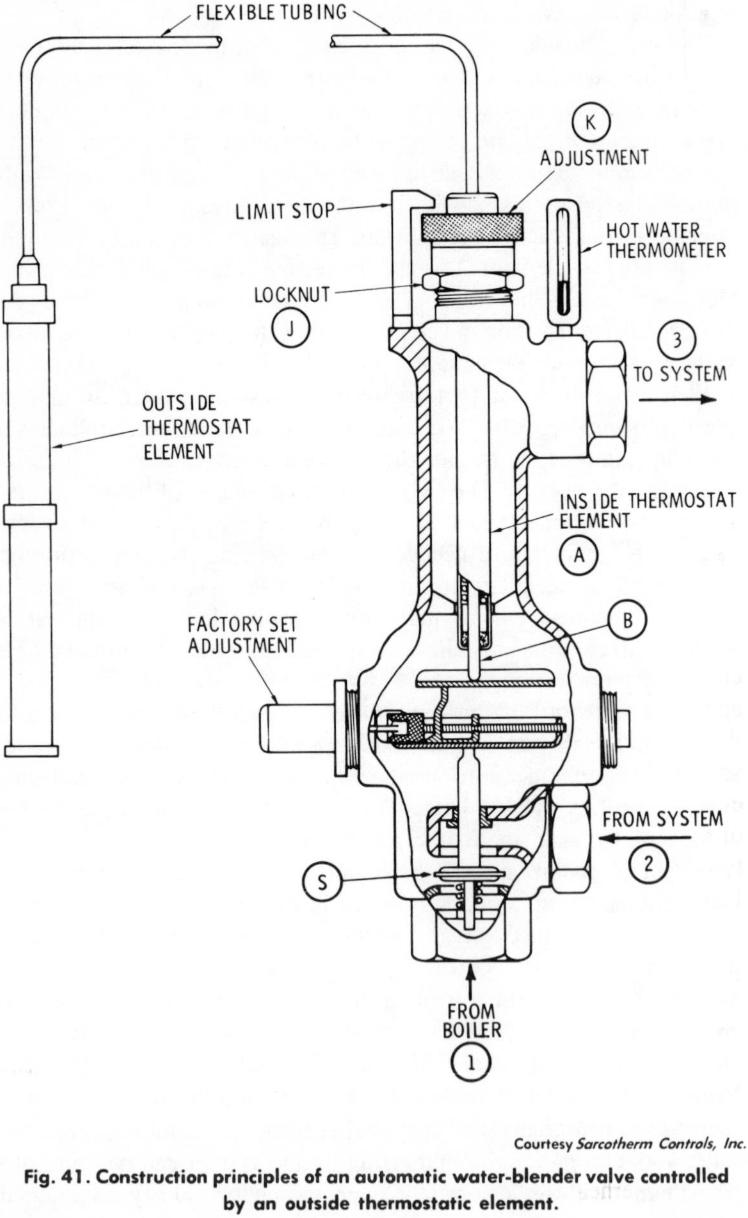
FLEXIBLE TUBING
K
ADJUSTMENT
LIMIT STOP
HOT WATER
THERMOMETER
LOCKNUT
J
3
TO SYSTEM
OUTSIDE
THERMOSTAT
ELEMENT
INSIDE THERMOSTAT
ELEMENT
A
B
FACTORY SET
ADJUSTMENT
FROM SYSTEM
2
S
FROM
BOILER
1

Courtesy *Sarcotherm Controls, Inc.*

Fig. 41. Construction principles of an automatic water-blender valve controlled by an outside thermostatic element.

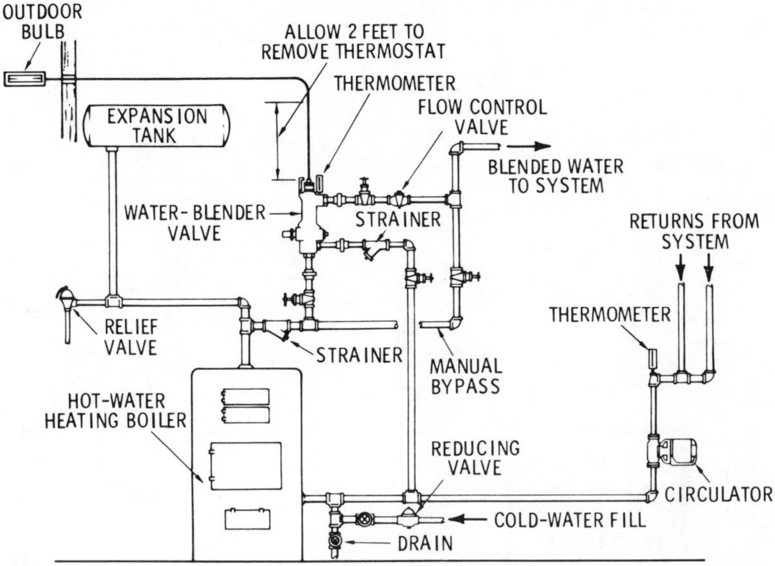
OUTDOOR
BULB
ALLOW 2 FEET TO
REMOVE THERMOSTAT
THERMOMETER
FLOW CONTROL
VALVE
EXPANSION
TANK
BLENDED WATER
TO SYSTEM
WATER-BLENDER
VALVE
STRAINER
RETURNS FROM
SYSTEM
RELIEF
VALVE
STRAINER
THERMOMETER
MANUAL
BYPASS
HOT-WATER
HEATING BOILER
REDUCING
VALVE
CIRCULATOR
COLD-WATER FILL
DRAIN

Courtesy *Sarcotherm Controls, Inc.*

Fig. 42. Piping diagram of a typical installation using an automatic water-blender valve controlled by an outside thermostat.

circuit to *H* when the desired room temperature is reached. This action applies heat to bulb *Q*, which will cause a greater pressure in the thermostatic flexible tube system, closing the hot—water inlet port and recirculating the cooler return water through the heating coils. This effectively checks the overheating while preserving the responsiveness of the system to outside temperature changes. The piping diagram of a typical system using this indoor—outdoor control feature is shown in Fig. 44.

An automatic heat-control system for use with heavy radiant-heat panels is shown in Fig. 45. In this system, the outdoor controller senses any change in the heating load as the outdoor temperature varies. This change is transmitted to the control panel, where a corresponding change is made in the setting of a temperature controller whose sensing bulb is located in the supply water line to the heating coils.

If a thermostat is used, it senses changes in the internal heat load caused by occupancy, solar gain, etc., and acts as a limit control to prevent overheating. The circulator operates continuously except when

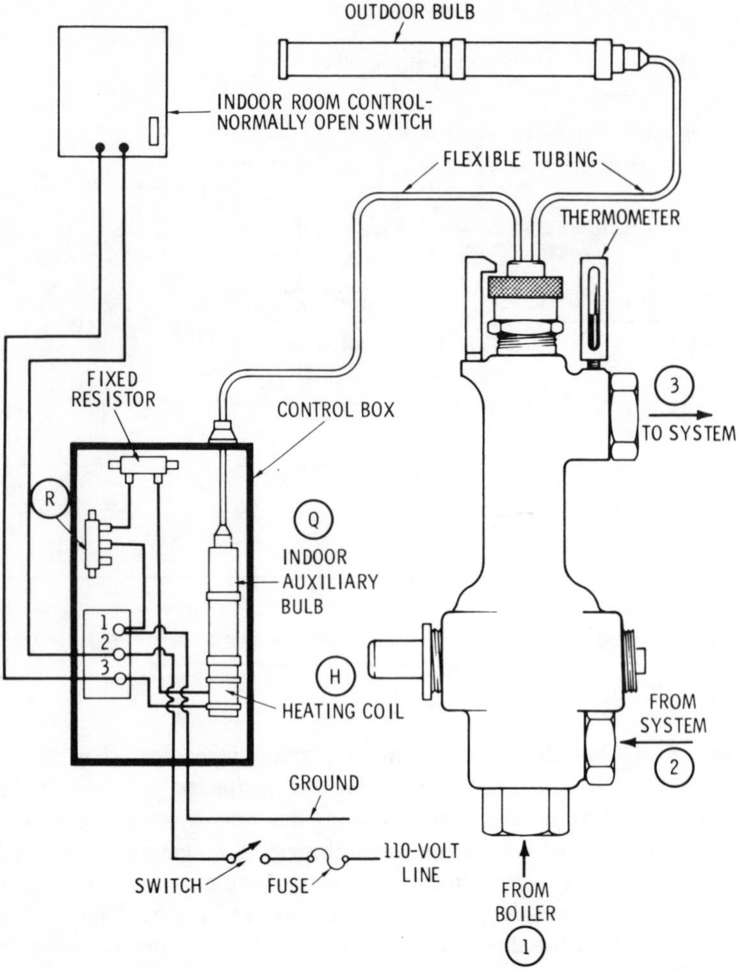
OUTDOOR BULB
INDOOR ROOM CONTROL-
NORMALLY OPEN SWITCH
FLEXIBLE TUBING
THERMOMETER
FIXED
RESISTOR
CONTROL BOX
3
TO SYSTEM
R
Q
INDOOR
AUXILIARY
BULB
1
2
3
H
HEATING COIL
FROM
SYSTEM
2
GROUND
110-VOLT
LINE
SWITCH
FUSE
FROM
BOILER
1

Courtesy *Sarcotherm Controls, Inc.*

Fig. 43. An automatic water-blender valve with indoor compensation feature.

no heat is required. The safety control prevents excessive temperature of the boiler water.

The automatic heat-control system in Fig. 46 is for use with light radiant-heat panels. This type has no outdoor sensing unit. Instead, the room thermostat senses any change in room temperature and transmits this change to the control panel, which adjusts the control point of a

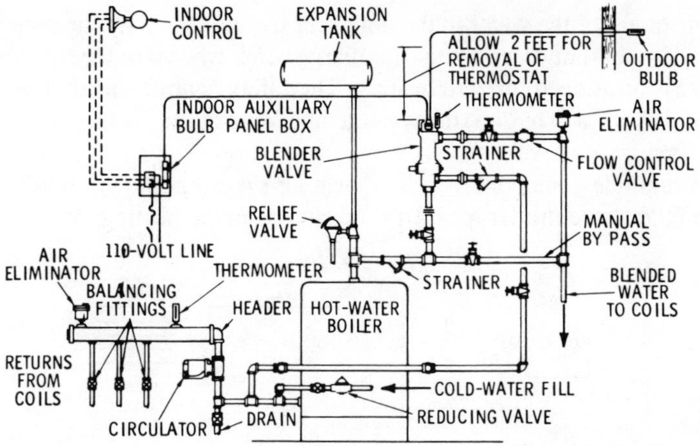
INDOOR CONTROL
EXPANSION TANK
ALLOW 2 FEET FOR REMOVAL OF THERMOSTAT
OUTDOOR BULB
THERMOMETER
AIR ELIMINATOR
INDOOR AUXILIARY BULB PANEL BOX
BLENDER VALVE
STRAINER
FLOW CONTROL VALVE
RELIEF VALVE
MANUAL BY PASS
110-VOLT LINE
AIR ELIMINATOR
THERMOMETER
STRAINER
BLENDED WATER TO COILS
BALANCING FITTINGS
HEADER
HOT-WATER BOILER
RETURNS FROM COILS
COLD-WATER FILL
DRAIN
REDUCING VALVE
CIRCULATOR

Courtesy *Sarcotherm Controls, Inc.*

Fig. 44. Diagram of heating system using indoor-outdoor controls.

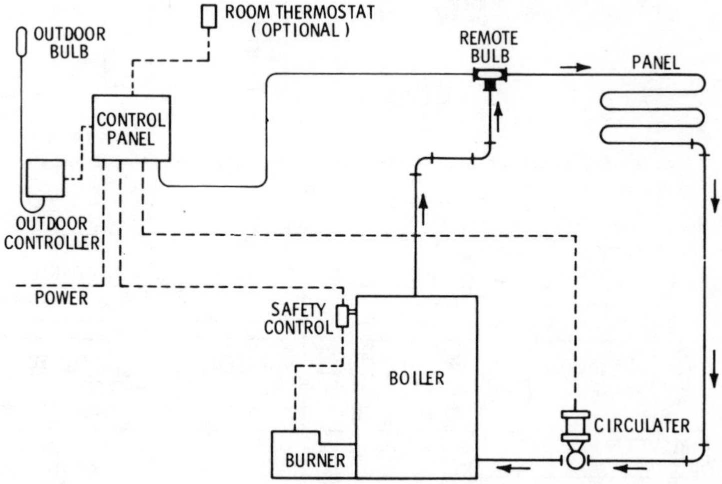
ROOM THERMOSTAT (OPTIONAL)
OUTDOOR BULB
REMOTE BULB
PANEL
CONTROL PANEL
OUTDOOR CONTROLLER
POWER
SAFETY CONTROL
BOILER
CIRCULATER
BURNER

Courtesy *Honeywell Inc.*

Fig. 45. An automatic heat-control system for heavy radiant-heat panels.

temperature controller whose sensing bulb is located in the supply water to the heating coil. The temperature controller repositions the three-way valve, which mixes the hot boiler water with the cooler return water from the bypass to provide a mixture of the proper temperature to satisfy the setting of the temperature controller. The burner is controlled by an aquastat (immersion temperature controller) to maintain the necessary temperature of the water in the boiler for the indirect water heater. The circulator is shut down by the auxiliary switch whenever the three-way valve is positioned for zero heating. The safety control will shut off the burner if, for any reason, the water temperature in the boiler reaches the maximum safe level.

A multiple-zone control radiant-heating system is shown in Fig. 47. The outdoor controller measures the change in the heating load caused

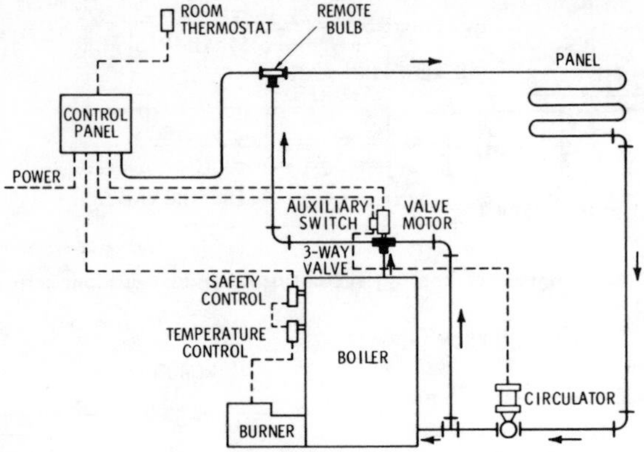
ROOM
THERMOSTAT
REMOTE
BULB
PANEL
CONTROL
PANEL
POWER
AUXILIARY
SWITCH
VALVE
MOTOR
3-WAY
VALVE
SAFETY
CONTROL
TEMPERATURE
CONTROL
BOILER
CIRCULATOR
BURNER

Courtesy *Honeywell Inc.*

Fig. 46. An automatic heat-control system for light radiant-heat panels.

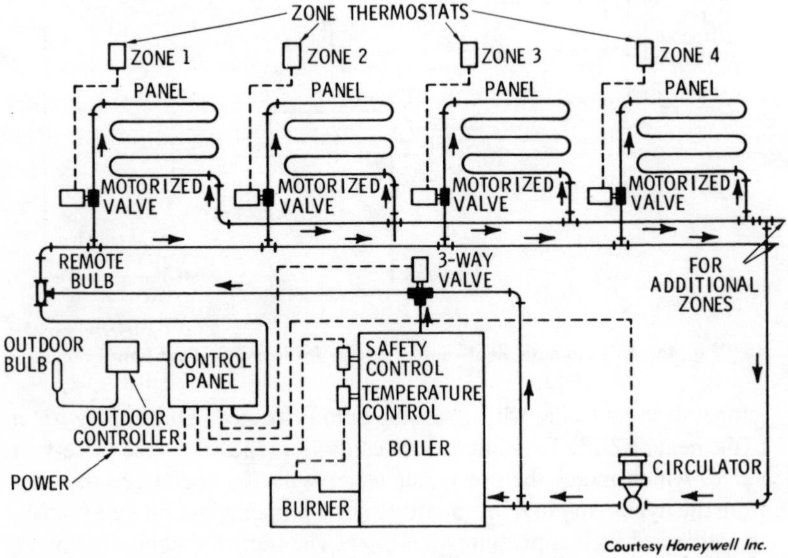
ZONE THERMOSTATS
ZONE 1
ZONE 2
ZONE 3
ZONE 4
PANEL
PANEL
PANEL
PANEL
MOTORIZED
VALVE
MOTORIZED
VALVE
MOTORIZED
VALVE
MOTORIZED
VALVE
REMOTE
BULB
3-WAY
VALVE
FOR
ADDITIONAL
ZONES
OUTDOOR
BULB
CONTROL
PANEL
SAFETY
CONTROL
TEMPERATURE
CONTROL
OUTDOOR
CONTROLLER
BOILER
CIRCULATOR
POWER
BURNER

Courtesy *Honeywell Inc.*

Fig. 47. A typical control system for a multiple-zone radiant-heating system.

by a change in the outdoor temperature. This change is transmitted to the control panel, causing a corresponding change in the setting of a temperature controller whose sensing bulb is located in the main distribution line, and downstream from the three-way valve. The temperature controller positions the three-way valve to mix the hot boiler water with the cooler return water in the proper proportions to furnish supply water to the heating coils at the proper temperature. A temperature controller in the boiler controls the burner so as to maintain the boiler water at the temperature required by the indirect heater. A room thermostat in each zone controls a motorized valve in its particular supply line. This valve regulates the panel-coil temperature in accordance with the individual requirements of the zone.

CHAPTER 7

Boiler Fittings

Certain fittings or devices that are usually mounted or connected to the boiler in steam and hot-water heating systems are essential to the proper operation of the boiler. Among these are:

1. Valves
   a. Safety.
   b. Stop.
   c. Check.
   d. Blow-off.
2. Water-gauge cocks.
3. Water gauges.
4. Steam gauges.
5. Injectors.
6. Fusible plugs.

All of these fittings will not necessarily be installed on all boilers (especially hot-water boilers), but the pipe fitter should be familiar with their purpose. Fig. 1 shows the piping diagram of a typical steam boiler with the control and indicating devices essential to its proper operation.

VALVES

The valves used in boiler installations are of several different types including the familiar globe and gate valves. Only the best-quality valves should be used because of the demands placed on them by the

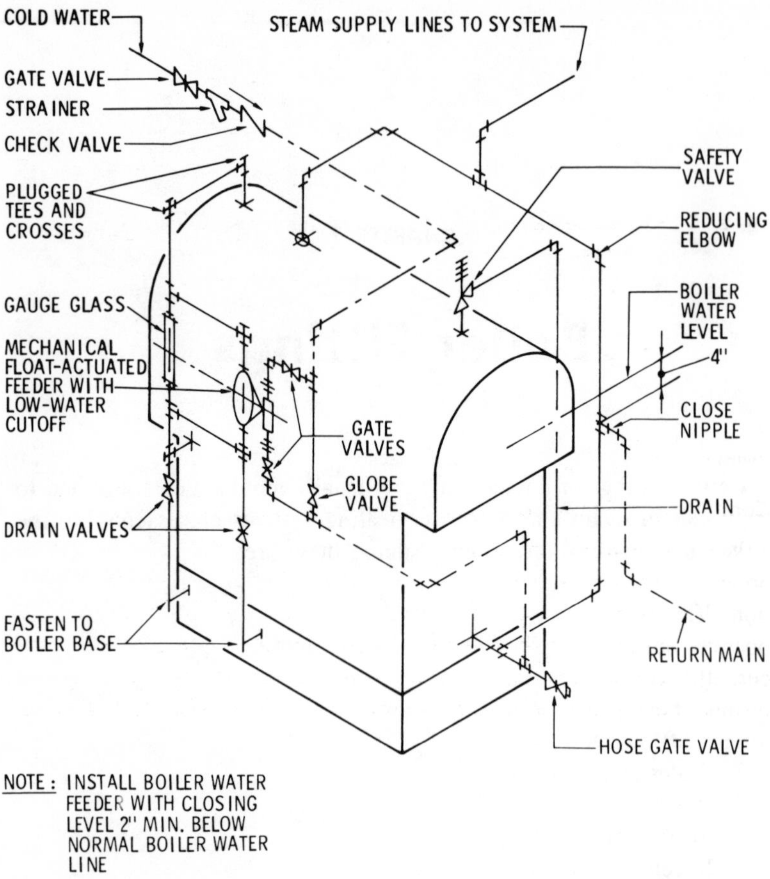
COLD WATER
STEAM SUPPLY LINES TO SYSTEM
GATE VALVE
STRAINER
CHECK VALVE
SAFETY VALVE
PLUGGED TEES AND CROSSES
REDUCING ELBOW
BOILER WATER LEVEL
4"
GAUGE GLASS
MECHANICAL FLOAT-ACTUATED FEEDER WITH LOW-WATER CUTOFF
CLOSE NIPPLE
GATE VALVES
GLOBE VALVE
DRAIN
DRAIN VALVES
FASTEN TO BOILER BASE
RETURN MAIN
HOSE GATE VALVE
NOTE: INSTALL BOILER WATER FEEDER WITH CLOSING LEVEL 2" MIN. BELOW NORMAL BOILER WATER LINE

Fig. 1. A typical boiler, showing the control and indicating devices essential to its proper and safe operation.

exposure of their working parts to the high temperatures of the steam or hot water they are controlling.

Safety Valves

Every boiler installation must have a *safety valve* incorporated to protect the boiler itself and the building occupants in case of malfunction. These valves are adjusted to open and relieve the internal pressure, should it rise above a safe predetermined level. A boiler water safety relief valve is shown in Fig. 2.

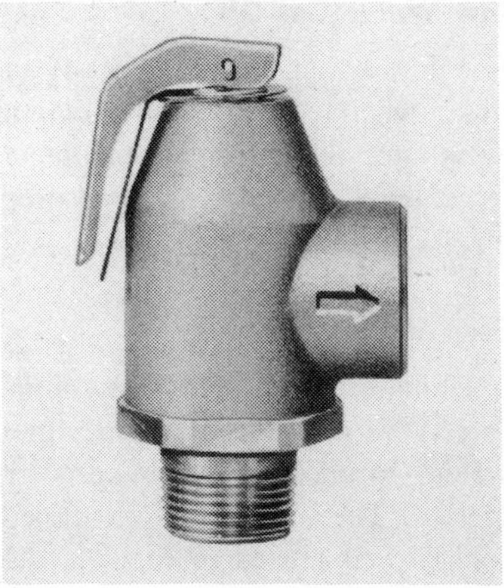

Fig. 2. A boiler water safety relief valve.

Courtesy *Watts Regulator Co.*

Combination Feeder and Low Water Cut-off

When a combination low water feeder and cut-off is used on a steam boiler the feeder adds make-up water when needed for normal operation. If, for any reason, an emergency occurs, the low water cut-off operates to shut down the firing mechanism by interrupting the electrical current to the burner. When the emergency has passed, the water feeder resumes normal operation. The combination feeder shown in Fig. 3 has fittings for a gauge glass connection.

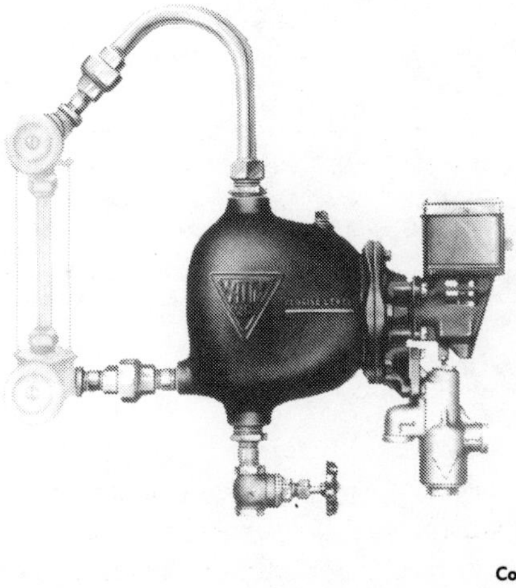

Courtesy *Watts Regulator Co.*

Fig. 3. A combination boiler water feeder and low water cutoff.

Low Water Cut-Off

A low water cut-off (Fig. 4) automatically shuts off the burner or stoker whenever the water level in the boiler falls below a safe operating minimum. The cut-off is mounted with the float level with the boiler water level. If the water level drops the float also drops, breaking the electrical circuit to the burner or stoker.

Stop Valves

Globe or gate valves are commonly used as *stop valves.* They are merely valves placed in pipe lines—either steam, hot water, or cold water—to allow the passage of steam or water, or to block it completely. They normally are turned completely on or completely off, seldom being used to regulate the flow in any intermediate manner.

Stop-Check Valves

Stop-check valves should not be confused with ordinary stop valves. A Stop-check valve is primarily designed to act as an automatic non-return valve by preventing a backflow of steam from the main steam header to the boiler to which it is connected, in the event of failure of that boiler. Stop-check valves are used when more than one boiler is connected to a common steam header.

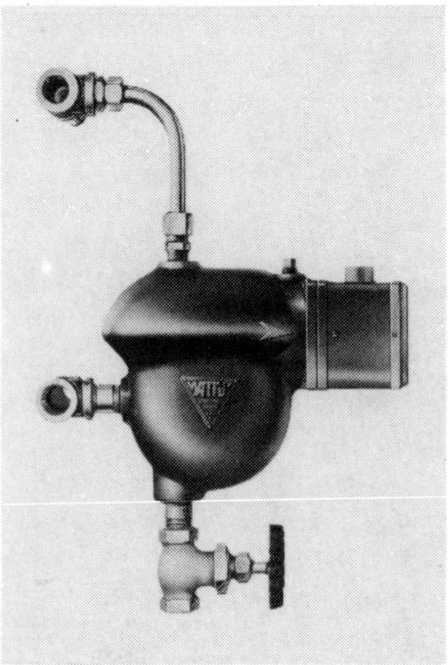

Fig. 4. A boiler low water cutoff.

Courtesy *Watts Regulator Co.*

Check Valves

It is sometimes necessary to allow steam or water to flow through a pipe in one direction but not in the other. An example of this is in the cold-water supply line to the boiler. A *check valve* is normally installed in this line to permit water to enter the boiler but prevent water (or steam) in the boiler from backing up into the supply line. These valves usually have a swinging gate and must be installed horizontally with the hinge at the top of the valve.

Blow-off Valves

A valve installed in a line connected to the lowest part of the water chamber in a boiler is called a *blow-off valve.* It is used to drain off sediment, scale, and other impurities that settle out of the water as it is heated. This settlement accumulates at the lowest point in the boiler, where it is removed by periodic opening of the blow-off valve.

BOILER WATER LEVELS

It is very important that the water level in the boiler be maintained within certain limits at all times. If the level is too high, inefficient operation results—if too low, there is danger of permanent damage to the boiler or the possibility of an explosion. Two principal methods are used to determine the water level.

Try Cocks

Small valves called *try cocks* are installed on steam boilers at the safe high-water level and at the safe low-water level. On larger boilers, a third try cock, halfway between the other two, is sometimes used. Fig. 5 shows a typical handwheel type of try cock. For large steam boilers

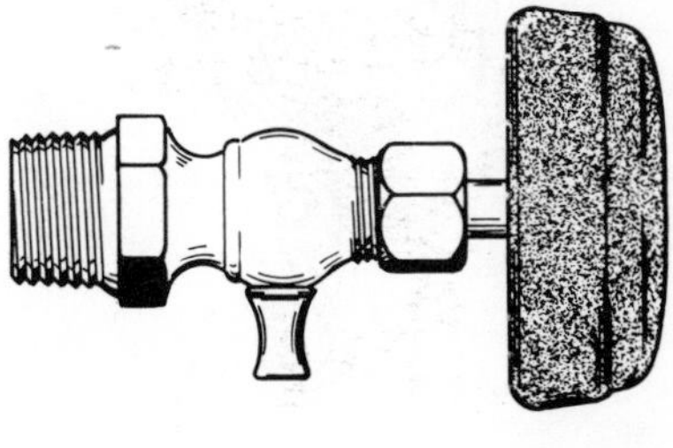

Fig. 5. A bronze handwheel try cock.

Courtesy *Eugene Ernst Products Co.*

where the try cocks are at a level so high as to be unhandy to reach, a lever (Fig. 6) is used which is opened and closed by means of a chain.

To ascertain the water level, first one cock and then the other is opened slightly and the presence of water or steam is determined by its appearance or sound. With a little experience it becomes easy to determine the approximate water level.

The correct method must be used to check the water level with try cocks. Fig. 7 shows the results of the right and wrong way when the water is at a particular level with respect to the try cock. By opening the try cock slightly, the escape of steam indicates the water level is below the fitting. If, however, the level is only slightly below the try cock and the cock is opened wide, the resulting violent agitation of the water will cause some of it to escape, giving the false impression that the water level is at or above the fitting. Therefore, try cocks should be opened only slightly to ensure that a false water level is not indicated.

Water Gauges

Water gauges are installed on steam boilers to give a visual indication of the boiler water level. When a low water cut-off is used as shown in Fig. 3 the gauge glass can be mounted on the low water cut-off.

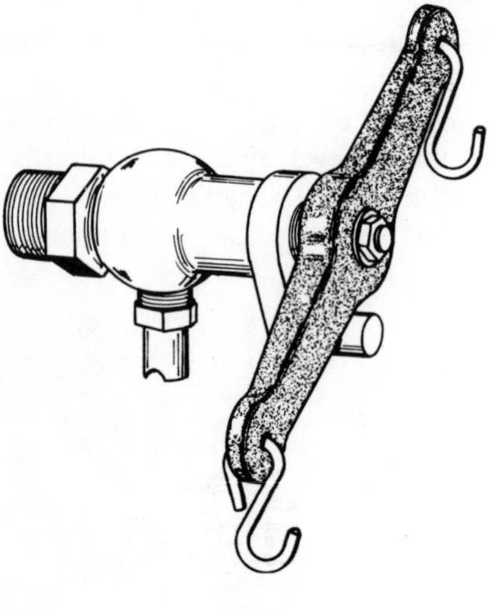

Fig. 6. A bronze lever-type try cock.

Courtesy *Eugene Ernst Products Co.*

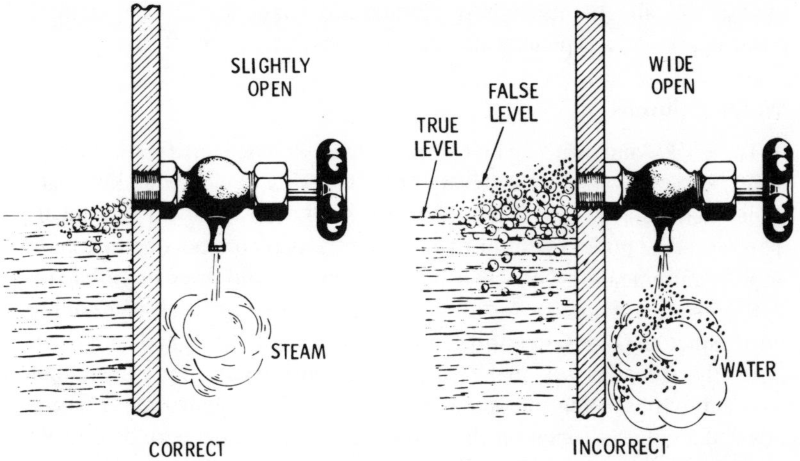
SLIGHTLY
OPEN
TRUE
LEVEL
FALSE
LEVEL
WIDE
OPEN
STEAM
WATER
CORRECT
INCORRECT

Fig. 7. Correct and incorrect method of testing the water level in a steam boiler.

A water gauge (Fig. 8) consists of a strong glass tube long enough to cover the safe range of water level and having the ends connected to the boiler interior by fittings. It is connected to the boiler at such an elevation that the lower and upper ends of the tube will be below the

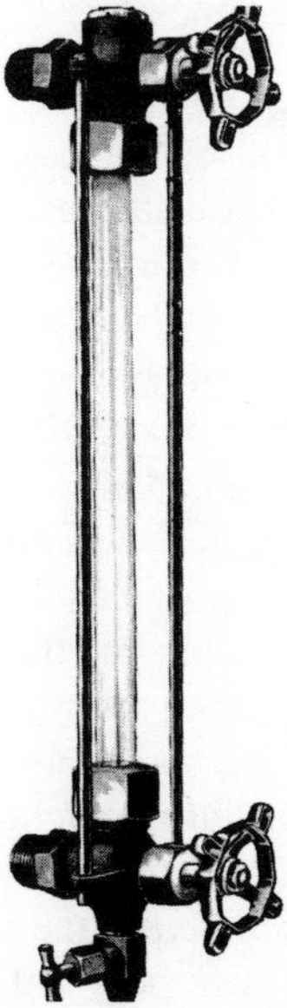

Fig. 8. A typical two-rod plain water gauge.

Courtesy *Lunkenheimer Co.*

lowest and above the highest permissible water level. The internal construction of a typical water gauge is shown in Fig. 9.

Water Columns

Try cocks and water gauges are frequently connected to a special fitting known as a *water column* (Fig. 10). This device contains floats which actuate a whistle when the water level drops below a safe level. This alerts the boiler operator to add water. It is recommended that local code requirements be checked before a water column is installed.

Provision is made for installing crosses with plugs at all right-angle turns, since the pipes and fittings often stop up with scale and sediment. After the steam is up, make sure the column connections are unobstructed by observing the action of the water in the gauge glass. Blow down the column (opening the valves fully for a short period of time) for a double check, which will also test the low-water alarm. Keep the gauge glass clean and properly lighted for good visibility. Operate the gauge cocks regularly to ensure their proper operation for emergency service.

Sight-Glass Maintenance

Good maintenance pays dividends in safety and trouble-free operation. A few pointers for sight-glass maintenance are:

1. Keep sight glass in its original container until ready to install.
2. Avoid bumping, chipping, or scratching.
3. Gasket seats must be flat, smooth, and free of metal or gasket bits.
4. Only approved gaskets should be used. These should be at least 1/16″, but not more than 1/4″ thick on the vessel sides. Clamp-ring gaskets should be less than 1/16″ thick. They should be of smooth material without wire inserts.
5. All bolts and nuts should be free of burrs, and should be well lubricated with graphite grease. They should be tightened to the value specified by the manufacturer (usually 10 to 20 lb.-ft., depending on the number of bolts and their size). A regular tightening sequence should be followed to avoid uneven loads on the glass. The use of conical spring washers will keep the joints tight during cycling.
6. A commercial glass cleaner should be used to keep the glass clean. Avoid the use of wire brushes, scrapers, and harsh abrasives. Inspect and replace the glass tubes on a regular schedule.

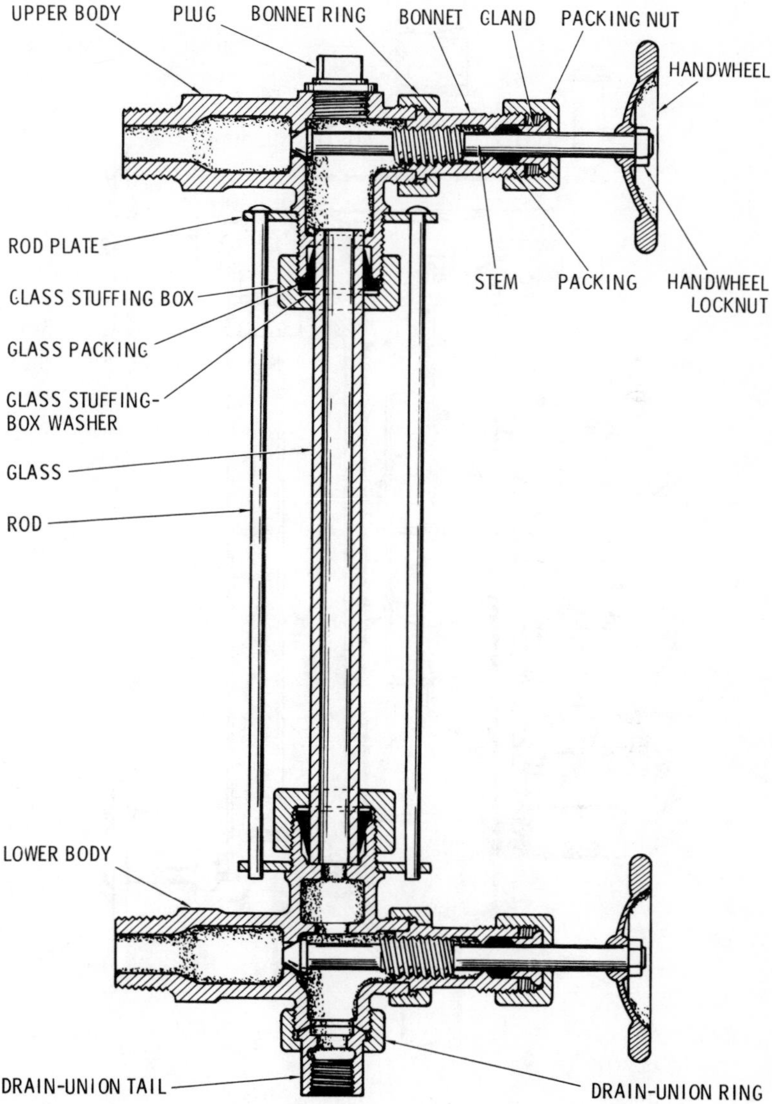
UPPER BODY
PLUG
BONNET RING
BONNET
GLAND
PACKING NUT
HANDWHEEL
ROD PLATE
STEM
PACKING
HANDWHEEL LOCKNUT
GLASS STUFFING BOX
GLASS PACKING
GLASS STUFFING-BOX WASHER
GLASS
ROD
LOWER BODY
DRAIN-UNION TAIL
DRAIN-UNION RING

Courtesy *Lunkenheimer Co.*

Fig. 9. Cross-sectional view of a typical water gauge.

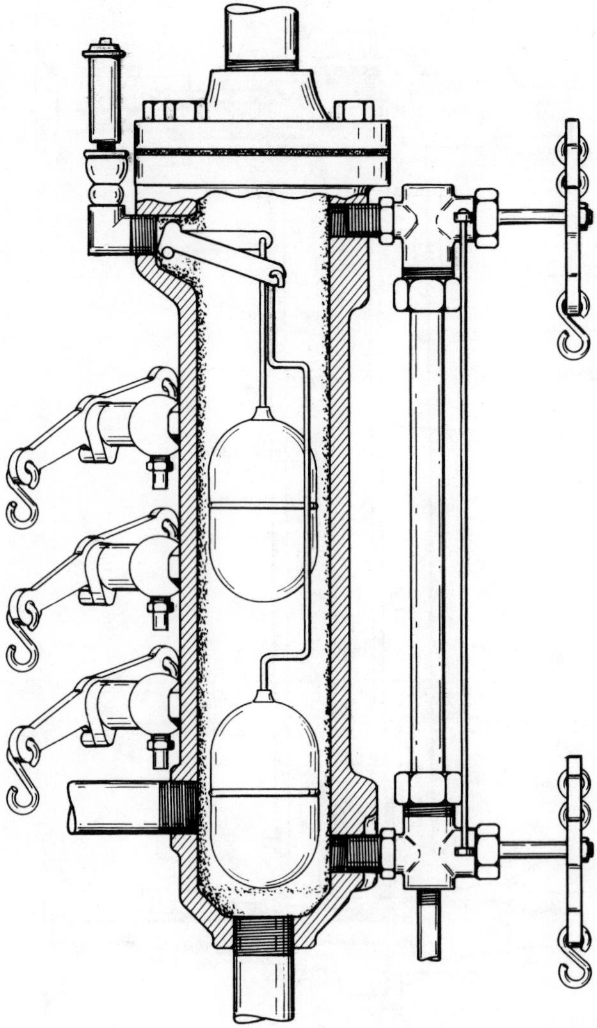

Courtesy *Eugene Ernst Products Co.*

Fig. 10. A water column with a vertical gauge.

7. Use a bright light to check for glistening marks in or on the glass, and for cloudy or roughened surfaces. If any of these signs is found, replace the glass regardless of the maintenance schedule.
8. A replacement glass must have a pressure and temperature rating higher than the service rating of the vessel. Be sure to replace all shields intended to protect the glass from falling objects or sudden blasts of cold air.

It is sometimes necessary or desirable to cut a sight glass from a long section of tube. This can be accomplished quite easily if a sight-glass cutter, such as the one in Fig. 11, is used. Follow these steps:

1. Score the inner surface of the tube at the correct point. It is important to score once around only—additional scoring will result in a jagged edge and will dull the cutter.
2. Heat the outside of the glass tube directly over the score for a few seconds with the tip of the hot blue flame of a Bunsen burner, propane torch, or oxyacetylene flame; the glass will crack audibly and visibly at the score. Heat only until the crack encircles the glass.
3. If only a short piece is to be cut off, tap the scored end lightly on a hard surface, preferably wood. If a long piece is to be cut off, grip the glass firmly on each side of the score and pull apart without bending.

Fig. 11. A boiler sight-glass cutter.

STEAM GAUGES

A steam gauge is a very important fixture which should be tested at frequent intervals to ascertain if it is indicating the correct steam pressure. This device indicates the difference in pressure between the inside and outside of the boiler. In other words, it indicates a *gauge pressure* as distinguished from *absolute pressure.*

A steam gauge works on one of two principles—the expansion of a corrugated diaphragm when pressure is applied, or the tendency of a curved tube to assume a straight position when under pressure.

The bent-tube type of gauge is now almost universally used. Since the movement of the free end is necessarily small, its motion is multiplied by means of a rack-and-pinion gear which causes a pointer to rotate. The principle of operation is shown in Fig. 12.

When a gauge is in good working order, the pointer will move easily with every change of pressure in the boiler. The pointer should return to zero when the steam is shut off.

A properly installed gauge should have one or more turns of tubing placed between it and the boiler, as in Fig. 13. This "pigtail" fills with condensate to prevent live steam from reaching the working parts of the gauge, thus preventing damage. The gauge will be ruined if such a "pigtail" is not used, especially if superheated steam is produced. In

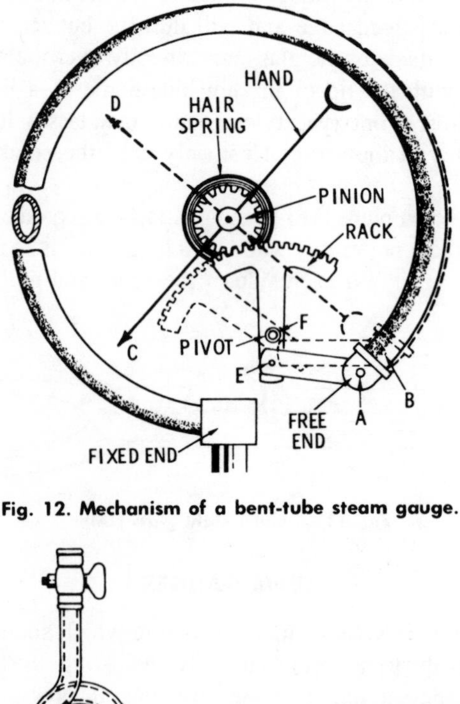
HAND
D
HAIR
SPRING
PINION
RACK
F
PIVOT
C
E
B
FREE
END
A
FIXED END

Fig. 12. Mechanism of a bent-tube steam gauge.

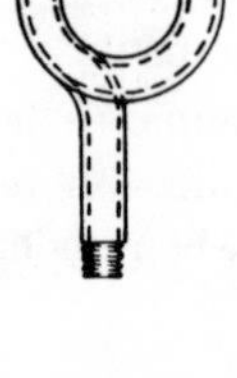

Fig. 13. Steam-gauge pigtail.

fact, before installing a gauge, *the "pigtail" should be filled with water.* Locate the gauge in a cool place and secure it to some substantial object where it will be free from vibration or shock.

INJECTORS

An *injector* is a device for forcing water into a boiler against the boiler pressure by means of a steam jet. The simplest form of an injector is shown in Fig. 14. Steam from the boiler enters the fixed steam nozzle and passes through the space between the nozzle and combining tube. It enters the combining tube and then passes into the overflow chamber and out through the overflow check valve. As it passes from the steam nozzle into the combining tube, a vacuum is formed which draws water in through the water inlet. The inrush of cold water condenses the steam in the combining tube, and the water thus formed is at first driven out through the overflow. As the velocity of the water jet increases, however, sufficient momentum is obtained to overcome the boiler pressure; and water enters the delivery tube, passes the main check valve, and enters the boiler. With the water now flowing into the boiler, a vacuum is produced in the overflow chamber; the vacuum causes the

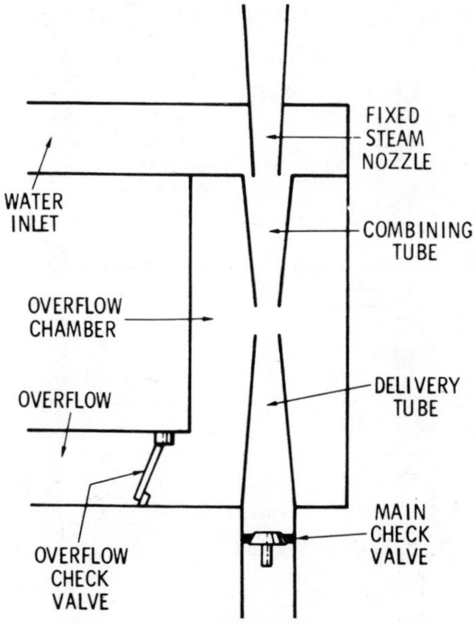
FIXED STEAM NOZZLE
WATER INLET
COMBINING TUBE
OVERFLOW CHAMBER
OVERFLOW
DELIVERY TUBE
MAIN CHECK VALVE
OVERFLOW CHECK VALVE

Fig. 14. Elementary principles of an injector.

overflow check valve to close, preventing air from entering the delivery tube. This sequence of events is shown in Fig. 15.

There are three pipe connections to be made to an injector. One is the steam pipe, which must be the same size as the injector connection. This pipe must be connected to the highest possible point on the boiler and be independent of any other pipe in order to secure the best results. It must be blown out with steam before the final connection to the injector is made.

The second pipe connected to the injector is the suction pipe. It must be at least as large as the injector connection. Where the lift is over ten feet, the suction pipe should be one or two sizes larger, reduced to the injector size as close as possible to the injector and with a globe valve the same size as the pipe installed as near the injector as possible. A second globe valve is sometimes installed in this line if the water source has a high pressure. It is used to reduce the pressure while the first valve regulates the amount of water.

A foot valve should be installed on the lower end of the suction pipe where a long lift is necessary. Without a foot valve, the air in the pipe must be exhausted each time water is called for, resulting in a considerable waste of steam.

The delivery pipe should have both a globe valve and a check valve installed between the injector and the boiler. A drain pipe to take care of

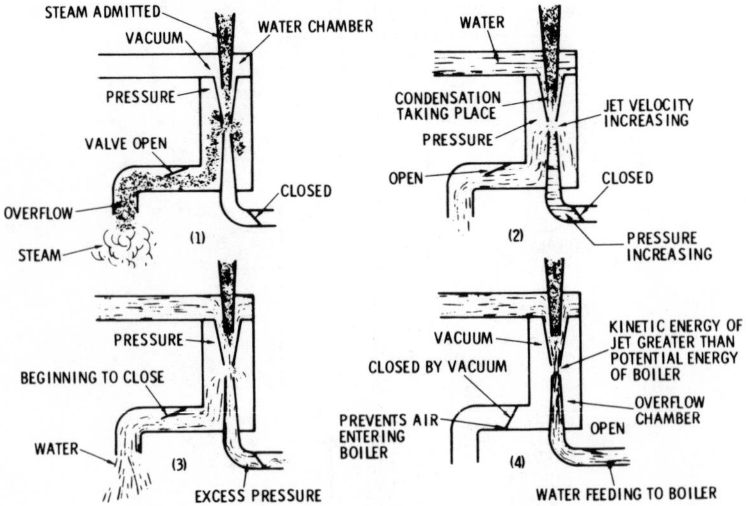
STEAM ADMITTED
VACUUM
WATER CHAMBER
PRESSURE
VALVE OPEN
CLOSED
OVERFLOW
STEAM
(1)
WATER
CONDENSATION TAKING PLACE
JET VELOCITY INCREASING
PRESSURE
OPEN
CLOSED
(2)
PRESSURE INCREASING
PRESSURE
BEGINNING TO CLOSE
WATER
(3)
EXCESS PRESSURE
VACUUM
CLOSED BY VACUUM
KINETIC ENERGY OF JET GREATER THAN POTENTIAL ENERGY OF BOILER
OVERFLOW CHAMBER
PREVENTS AIR ENTERING BOILER
OPEN
(4)
WATER FEEDING TO BOILER

Fig. 15. Starting cycle sequence of an injector.

the overflow completes the installation. A cross-sectional view of a typical injector is shown in Fig. 16.

FUSIBLE PLUGS

A *fusible plug* is a safety device installed on some boilers to protect it in case of dangerously low water. These plugs (Fig. 17) are commonly made of bronze and filled with pure tin. The melting temperature of the tin is approximately 450°F, at which point the plug will relieve the pressure within the boiler. Fusible plugs are normally used only on boilers having a steam pressure of less than 250 psi.

PRESSURE RELIEF VALVES

Steam and hot-water boilers must be equipped with an approved *relief valve* installed in accordance with local or A.S.M.E. codes. Such a valve is shown in cross section in Fig. 18. When pressure within the boiler reaches a certain level, the relief-valve mechanism actuates to relieve that pressure. After the pressure is reduced to a safe level, the valve closes.

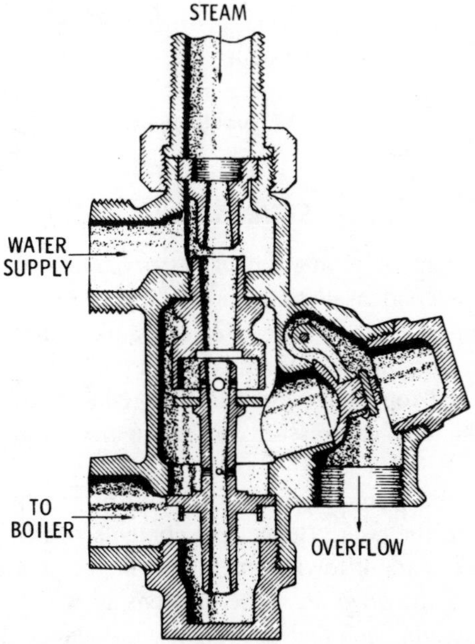
STEAM
WATER SUPPLY
TO BOILER
OVERFLOW

Fig. 16. Cross-sectional view of a typical injector.

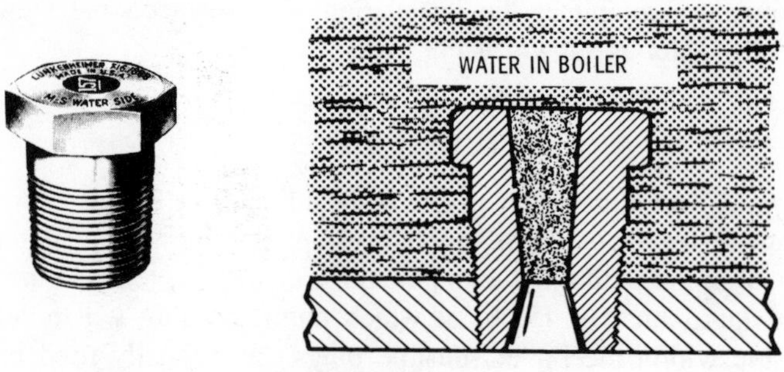
WATER IN BOILER

(A) INSIDE TYPE

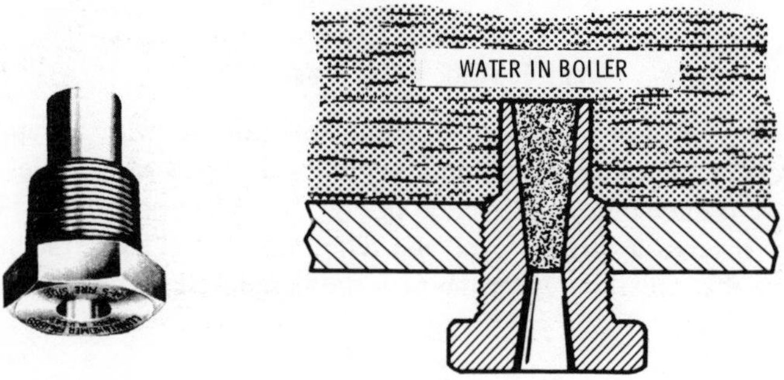
WATER IN BOILER

(B) OUTSIDE TYPE

Courtesy *Lunkenheimer Co.*

Fig. 17. Fusible plugs.

STEAM LOOP

An arrangement of piping called a *steam loop* is used in a steam system to return condensate to the boiler. The essential parts of a steam loop are shown in Fig. 19. The *riser* contains a mixture of water and steam. The steam portion of the mixture is condensed by means of the *condenser* at the top. This condensation reduces the pressure in the system, and the reduced pressure causes the mixture in the riser to flow upward.

As soon as the water mixed with steam passes the *goose neck,* it cannot return to the riser. Hence, the contents of the pipes constantly work from the separator toward the boiler, the condenser being slightly inclined toward the *drop leg.* The condensate will accumulate in the drop leg to a height such that its weight will balance the weight of the mixture in the riser.

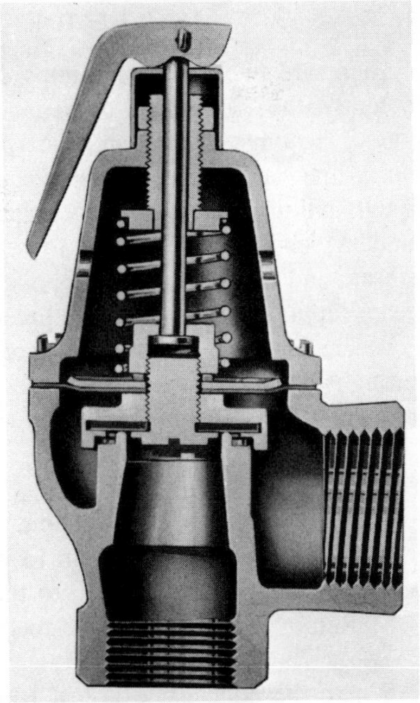

Fig. 18. A typical water pressure-relief valve.

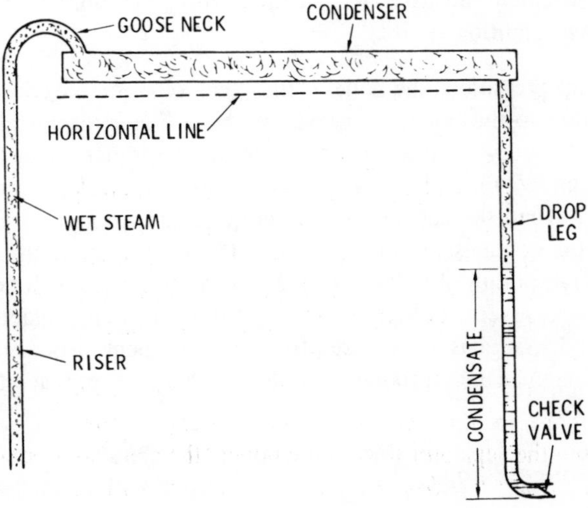
GOOSE NECK
CONDENSER
HORIZONTAL LINE
WET STEAM
RISER
DROP LEG
CONDENSATE
CHECK VALVE

Fig. 19. Essential parts of a steam loop.

HEATING PUMPS

There are two types of pumps designed for use in steam heating systems—the *condensation pump,* which is used with a two-pipe gravity heating system, and the *vacuum pump,* which is used with return-line vacuum and variable-vacuum heating systems. Although the design and ultimate operation of these two types differ, they have one fundamental feature common to both—the accumulation of the condensate and its delivery to the boiler. The water condensed from the steam supplied to the steam heating system is the source of condensate.

Wherever the condensate is not wasted to the sewer, the cycle through which the medium passes in steam heating systems is as follows:

1. Vaporization into steam.
2. Conveyance to the radiation.
3. Condensation from steam to water.
4. Conveying condensation to the pump.
5. Return of the condensate to the boiler.

Where the condensate is lost from the system or is wasted to the sewer, the fourth and fifth phases of the cycle may include wasting the condensate and supplying water from a source of "make-up."

Pumps should be used for steam heating systems if any of the following conditions exist:

1. If the pressure at the boiler is too great to permit the condensate to return to the boiler by gravity or by a boiler return trap.
2. If the "dry" return mains are below the boiler water line.
3. If the size of the heating system, in terms of condensing capacity, is beyond the capacity of the largest return trap.
4. If the condensate must be forced from a lower elevation over a "high-point" or to the floor level of the boiler room; for example, where several buildings are fed from one boiler plant and the condensate has to be raised from the basement of one and forced through an underground return to a higher level to reach the boiler room.

Condensation Pumps

Condensation pumps are available as complete, compact assemblies (packages) for returning water to low- and medium-pressure boilers in gravity heating systems, steam process equipment, or combinations of both; and are available as components which may be assembled (installed) on the job to fit special conditions. They are also used in low-pressure systems where the return mains are located at elevations which do not permit a gravity flow of the condensate to the boiler.

These pumps are operated in response to a float switch, and will quickly elevate the condensate from a low to a high return line, or will automatically pump the condensate from auxiliary apparatus to storage tanks. Their use permits the increase of usable heated space in a building by allowing the radiation to be installed at a level lower than the water level in the boiler.

These pumps are available as a single pump unit (Fig. 20) or a duplex (Fig. 21). The duplex unit comprises two pumps with a receiver to which the suction of each pump connects. Duplex pumps usually employ an alternator which transfers the operation from one pump to the other, in sequence, as the water level varies between the cut-in and cut-out points. In addition, they provide automatic standby service so that if one pump fails to start or cannot handle the load, the second pump starts automatically. Fig. 22 illustrates a typical piping and wiring arrangement for controlling a single condensation pump to maintain a uniform water level in a single boiler.

Courtesy *Dunham-Bush, Inc.*

Fig. 20. A heavy-duty single-unit condensation pump.

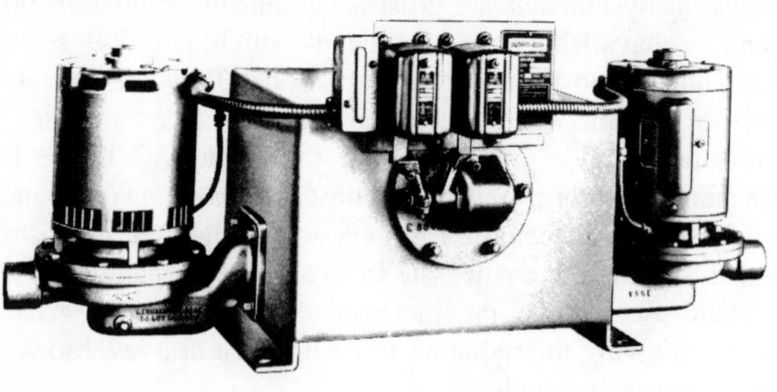

Courtesy *Dunham-Bush, Inc.*

Fig. 21. A duplex condensation pump.

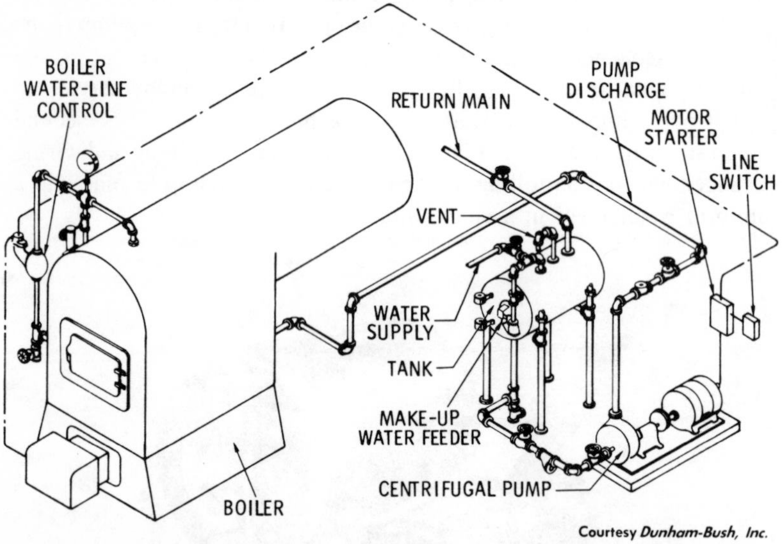
BOILER
WATER-LINE
CONTROL
RETURN MAIN
PUMP
DISCHARGE
MOTOR
STARTER
LINE
SWITCH
VENT
WATER
SUPPLY
TANK
MAKE-UP
WATER FEEDER
CENTRIFUGAL PUMP
BOILER

Courtesy *Dunham-Bush, Inc.*

Fig. 22. Typical piping and wiring arrangement for controlling a single condensation pump to maintain a uniform water level in a single boiler.

Vacuum Pumps

Vacuum pumps are similar to condensation pumps in one of their functions—they receive system condensate and pump it to the boiler. However, a vacuum pump has an added function. It produces a vacuum by removing air, vapor, and condensate from the system. The vacuum thus created causes circulation in the heating system before the pressure has been raised in the boiler. By quickly exhausting air (noncondensate gases) and condensate from the system, the vacuum pump induces the steam to circulate rapidly, reducing the warm-up time and allowing the heating system to function quietly. Fig. 23 shows a typical vacuum pump. Vacuum pumps are rated so that each standard unit will easily handle the requirements of any heating system up to the capacity of the pump without special allowance for covered piping.

Duplex units have the advantage over single pumps due to the provision of automatic standby service. Should one pump fail to start, or is unable to handle the load, the other pump is cut in and picks up the load since it is controlled by vacuum or float switches independent of the other pump.

Pump Selection

Whether a condensation pump or a vacuum pump is necessary, in most instances, depends on whether a lowest first cost or the best investment and maximum system performance is the major consideration. Some of the advantages of the vacuum pump over the condensation pump are as follows:

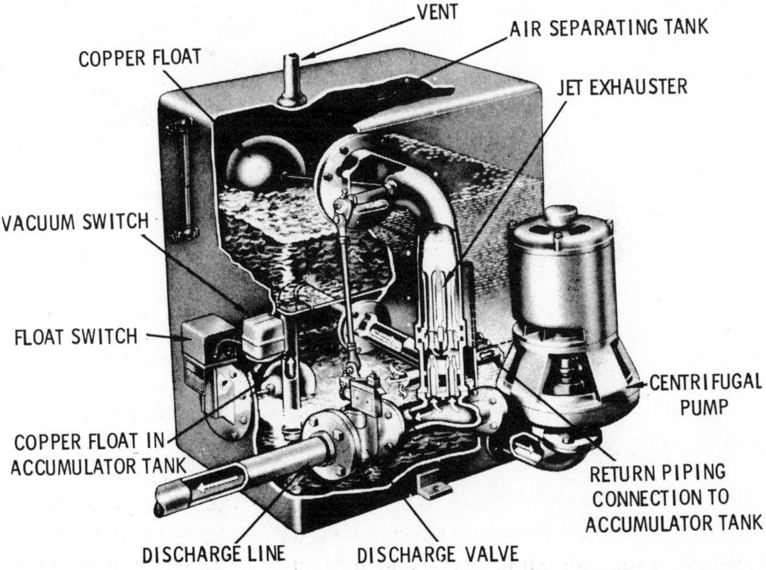
VENT
AIR SEPARATING TANK
COPPER FLOAT
JET EXHAUSTER
VACUUM SWITCH
FLOAT SWITCH
CENTRIFUGAL PUMP
COPPER FLOAT IN ACCUMULATOR TANK
RETURN PIPING CONNECTION TO ACCUMULATOR TANK
DISCHARGE LINE
DISCHARGE VALVE

Courtesy *Dunham-Bush, Inc.*

Fig. 23. A single-unit vacuum pump.

1. Rapid heating, when the system first starts up, is achieved by the addition of a vacuum pump. Air is evacuated from the system ahead of the steam, so that the steam is distributed while pressures are still low. Circulation to all radiators is assured because a vacuum in the return line continually removes any air released by the thermostatic traps. This results in savings in fuel costs because the boiler will operate under vacuum in mild weather.
2. If the return lines are too low to reach the pump inlet by gravity, a separate accumulator tank may be placed at a lower level than the pump and the condensate raised from it by the vacuum the pump produces. This, in some installations, the need for a large pump pit may be eliminated.
3. Pipe sizes may be reduced in systems which have vacuum pumps. This results in substantial savings on installation costs. The cost of a vacuum pump is greater than that of a condensation pump, but the advantages of the vacuum pump usually will offset this cost.

CHAPTER 8

Fuel-Oil Tank Installation

Fuel-oil tanks must conform in construction and installation to the regulations of the National Board of Fire Underwriters and to local ordinances. Vents, fill pipes, outlet pipes, and gauges are also governed by the foregoing regulations. Fuel-oil tanks may be classified with respect to their location, as:

1. Gravity tanks.
2. Subgravity tanks.
3. Inside tanks (basement).
4. Outside tanks (underground).

They may also be classified with respect to the number of pipe connections between the fuel tank and the oil burner, as:

5. Single-pipe installations.
6. Two-pipe installations.

Gravity tanks, which are the more common, are those which are placed at a higher level than the burner, so that the fuel is delivered by gravity to the burner. Fig. 1 shows a typical gravity installation of a tank installed outside the building.

Subgravity tanks are those from which the fuel is drawn by a pump which delivers it to the burner. Most burner pumps in installations of this type will lift the oil 12 feet or more while delivering it to the burner at the required pressure. Fig. 2 shows a typical subgravity installation of a tank installed outside the building.

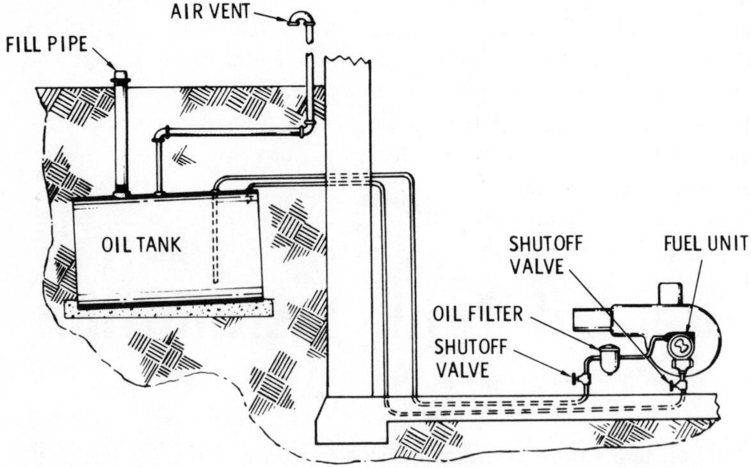
AIR VENT
FILL PIPE
OIL TANK
SHUTOFF
VALVE
FUEL UNIT
OIL FILTER
SHUTOFF
VALVE

Fig. 1. A typical underground, two-pipe tank installation with the oil burner below the level of the tank unit

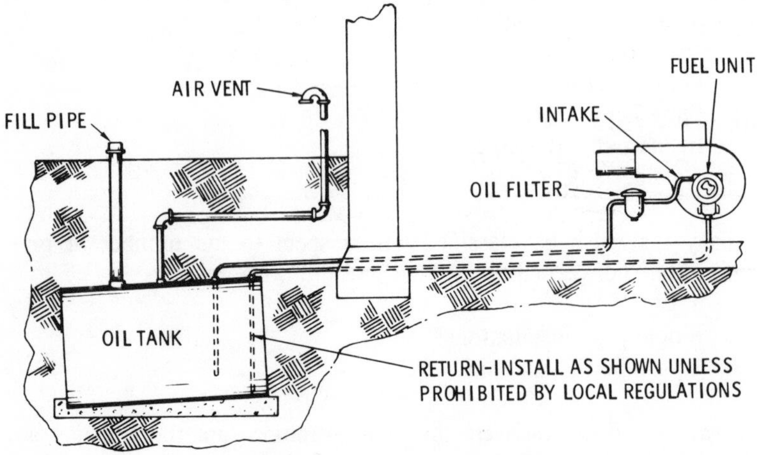
FUEL UNIT
AIR VENT
FILL PIPE
INTAKE
OIL FILTER
OIL TANK
RETURN-INSTALL AS SHOWN UNLESS
PROHIBITED BY LOCAL REGULATIONS

Fig. 2. A typical underground, two-pipe tank installation with the oil burner above the level of the tank unit.

Single-pipe installations are those in which only a suction line is required between the oil tank and the burner, whereas in two-pipe installations, two fuel lines are employed, one a suction line and the other an oil-return line.

A single-pipe installation, such as shown in Fig. 3, is used only where the bottom of the oil tank is above the burner pump. If there is any oil in the tank, there will then be a gravity flow to the pump. The suction or supply line is connected to the bottom end of the tank. The oil that is pumped in excess of that used is recirculated or bypassed at the burner, from the pressure-valve discharge into the strainer or filter housing on the inlet side of the pump.

A two-pipe or suction-and-return method of installation is used on inside tanks when the bottom of the tank cannot be set above the pump, and on all outside underground tank installations.

TANK CAPACITY

If space is available in the basement, a 275-gallon inside oil storage tank is generally preferred. The customary dimensions of such a vertical-oval type tank are approximately 27 inches by 42 inches on the oval end, by about 64 inches in length. Oil supply tanks located inside buildings shall not exceed 550 gallons individual capacity, or 550 gallons aggregate capacity.

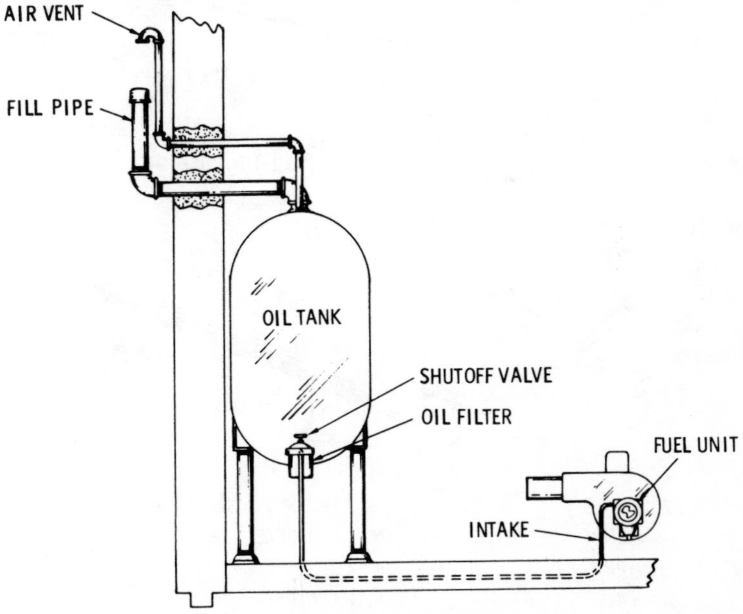
AIR VENT
FILL PIPE
OIL TANK
SHUTOFF VALVE
OIL FILTER
FUEL UNIT
INTAKE

Fig. 3. An inside, single-pipe tank installation with the oil burner below the level of the tank unit.

Fuel-Oil Tank Installation

A typical piping connection diagram for a double inside tank installation is given in Fig. 4. In locations where available basement space is at a minimum, or where it is desirable to store larger quantities of oil, an underground tank outside the building is used. Outside tanks are usually of 1000-gallon capacity or more.

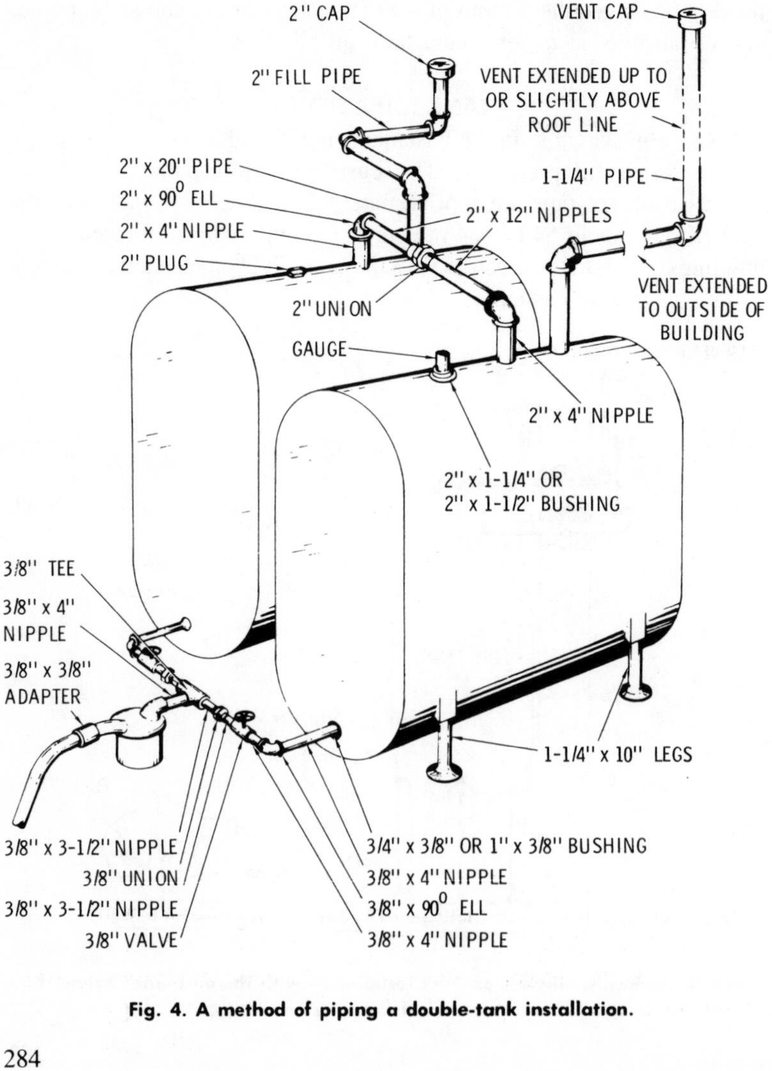

2" CAP
VENT CAP
2" FILL PIPE
VENT EXTENDED UP TO OR SLIGHTLY ABOVE ROOF LINE
2" x 20" PIPE
1-1/4" PIPE
2" x 90° ELL
2" x 12" NIPPLES
2" x 4" NIPPLE
2" PLUG
VENT EXTENDED TO OUTSIDE OF BUILDING
2" UNION
GAUGE
2" x 4" NIPPLE
2" x 1-1/4" OR 2" x 1-1/2" BUSHING
3/8" TEE
3/8" x 4" NIPPLE
3/8" x 3/8" ADAPTER
1-1/4" x 10" LEGS
3/8" x 3-1/2" NIPPLE
3/8" UNION
3/8" x 3-1/2" NIPPLE
3/8" VALVE
3/4" x 3/8" OR 1" x 3/8" BUSHING
3/8" x 4" NIPPLE
3/8" x 90° ELL
3/8" x 4" NIPPLE

Fig. 4. A method of piping a double-tank installation.

INSTALLATION

Where there are no regulations governing the installation of tank and burner, the following recommendations should be followed.

1. In selecting a location for the tank, whether in the basement (Fig. 5) or buried outside (Fig. 6), keep in mind the accessibility for filling, and locate the tank the proper distance from the heating plant as required by the Underwriter's or local ordinances governing its installation. In the absence of these regulations or ordinances, it is recommended that inside tanks be at least 7 feet from the heating plant, that they be properly vented to the outside of the building, and that the fill line be installed from the outside. The vent line should not be smaller than 1-1/4 inches, and the fill line should be from 1-1/4 inches to 3 inches. Outside tanks should be buried below the frost line and connected with the proper fill and vent pipes as well as the suction and return lines to the burner.
2. The suction line, where possible, should be 3/8 inch O.D. copper tubing. One-half inch O.D. copper tubing should be used for the return line. Make sure there are no leaky connections between the tank and the fuel unit. Continuous lengths of tubing should be used for the return and suction lines.
3. It is always advisable to run the return line to the same depth in the tank as the suction line on outside tank installations. This will avoid the possibility of air getting into the pump during the shutdown period of the burner.
4. Keep in mind that when a return line is used, a bypass plug must be inserted if a *Sundstrand* pump is used, or the top cover must be reversed if a *Webster* pump is used.
5. Whenever the top of the tank is above the burner level, an approved make of syphon-breaking or antisyphon valve should be installed in the suction line. The purpose of this valve is to prevent the contents of the tank from syphoning into the basement, should a pipeline break. A horizontal swing check valve should be installed in the suction line at the point where the line enters the basement.

It is recommended that inside storage tanks be provided with a drain opening. When such a drain opening is provided, the tank should be installed with the bottom slanted toward the drain opening, with a slope of not less than 1/4 inch per foot of length. The drain opening should be

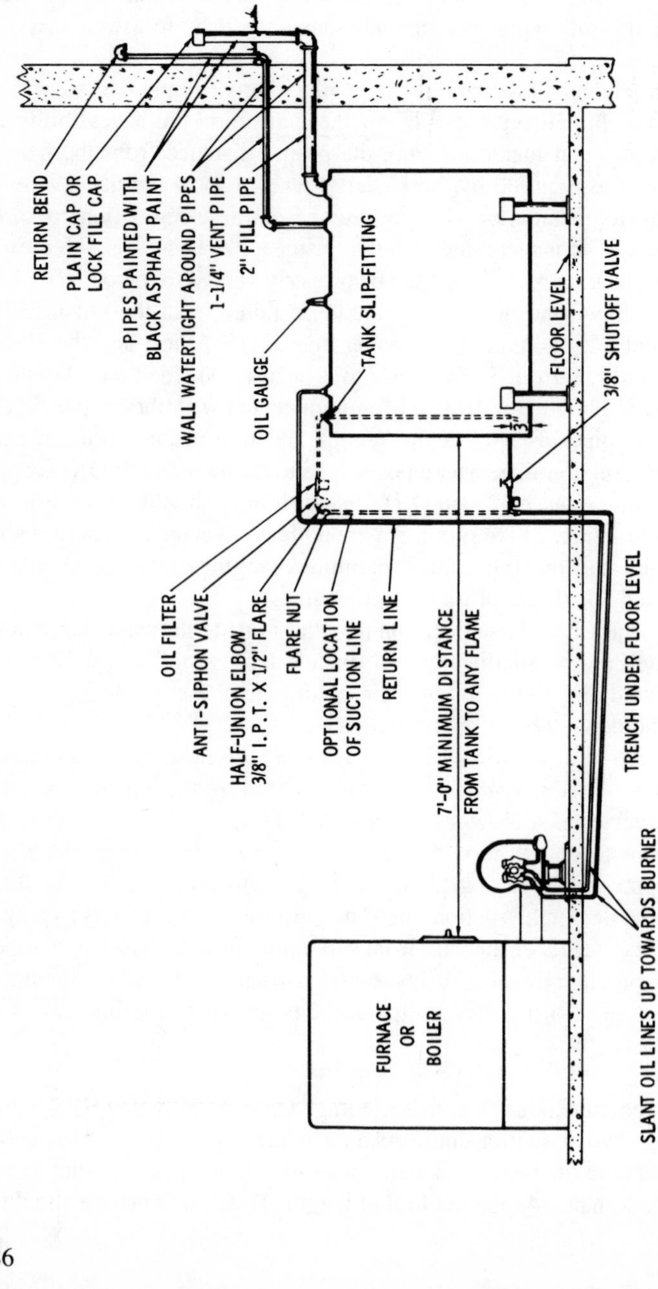
RETURN BEND
PLAIN CAP OR
LOCK FILL CAP
PIPES PAINTED WITH
BLACK ASPHALT PAINT
WALL WATERTIGHT AROUND PIPES
1-1/4" VENT PIPE
2" FILL PIPE
TANK SLIP-FITTING
OIL GAUGE
FLOOR LEVEL
3/8" SHUTOFF VALVE
3"
OIL FILTER
ANTI-SIPHON VALVE
HALF-UNION ELBOW
3/8" I.P.T. X 1/2" FLARE
FLARE NUT
OPTIONAL LOCATION
OF SUCTION LINE
RETURN LINE
7'-0" MINIMUM DISTANCE
FROM TANK TO ANY FLAME
TRENCH UNDER FLOOR LEVEL
SLANT OIL LINES UP TOWARDS BURNER
FURNACE
OR
BOILER

Fig. 5. An inside tank installation showing the piping and other details.

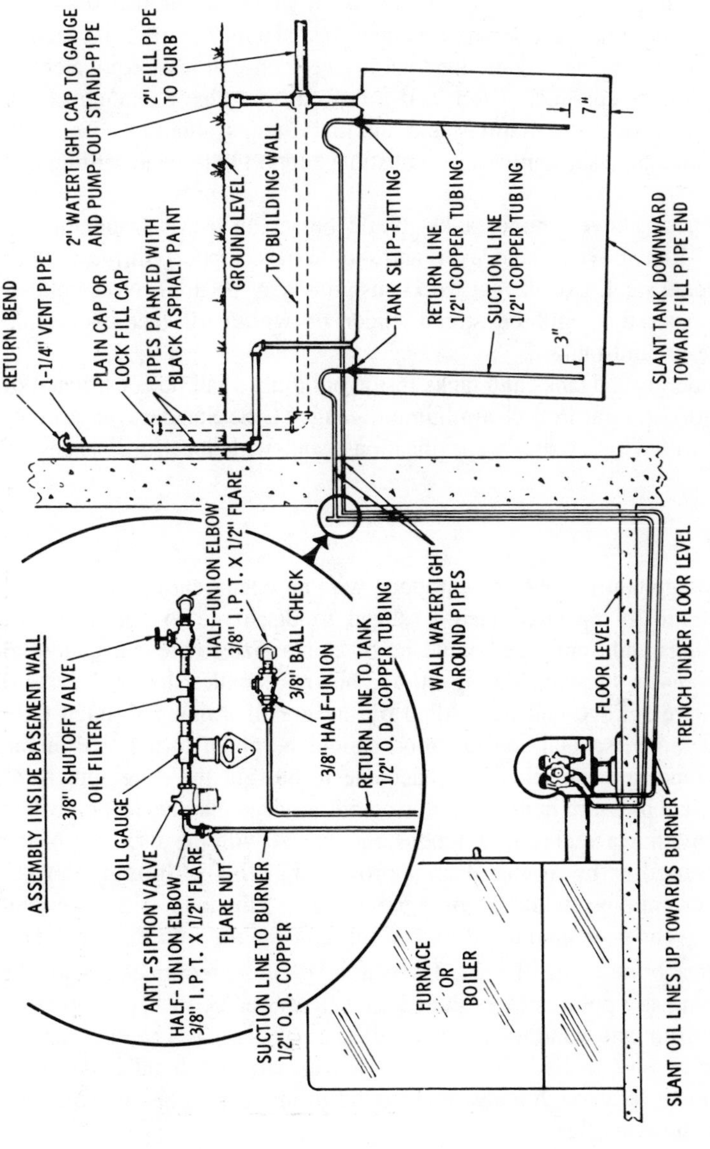
RETURN BEND
1-1/4" VENT PIPE
PLAIN CAP OR
LOCK FILL CAP
2" WATERTIGHT CAP TO GAUGE
AND PUMP-OUT STAND-PIPE
2" FILL PIPE
TO CURB
PIPES PAINTED WITH
BLACK ASPHALT PAINT
GROUND LEVEL
TO BUILDING WALL
TANK SLIP-FITTING
RETURN LINE
1/2" COPPER TUBING
SUSCTION LINE
1/2" COPPER TUBING
3"
7"
SLANT TANK DOWNWARD
TOWARD FILL PIPE END
ASSEMBLY INSIDE BASEMENT WALL
3/8" SHUTOFF VALVE
OIL FILTER
OIL GAUGE
ANTI-SIPHON VALVE
HALF- UNION ELBOW
3/8" I. P. T. X 1/2" FLARE
FLARE NUT
SUCTION LINE TO BURNER
1/2" O. D. COPPER
HALF-UNION ELBOW
3/8" I. P. T. X 1/2" FLARE
3/8" BALL CHECK
3/8" HALF-UNION
RETURN LINE TO TANK
1/2" O. D. COPPER TUBING
WALL WATERTIGHT
AROUND PIPES
FLOOR LEVEL
TRENCH UNDER FLOOR LEVEL
FURNACE
OR
BOILER
SLANT OIL LINES UP TOWARDS BURNER

Fig. 6. An outside tank installation showing the piping and other details.

provided with a suitable pipe connection in a form to provide a sump from which water or sediment can be readily drained at intervals.

Prior to setting an underground tank, it should be protected against erosion on the outside in a manner satisfactory to the inspection department having jurisdiction, but in every case at least equivalent to two primary coatings of red lead followed by a heavy coating of hot asphalt. Inside installations and auxiliary tanks should be securely supported by a substantial incombustible support to prevent settling and sliding.

Outside above-ground tanks should be set on a firm foundation at least 1 foot above ground level, and should have supports of masonry or protected steel, except that wood cushions may be used. No combustible material should be stored under or within 10 feet of outside above-ground tanks.

Underground tanks and tanks inside buildings shall be constructed of steel or wrought iron of a minimum gauge (U.S. Standard) in accordance with Underwriter's specifications, and with local ordinances, if any.

Tank Vents

Storage tanks should be equipped with an open vent or an approved automatically operated vent arranged to discharge to the open air. Openings and vent pipe should be of ample size to prevent abnormal pressures in the tank during filling, but not smaller than 1-1/2 inch.

Where storage tanks are filled by the use of a pump through tight connections, special consideration should be given to the size of the vent pipe to ensure that it is adequate to prevent the development of abnormal pressure in the tank during filling. This may be accomplished by providing a vent pipe commensurate in size with the discharge of the pump used or by means of an approved signaling or limiting device.

The vent pipe should be arranged to drain to the tank. The lower end of the vent pipe should not extend through the top into the tank for a distance of more than 1 inch. The vent pipe should terminate outside the building at a point not less than 2 feet, measured vertically or horizontally, from any window or other building or opening. The outer end of the pipe should be provided with a weatherproof hood and should terminate far enough above the ground to prevent its being obstructed with snow and ice.

The vent pipe should not be cross connected with the fill pipe or return line from the burner. The vent opening on outside above-ground tanks should be provided with approved flame arresters. Local building regulations must be followed when installing the oil tank vent pipe. Some areas require that an oil tank vent be extended to or above the roof line.

Fill and Overflow Pipes

Storage tanks, other than outside above-ground tanks, should be filled only through fill pipes terminating outside the building at a point at least 5 feet from any building opening at the same or lower level. When not in use fill terminals should be closed tightly by a metal cover designed to prevent tampering. Auxiliary tanks should be filled by pumping from storage tanks. Cross connections permitting gravity flow from one tank to another should not be made. This, however, should not be construed as prohibiting filling of an outside tank by gravity.

All fill pipes should be located near a driveway or street curb to avoid the necessity of walking on soft or soggy lawns with a fill hose during spring and fall. In the event that a furnace is to be installed during cold weather, it is desirable to have the tank installed and fuel delivered before starting to install the burner.

Valves

A readily accessible shutoff valve of an approved type should be installed in the oil supply line near the burner. A shutoff valve of an approved type should be installed on each side of an oil strainer that is not a part of the oil-burner unit, and on the discharge and suction line of an oil pump directly to the burner, but which is not a part of the burner unit.

TESTING

After installation and before being covered, tanks and piping should be tested hydrostatically or with air, at a pressure from 5 to 10 pounds per square inch, but not less than 5 pounds per square inch at the highest point of the system. Instead of a pressure test, suction lines may be tested under a vacuum of not less than 20 inches of mercury. Tests should continue for at least 30 minutes without a noticeable drop in pressure or vacuum.

CHAPTER 9

Gas Piping

The word "piping," as used in reference to the installation of gas appliances, relates only to rigid conduit of steel, iron, brass, or copper that is acceptable to the authorities having jursidiction. Reference should also be made to manufacture's instructions, gas supplier regulations, and local building, heating, plumbing, or other codes in effect in the area in which the installation is made.

Installation and replacement of either gas piping or gas appliances shall be performed only by a qualified installing agency. The term "qualified installing agency" means any individual, firm, corporation, or company which either in person or through a representative is engaged in and is responsible for the installation or replacement of gas piping on the outlet side of the meter, or for the connection, installation, or repair of gas appliances; and those who are experienced in such work, who are familiar with all the precautions required, and who have complied with all the requirements of the authority having jurdisdiction.

GENERAL PRECAUTIONS

When installing gas piping or gas appliances, all gas supplies shall be turned off in order to eliminate hazards from leakage of gas. Before turning off the gas to the premises for the purpose of installation, repair, replacement,or maintenance of gas piping or appliances, all burners shall be turned off. When two or more consumers are served from the same supply system, precautions shall be exercised to assure that only service to the proper consumer is turned off.

When checking for gas leaks, only a soap and water solution, or other material accepted for that purpose, shall be used. Matches, candles, flames or other sources of ignition shall not be used for this purpose. Artificial illumination shall be restricted to battery-powered flashlights (preferably of the safety type) or approved safety lamps. In searching for leaks, electric switches should not be operated. If electric lights are already on, they should not be turned off. Smoking shall not be permitted.

INSTALLATION OF GAS PIPING

Before proceeding with the installation, a piping sketch or piping plan should be made of the gas piping system. This plan should show the proposed location of the piping as well as the size of the various branches. Adequate provisions should be made for additional appliances which might be installed later.

Prior to piping installation, however, the proposed piping layout or specifications should be submitted for approval to the gas supplier or the authorities having jurisdiction. When additional appliances are to be served through any existing piping system, the capacity of the system in service should be checked for adequacy and replaced with larger piping if necessary.

Location of the Gas Meter

Meters shall be located as near as practicable to the point of entrance of the gas service and where the connections are readily accessible for servicing. Location, dimensions, space requirements, and type of installation shall be acceptable to the gas supplier.

Gas piping shall be of such a size and so installed as to provide an uninterrupted supply of gas sufficient to meet the maximum demand without undue loss of pressure between the meter (or service regulator when a meter is not provided) and the appliance or appliances. The size of gas piping depends on the following factors:

1. Allowable loss in pressure from the meter (or a service regulator when a meter is not provided) to the appliance.
2. Maximum gas consumption to be provided.
3. Length of piping and number of fittings.
4. Specific gravity of the gas.
5. Diversity factor.

The specific gravity of most gases varies from 0.45 to 0.65, and since the capacity of the piping is affected by this factor inversely as the square root, it is usually sufficiently accurate to use an assumed gravity factor of 0.60 for all calculations.

Gas Consumption

The quantity of gas to be provided at each outlet shall be determined directly from the manufacturer's Btu rating of the appliance to be installed. If the ratings of the appliances to be installed are not known, Table 1 lists the approximate consumption of the average appliance in Btu per hour.

Size of Piping

Capacities for different sizes and lengths of pipe in cubic feet per hour for gas of 0.60 specific gravity, based on a pressure drop of 0.3 inch of

Table 1. Approximate Gas Consumption of Typical Appliances.

Appliance	Input Btu per hour (approx.)
Boiler or furnace (domestic)	100,000 to 250,000
Range (free-standing, domestic)	65,000
Built-in oven or broiler unit (domestic)	25,000
Built-in top unit (domestic)	40,000
Water heater, automatic storage (50-gal. tank)	55,000
Water heater, automatic instantaneous	
2 gal. per minute	142,800
4 gal. per minute	285,000
6 gal. per minute	428,400
Water heater, domestic, circulating or side-arm	35,000
Refrigerator	3,000
Clothes dryer, domestic	35,000

water column are shown in Table 2. In adopting a 0.3-inch pressure drop, due allowance for an ordinary number of fittings should be made. Table 2 is based on a gas of 0.60 specific gravity, but if it is desired to use an exact specific gravity for a particular condition, the values in the table can be corrected by multiplying by $\sqrt{0.6/\text{sp.gr.}}$

To obtain the size of piping required for a certain unit, it is first necessary to determine the number of cubic feet of gas per hour consumed by the unit as follows:

$$\text{cu. ft. of gas/hr.} = \frac{\text{total Btu/hr. required by unit}}{\text{Btu/cu. ft. of gas}}$$

Example: It is required to determine the size of the piping in a gas-fired boiler installation when the burner input is 155,000 Btu per hour, and the heating value per cubic foot of gas is 1000 Btu. Assume the distance between the gas meter and the burner to be 75 feet.

Table 2. Capacity of Pipe of Different Diameters and Lengths in Cu. Ft. Per Hour With Pressure Drop of 9.3 In. and Specific Gravity of 0.60.

Length of Pipe (ft.)	Iron-pipe sizes (IPS) (inch)				
	½	¾	1	1¼	1½
15	76	172	345	750	1220
30	52	120	241	535	850
45	43	99	199	435	700
60	38	86	173	380	610
75		77	155	345	545
90		70	141	310	490
105		65	131	285	450
120			120	270	420
150			109	242	380
180			100	225	350
210			92	205	320
240				190	300
270				178	285
300				170	270
450				140	226
600				119	192

By a substitution of values in the foregoing formula, we obtain:

$$\text{cu. ft. of gas/hr.} = \frac{155{,}000}{1000} = 155$$

From Table 2 it will be noted that a 1-inch pipe 75 feet long will handle 155 cu. ft./hr. with 0.3-inch pressure drop. The determination of gas-pipe sizes for other heating appliance units can be made in a similar manner.

Example: It is required to determine the pipe size of each section and outlet of the piping system, with a designated pressure drop of 0.3-inch water column. The gas to be used has a specific gravity of 0.60 and a heating value of 1000 Btu per cubic foot.

Reference to the piping layout and a substitution of values for maximum gas demand for the various gas outlets (Fig. 1) will be as follows:

$$\text{Gas demand for outlet A} = \frac{55{,}000}{1000} = 55 \text{ cu. ft./hr.}$$

$$\text{Gas demand for outlet B} = \frac{3{,}000}{1000} = 3 \text{ cu. ft./hr.}$$

$$\text{Gas demand for outlet C} = \frac{65{,}000}{1000} = 65 \text{ cu. ft./hr.}$$

$$\text{Gas demand for outlet D} = \frac{116{,}000}{1000} = 116 \text{ cu. ft./hr.}$$

From the piping layout in Fig. 1 it will be noted that the length of gas pipe from the meter to the most remote outlet (A) is 105 feet. This is the only distance used. Reference to Table 2 indicates that:

outlet A supplying 55 cu. ft./hr. requires 1/2-inch pipe;
outlet B supplying 3 cu. ft./hr. requires 1/2-inch pipe;
outlet C supplying 65 cu. ft./hr. requires 3/4-inch pipe;
outlet D supplying 116 cu. ft./hr. requires 3/4-inch pipe;
section 3 supplying outlets A and B, or 58 cu. ft./hr., requires 1/2-inch pipe;
section 2 supplying outlets A, B, and C, or 123 cu. ft./hr., requires 3/4-inch pipe;
section 1 supplying outlets A, B, C, and D, or 239 cu. ft./hr., requires 1-inch pipe.

The determination of sizes of gas pipe for other piping layouts may be made in a similar manner.

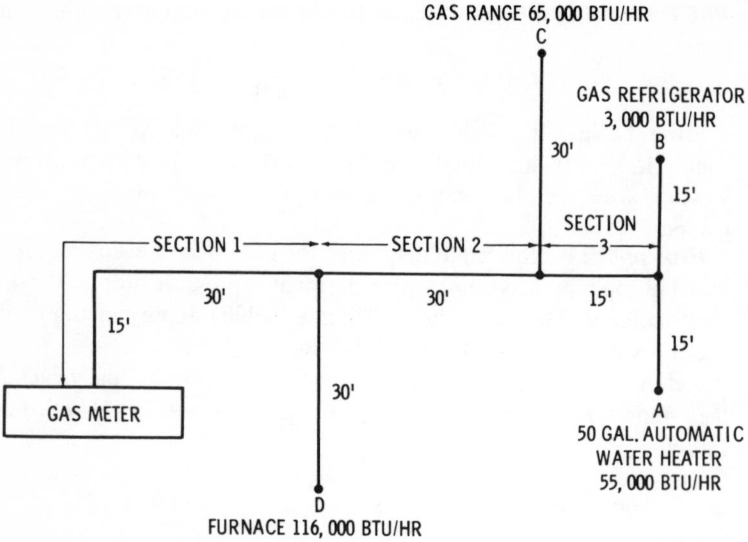
GAS RANGE 65,000 BTU/HR
C
GAS REFRIGERATOR
3,000 BTU/HR
B
30'
15'
SECTION
SECTION 1
SECTION 2
3
30'
30'
15'
15'
15'
30'
GAS METER
A
50 GAL. AUTOMATIC
WATER HEATER
55,000 BTU/HR
D
FURNACE 116,000 BTU/HR

Fig. 1. Diagram of the gas-piping layout for a typical domestic installation.

Diversity Factor

The diversity factor, or ratio of the maximum probable demand to the maximum possible demand, is an important factor in determining the correct size of gas piping to be used in multiple-family dwellings. It is dependent on the number and kinds of gas appliances being installed. Consult the servicing gas supplier or authority having jurisdiction for the diversity factor to be used.

Piping Material

Standard weight wrought-steel and wrought-iron pipe shall be used in installation of appliances supplied with utility gases. Threaded copper or brass pipe in iron-pipe sizes may be used with gases not corrosive to such material. Gas pipe shall not be bent. Fittings shall be used when making turns in gas pipe.

The connection of steel or wrought-iron pipe by welding is permissible. Threaded pipe fittings (except stopcocks or valves) shall be of malleable iron or steel when used with steel or wrought-iron pipe, and shall be copper or brass when used with copper or brass pipe. When approved by the authority having jurisdiction, special fittings may be used to connect steel or wrought-iron pipe.

Piping Material for Liquefied Petroleum (LP) Gases

Gas piping for use with undiluted liquefied petroleum (LP) gases shall be of steel or wrought-iron pipe complying with the American Standard for wrought-steel and wrought-iron pipe, and brass or copper pipe, or seamless copper, brass, steel, or aluminum tubing. All pipe or tubing shall be suitable for a working pressure of not less than 125 pounds per square inch. Copper tubing may may be of the standard K or L grade, or equivalent, having a minimum wall thickness of 0.032 inch. Aluminum tubing shall not be used in exterior locations or where it is in contact with masonry or plaster walls, or with insulation.

Defective Material

Gas pipe or tubing and fittings shall be clear and free from burrs and defects in structure or threading and shall be thoroughly brushed, chipped, and scale blown. Defects in pipe, tubing, or fittings shall not be repaired. When defective pipe, tubing, or fittings are located in a system, the defective material shall be replaced.

Gas pipe, tubing, fittings, and valves removed from any existing installation shall not be used again until they have been thoroughly cleaned, inspected, and ascertained to be equivalent to new material. Joint compounds shall be applied sparingly and only to the male threads of pipe joints. Such compounds shall be resistant to the action of liquefied petroleum (LP) gases.

Pipe Threads

Pipe and fittings shall comply with the American Standard for pipe threads. Pipe with damaged threads shall not be used. All gas piping shall be threaded in accordance with specifications in Table 3.

Branch Connections

All branches should be taken from the top or side of horizontal pipes and not from the bottom. When ceiling outlets are taken from horizontal piping, the branch should be taken from the side and carried in a horizontal direction for a distance of not less than six inches. Fig. 2 shows the right and wrong way to connect a drop branch to a horizontal run.

Table 3. Approximate Length of Thread Required for Various Sizes of Gas Pipe.

Size of pipe, in inches	Approximate length of threaded portion, in inches	Approximate number of threads to be cut
3/8	5/16	10
1/2	3/4	10
3/4	3/4	10
1	7/8	10
1 1/4	1	11
1 1/2	1	11
2	1	11
2 1/2	1 1/2	12
3	1 1/2	12
4	1 3/4	13

Pipe Supports

Care must be taken to properly support the pipe and not subject it to any unnecessary strain. Fig. 3 shows a pipe supported by pipe clips or straps fastened to the building joists. The maximum distance these supports should be spaced depends on the pipe size. The following distances between supports should never be exceeded, with closer spacing being preferable.

3/8″ and 1/2″ pipe . . . . .6 feet
3/4″ and 1″ pipe . . . . .8 feet
1-1/4″ and larger (horizontal) . . . . .10 feet
1-1/4″ and larger (vertical) . . . . .every floor level

Pipe shorter than the support spacing listed should still be adequately supported. Wherever a branch fitting is used, or wherever there is a change of direction of 45° or more, a support should be provided within six inches of at least one side of the fitting or bend.

Pipe straps or iron hooks should not be used for fastening pipe larger than two inches. Beyond this size, when the pipe is horizontal and is to be fastened to floor joists or beams, pipe hangers should be used; when the pipe is horizontal and is to be fastened to a wall, hook plates should be used.

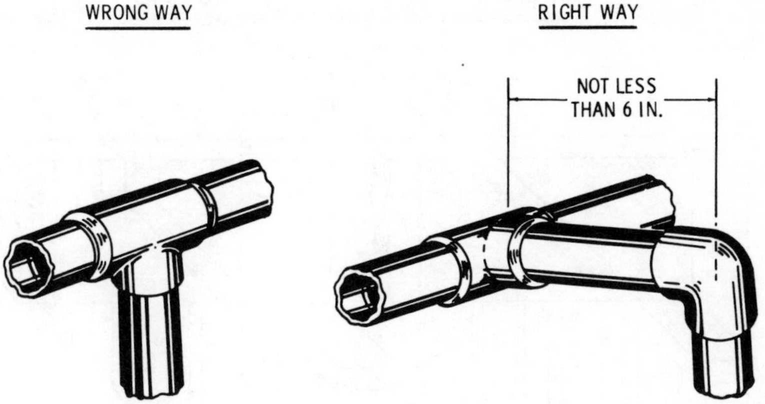
WRONG WAY
RIGHT WAY
NOT LESS THAN 6 IN.

Fig. 2. Connection of a drop branch to a horizontal pipe.

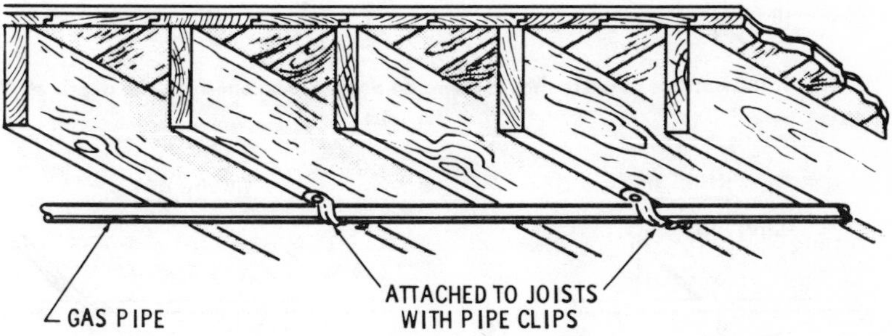
GAS PIPE
ATTACHED TO JOISTS WITH PIPE CLIPS

Fig. 3. Gas piping should be supported from the joists instead of on the wall.

When pipes run crosswise to beams, do not cut the beams to a depth of more than one-fifth of the depth of the timber (Fig. 4). This cutting should be as near the support of the beam as possible (Fig. 5), but in no case should it be farther from a support than one-sixth of the span. When possible, pipes should be run parallel to the beams to avoid cutting and weakening of the beams. Horizontal lines should have some pitch, as shown in Fig. 6, to provide for drainage of any condensed liquid, especially where pipes are exposed to cold. A T with a capped suspended nipple should be provided at the lowest point to allow removal of the liquid. The size and length of the nipple should be in proportion to the amount of piping which drains into it. A cross instead of a T can be used to advantage in some installations. The use of a cross permits cleaning the horizontal and vertical runs, should the need ever arise. Fig. 7 shows a typical installation using a cross.

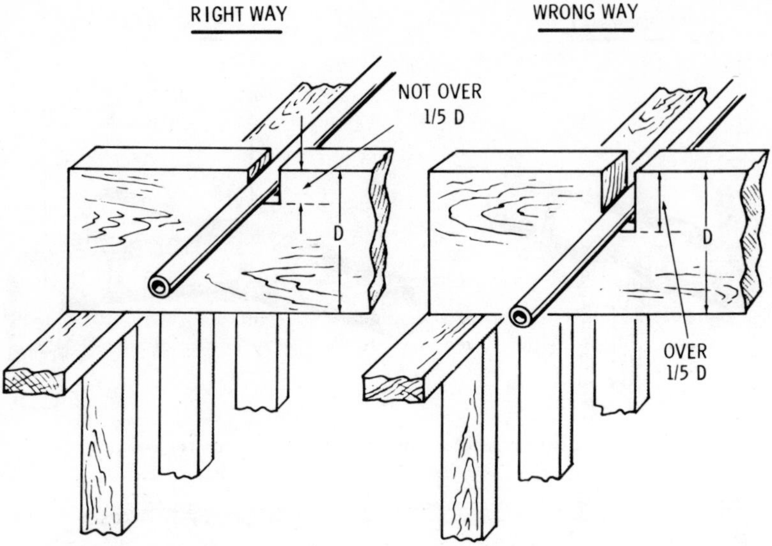
RIGHT WAY
WRONG WAY
NOT OVER
1/5 D
D
D
OVER
1/5 D

Fig. 4. The notches in joists should never be deeper than one-fifth the depth of the joist.

RIGHT WAY
WRONG WAY
SUPPORT
NEAR
SUPPORT
SUPPORT NEAR
CENTER OF SPAN

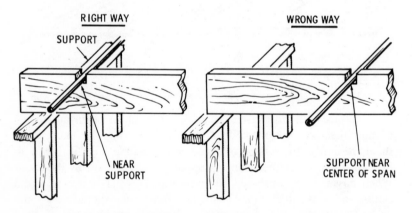

Fig. 5. Joists should be notched near where they are supported when it is necessary to run pipe across them.

Offsets

Where an offset in the line is necessary, such as a projection in a wall, the offset should be made with 45° elbows instead of 90° fittings. This procedure reduces the friction to the flow of gas, and also reduces the

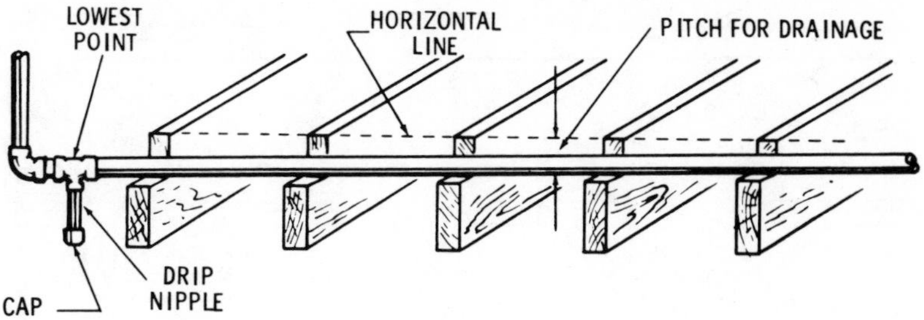
LOWEST POINT
HORIZONTAL LINE
PITCH FOR DRAINAGE
CAP
DRIP NIPPLE

Fig. 6. Horizontal runs of gas piping should be slanted slightly, with a drip nipple installed at the lowest point to collect any condensate moisture.

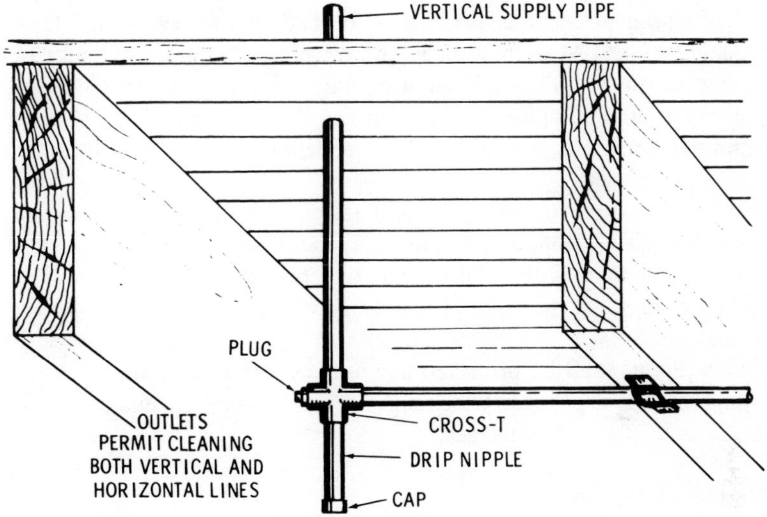
VERTICAL SUPPLY PIPE
PLUG
OUTLETS PERMIT CLEANING BOTH VERTICAL AND HORIZONTAL LINES
CROSS-T
DRIP NIPPLE
CAP

Fig. 7. The use of a cross at the junction of a horizontal and vertical run permits cleaning either pipe as well as permitting the installation of a drip nipple.

likelihood of stoppage in the line. Correct and incorrect offset installations are illustrated in Fig. 8.

Concealed Piping

Gas piping smaller in size than standard 1/2-inch pipe shall not be used in any concealed location. Concealed piping should be located in hollow portions rather than in solid portions. Tubing used with undiluted liquefied petroleum (LP) gases shall not be run inside any walls or partitions unless they are protected against physical damage. This rule does not apply to tubing which passes through walls or partitions.

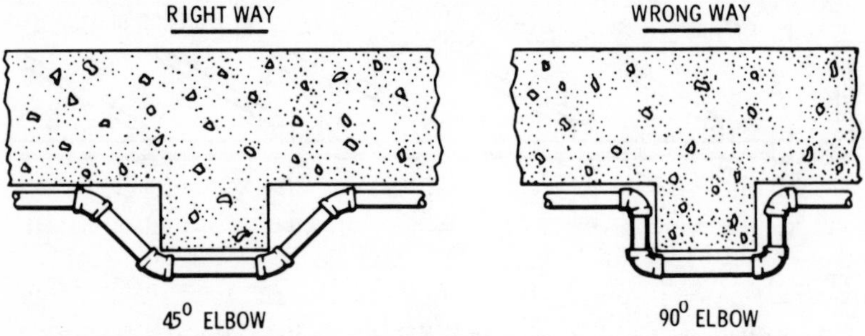
RIGHT WAY
WRONG WAY
45° ELBOW
90° ELBOW

Fig. 8. Offsets should be installed with 45° elbows instead of 90° fittings.

Gas piping in concrete floors shall be laid in channels in the floor suitably covered to permit access to the piping with a minimum of damage to the building. When installing gas piping which is to be concealed, unions, tubing fittings, running threads, right- and left-hand couplings, bushings, and swing joints made by combinations of fittings *shall not* be used

When necessary to insert fittings in gas pipe which has been installed in a concealed location, the pipe may be reconnected by use of a ground-joint union with the nut center punched to prevent loosening by vibration. Reconnection of tubing to be installed in a concealed location is prohibited.

Gas piping shall be protected against freezing temperatures. When piping must be exposed to wide ranges in temperature, special care shall be taken to prevent stoppages.

Closing of Outlets

Each outlet, including a valve or cock outlet, shall be securely closed and made gas tight with a threaded plug or cap immediately after installation; it shall be left closed until an appliance is connected. Similarly, when an appliance is disconnected from an outlet, the outlet shall be securely closed and made gas tight to prevent leaks.

Outlets shall be far enough from floors, walls, and ceilings to permit the use of proper wrenches without straining, bending, or damaging the piping. Outlets shall not be placed behind doors.

Electrical Grounding

A gas piping system within a building shall be electrically continuous and bonded to any grounding electrode, as defined by the National

Electrical Code. The point of attachment of a grounding conductor to gas piping shall always be on the street side of the gas meter and shall be accessible where practicable.

Gas Shutoff Valves

Manual shutoff valves controlling one or several gas piping systems shall, in multiple systems, be placed at an adequate distance from each other so that they will be easily accessible for operation and shall be installed so as to be protected from physical damage. Each valve should be plainly marked with a metal tag attached by the installing agency so that the gas piping systems supplied through them can be readily identified. It is advisable to place shutoff valves at every point where safety, convenience of operation, and maintenance demand them.

Gas Regulator and Gas Regulator-Vent Outlets

Each and every existing gas service connection supplying gas into any building or premises at a pressure in excess of one pound per square inch shall be provided with a regulator which will reduce the pressure of such gas prior to entering the meter in the said building or premises to not more than one-half pound per square inch, except that the authorities may permit a higher pressure for commercial or industrial use.

Each gas regulator shall be provided with a ventilating pipe which shall lead directly to the outer air, and said outlet, where practicable, shall not be located under a window or any opening leading back into the premises. It shall be unlawful for any person to cover over, plug up, or otherwise obstruct any gas regulator-vent outlet. A gas vent identified by suitable marking shall be attached to the outlet on the outside of the building.

Gas Service Connection

Each and every gas service connection which is hereafter brought into a structure shall be fitted with an approved lubricated-type, straightway shutoff stopcock or shutoff valve, or equivalent, so designed and constructed as to preclude the core from being blown out by the pressure of the gas in such pipe. Such stopcock or shutoff valve shall be placed in an accessible position immediately inside the street side of the gas meter and of the gas regulator, if any.

In all high-pressure areas, the gas company concerned shall at least once each year inspect the stopcock or shutoff valves to insure that they are all in good working order and ready for immediate use. All materials used in the installation of stopcock or shutoff valves shall be approved by the authorities. All matters in relation to stopcock or shutoff valves not covered by this section shall be determined by the authorities.

Test Methods

Before appliances are connected, piping systems shall stand a pressure of at least six inches of mercury or three pounds of gauge pressure for a period of not less than ten minutes without showing any drop in pressure. Pressure shall be measured with a water manometer or an equivalent device so calibrated as to read in increments of not greater than one-tenth inch water column for a period of not less than ten minutes without showing any drop in pressure. The source of the pressure shall be isolated before the pressure tests are made.

Purging

After gas piping has been checked, the piping receiving the gas shall be fully purged. One method of purging gas appliance piping is to disconnect the pilot piping at the outlet of the pilot valve. Under no circumstances shall piping be purged into the combustion chamber of an appliance.

LIQUID PETROLEUM GAS

Liquefied petroleum gas is a fuel made up principally of propane or butane, or a mixture of these products. At atmospheric pressure pure butane returns to gas, or vaporizes, at a temperature of 32° F. Pure propane vaporizes or returns to gas at 44° F. Butane is more widely used in mild climates, propane can be used in much colder climates. L.P. gas is used in many areas which are not served by natural gas mains. It is also used in many portable applications such as campers and service trucks. L.P. gas has some advantages over natural or manufactured gas and also some disadvantages. The principal advantage is the higher Btu content of L.P. gas. The heat content of L.P. gas is roughly three times greater than natural gas and six times greater than manufactured gas. The principal disadvantage is that L.P. gas is heavier than air and therefore extreme care must be taken when installing tanks and piping for L.P. gas.

Natural gas, which is lighter than air will escape upward in the event of leakage. L.P. gas, which is heavier than air, will drop from the point of leakage and may collect in a low place or pocket and remain there for a considerable length of time. If a spark or flame should come into contact with the gas an explosion could occur. The explosion in the Coliseum at the Indiana State Fairgrounds several years ago, resulting in many casualties and considerable property damage, was traced to a leakage of L.P. gas stored and used in the building.

L.P. gas tanks and cylinders must be installed or stored outside of any building used for human occupancy, in a well ventilated area and so placed that they are readily accessible at all times for inspection, testing and shutting off the gas supply. All service piping and main supply shut-off valves should be outside of the building and readily accessible. Special pipe joint compounds which will not deteriorate or dissolve when in contact with L.P. gas must be used on all threaded connections. All L.P. gas installations must comply with local codes or regulations.

USING A MANOMETER

Although manometers can be used for several purposes, the plumber and pipefitter will find the manometer useful for testing gas piping and for measuring the gas pressure delivered to the appliances in a building. In almost all areas using natural gas, the gas is delivered to a regulator located on the main side of the meter at a higher pressure than the appliances in the building are designed for. The higher pressure is reduced at the regulator and then passed on through the meter to the appliances. The pressure may vary somewhat from one area to another, in general usage the pressure on the downstream side of the regulator will average from 5 to 6 inches of water column. In order to check the gas pressure it will be necessary to connect the hose from one side of the manometer to the gas piping. This can be done at a gas stove by removing a burner and sliding the manometer hose on the burner connection or by connecting the manometer hose to the piping at any other desirable point. The manometer is prepared for use by first pouring water into the tube on one side until it equalizes at the zero point on the manometer gauge. To insure an accurate reading this should be done carefully; it may be necessary to add water two or three times in order to equalize the water in the tube. When the water level is equalized at the zero point, connect the manometer to the test point. Turn the gas on at the test point and observe the reading on each side of the manometer tube. Fig. 9A shows a manometer with the water

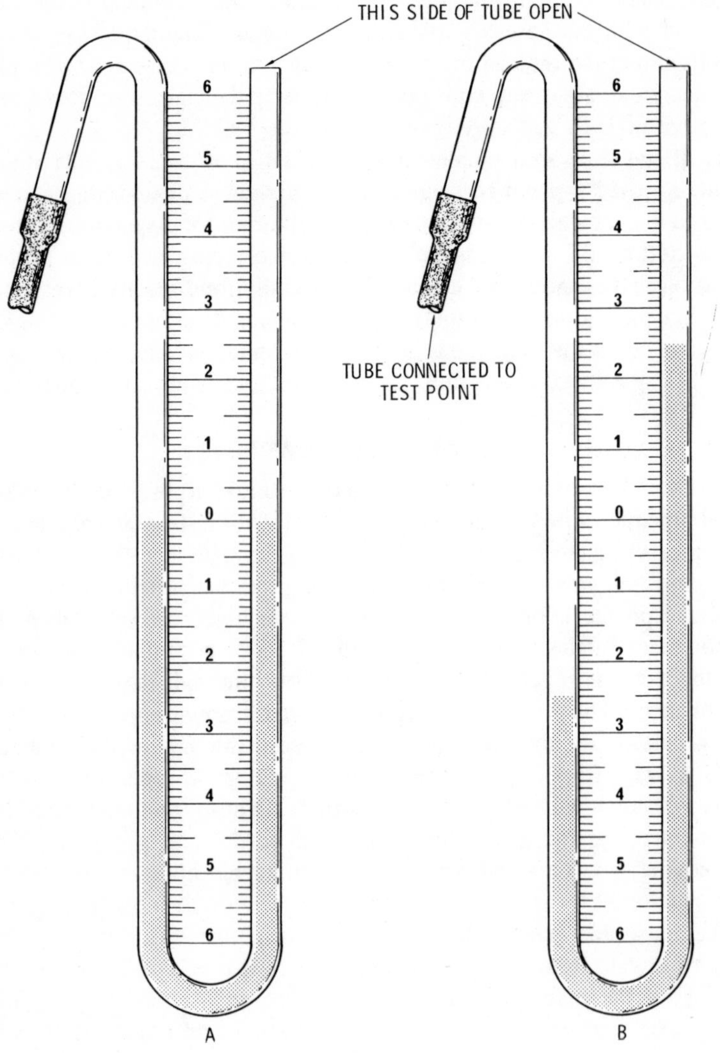
THIS SIDE OF TUBE OPEN
TUBE CONNECTED TO TEST POINT
6
5
4
3
2
1
0
1
2
3
4
5
6
A
B

Fig. 9. Using a manometer.

equalized at the zero point. 9B shows the water column after the gas has been turned on at the test point. The reading at the left side of the tube is 2-1/2 below the zero point. The reading on the right side of the tube shows 2-1/2 above the zero point. The two readings are added together, 2-1/2 + 2-1/2 = 5. The pressure at the test point is 5 inches of water column pressure. The readings of both sides of the manometer tube are always added together to obtain the pressure (2 + 2 = 4) (3-1/2 + 3-1/2 = 7). When a manometer is used for testing pressure one side of the tube is always left open, the other side is connected to a convenient point to make the test. Gas pressures may vary somewhat due to the usage at a particular time or season, but if the manometer shows a greater deviation from normal than should be expected, the utility or supplier should be notified so that corrective action may be taken.

One of the causes of low gas pressure is a stopped-up vent opening from a regulator. A regulator, to work properly, must "breathe", the diaphragm flexes or expands outward each time gas flows through the meter, and contracts inward when the gas flow ceases. If the vent opening becomes plugged by snow, ice, or any other cause, the diaphragm will not be able to flex, or open, and little or no gas will be able to enter the piping. In the event of low gas pressure, the first item to check is the regulator.

CHAPTER 10

Air Conditioning

Year round air conditioning is designed to control the quality of indoor air during the entire year. Summer air conditioning lowers the air temperature, filters, and cleans the air and de-humidifies it. Winter air conditioning heats the air, cleans and filters it and adds humidity.

Three factors are important in summer air conditioning; temperature, relative humidity, and air circulation. Relative humidity is the percentage of moisture in the air, compared to the greatest amount, (100%) which the air can hold. The warmer the air is the more moisture it can hold.

If the temperature of the air in a building is 80°, the air is *capable* of holding 12 grains of moisture per cubic foot. If the moisture content of the air is measured and found to be only 6 grains per cubic foot, the relative humidity is 50%. If the same air is passed through a cooling coil and cooled to a temperature of 60° with 6 grains per cubic foot, the relative humidity of the air is 100% since air at 60° is capable of holding only 6 grains per cubic foot.

When air is cooled to the point where its relative humidity is 100% it is at the dew point.

If the 60° air is cooled to 50° as it passes through the cooling coil, it can hold only 4 grains of moisture per cubic foot. Two grains of moisture will be condensed into water as it passes over and through the cooling coil. In a typical air-conditioned building in the summer, cooled, dried, and cleaned air is circulated, through the supply ducts, into the room spaces. As the cool dry air circulates it absorbs heat and moisture and is returned through the return ducts to the filters and then to the cooling coil, or evaporator. When the warmed moist air passes over and through the cooling coil it is cooled and the excess moisture in the air is condensed into water which drips off of the coil and is drained away.

All refrigeration systems, whether operated by gas or electricity, depend on two basic principles:

1. *A liquid absorbs heat when it boils or evaporates to a gas.* Examples:
   A. Freon 12 boils to vapor at -21° F. at atmospheric pressure.
   B. Water boils to a gas (steam) at 212° at atmospheric pressure.
   C. Liquid ammonia boils to vapor at -28° F. at atmospheric pressure.
2. *As vapor or gas condenses to a liquid form, heat is released.* One of the important rules of refrigeration is: The boiling point of a liquid is changed by changing the pressure the liquid is under. *Raising* the pressure *raises* the *boiling point. Lowering* the pressure *lowers* the *boiling point.*

COOLING TOWERS

Water is used as a cooling medium in some types of air conditioning systems. It is most practical and economical to reuse this water and since the water absorbs heat in the condenser, it must be cooled in order to re-use it. Fig. 1 shows a block diagram of a water air conditioning system using a cooling tower. The cycle of operation is basically simple; cooled water from the cooling tower basin is pumped through the water jacket of the condenser. The cooled water absorbs heat from the hot gas in the refrigerant piping coils in the condenser. The water,

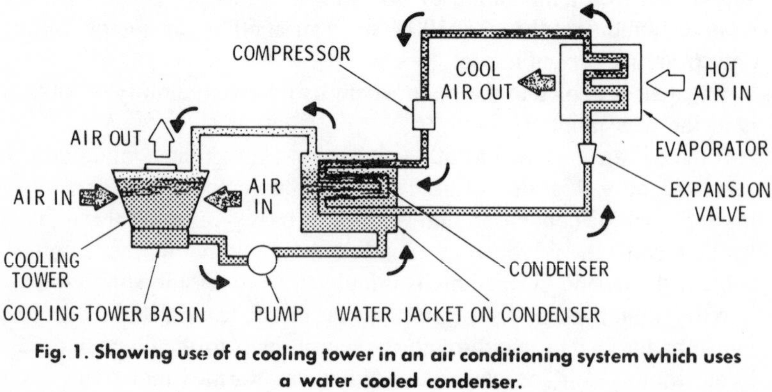

COMPRESSOR
COOL AIR OUT
HOT AIR IN
EVAPORATOR
EXPANSION VALVE
AIR OUT
AIR IN
AIR IN
COOLING TOWER
CONDENSER
COOLING TOWER BASIN
PUMP
WATER JACKET ON CONDENSER

Fig. 1. Showing use of a cooling tower in an air conditioning system which uses a water cooled condenser.

now hot from the absorbed heat, is circulated to the cooling tower where it runs into a distribution trough and drips down through the baffles of the tower. A fan pulls air into the tower, circulating the air into, through and around the baffles. The water dripping through the baffles gives off heat into the air and the heated air is pushed out of the exhaust opening in the tower. The cooled water drips into the basin in the base of the tower, ready to be circulated back to the condenser to again absorb heat. A small amount of water is lost by evaporation in the tower and a make-up water valve is installed in the tower basin to replace the lost water. Fig. 2 shows a cut-away drawing of a cooling tower, and Fig. 3 shows a typical cooling tower installation.

ELECTRICAL AIR CONDITIONING

The operation of an electrical (compressor) type of air conditioning system using an air cooled condenser is essentially the same as one using a water cooled condenser, the only difference being that air, pulled through the condenser by a fan, is used as the condenser cooling medium. The path of the refrigerant, and the changes which take place in the refrigerant as it passes through the compressor, condenser,

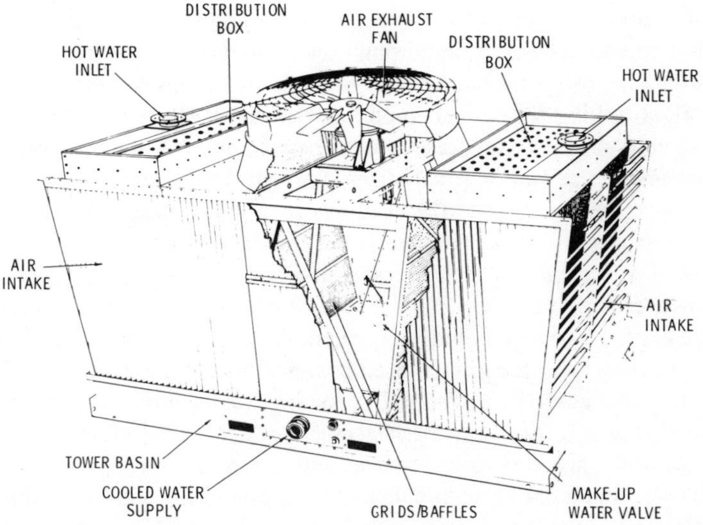
DISTRIBUTION BOX
AIR EXHAUST FAN
DISTRIBUTION BOX
HOT WATER INLET
HOT WATER INLET
AIR INTAKE
AIR INTAKE
TOWER BASIN
COOLED WATER SUPPLY
GRIDS/BAFFLES
MAKE-UP WATER VALVE

Courtesy *The Marley Cooling Tower Co.*

Fig. 2. A cutaway view of a cooling tower.

Courtesy *The Marley Cooling Tower Co.*

Fig. 3. Cooling towers used with water cooled condensers.

expansion valve and evaporator are explained elsewhere in this chapter. Fig. 4 shows a block diagram of an electrical type of air conditioning system.

A typical electric air conditioner operates by passing a cold liquid-plus-vapor mixture, at a lowered pressure, through an evaporator, or cooling coil. The cold liquid-plus-vapor mixture absorbs heat from the air passing over the cooling coil. As the liquid-vapor passes through the coil, absorbing heat, it is completely vaporized, changing from a liquid to a cold gas. The compressor pulls the cold gas from the evaporator or cooling coil as fast as it vaporizes, through the *suction line.* Next, the compressor compresses the cold gas to a higher pressure and in the process also raises the temperature of the cold gas, changing it to a hot gas. The increased pressure forces the hot gas through the condenser where either air or water removes the heat from the refrigerant gas. The hot gas must be at a higher temperature than the air or water in order for heat to flow from the gas to the air or water. As the hot gas gives off heat in the condenser it *condenses*, becoming a hot *liquid*, and still under high pressure. The liquid refrigerant flows from the condenser to an expansion valve or a capillary tube, and then passes through the valve to the evaporator. The pressure up to the expansion valve is higher than the pressure in the evaporator, therefore as the liquid refrigerant passes through the expansion valve the refrigerant expands rapidly and partially vaporizes as it enters the evaporator. Because the pressure is

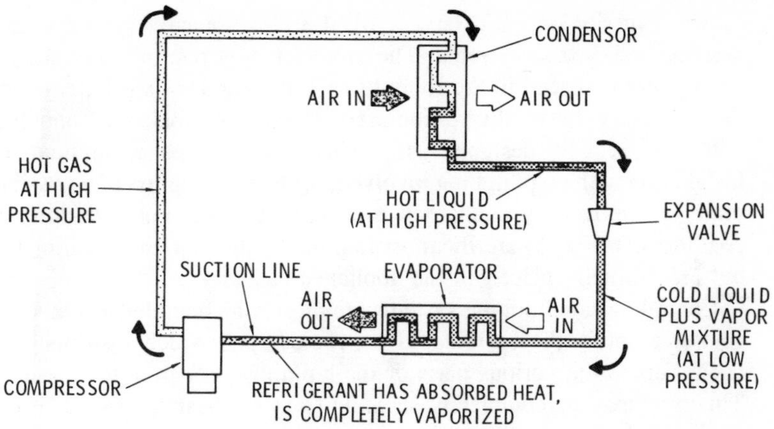
CONDENSOR
AIR IN
AIR OUT
HOT GAS AT HIGH PRESSURE
HOT LIQUID (AT HIGH PRESSURE)
EXPANSION VALVE
SUCTION LINE
EVAPORATOR
AIR OUT
AIR IN
COLD LIQUID PLUS VAPOR MIXTURE (AT LOW PRESSURE)
COMPRESSOR
REFRIGERANT HAS ABSORBED HEAT, IS COMPLETELY VAPORIZED

Fig. 4. A block diagram of an electrical type of air conditioning system.

reduced at the evaporator, the temperature is also reduced, changing the refrigerant from a hot liquid to a cold liquid-plus-vapor mixture. The cold liquid-plus-vapor mixture can now absorb the heat from the air passing over and through the evaporator (cooling coil) and the process of air conditioning continues. The block diagram of an electric type air conditioning unit (Fig. 4) explains the steps in the cycle of operation and the changes which take place in the refrigerant as it goes through the cycle.

It is important to remember that an air cooled condenser operates because the refrigerant is at a higher temperature than the air passing through the condenser and thus the heat which is absorbed by the refrigerant in the evaporator can flow to the cooler air at the condenser.

UNIT AND CENTRAL AIR-CONDITIONING SYSTEMS

The installation, maintenance, and repair of air-conditioning systems has become an important and profitable addition to the heating and plumbing trade in most parts of the country. The relatively recent demand for central, full-house air conditioning that is installed as an integral part of the heating system has been largely responsible for the acceptance of air conditioning as a natural part of the heating and plumbing trade. The installation of commercial air conditioning also involves the use of many plumbing techniques, again emphasizing how closely air conditioning is allied with plumbing and pipe fitting.

Air conditioning may be classified in two general groups—*unit* systems and *central* systems. The unit system is one in which the air conditioner is placed in the room to be cooled and, as a rule, contains the necessary fans, filters, compressor, and controls, all contained within a specially designed case. This type requires a minimum of installation with no plumbing involved, with the exception of a possible drain to remove any condensed moisture from the unit. Window air conditioners (Fig. 5) are the most familiar of the unit air conditioners, but are normally placed in the appliance category.

Central air-conditioning systems are those which are located at some central point in the building to be cooled, and which distribute the cooled air to the various parts of the building through a duct system. This type may be completely separate from the heating system, especially in the case of commercial buildings. The majority of central air-conditioning systems designed for homes, however, utilize the existing duct-work of warm-air furnaces to cool the home. Where steam, hot-water, or radiant heating systems are used, a separate cooling system is necessary which will require the installation of ducts to carry the cool air to the various rooms.

Combination heating and cooling systems are quite popular and offer the most economical means of year-round comfort in the home. Many of these units automatically switch from heating to cooling as required

GIBSON

Courtesy *Gibson Refrigerator Sales Corporation*

Fig. 5. A typical room air conditioner.

without any attention from the occupants other than setting a thermostat to the temperature desired. Some units do require a switch to be thrown to select either heating or cooling.

CAPACITY REQUIREMENTS

The important variables to keep in mind when estimating the Btu requirements for an air conditioning installation are:

1. The total square feet of floor area.
2. The type of wall and ceiling construction.
3. The proportion of the outside wall area that is glass.
4. The walls of the space to be air-conditioned that are exposed to the sun.

Additional factors to be taken into account are ceiling heights, number of persons using the space, difference in temperature between inside and outside, and miscellaneous heat loads such as lamps, radio and television sets, etc.

The cooling capacity of an air-conditioning system is its ability to remove heat from the building, and is usually measured in Btu's per hour for the smaller units. The higher the Btu rating, the more heat can be removed. The capacity rating of a unit is normally given on the nameplate, together with such other necessary information as voltage and wattage requirements. It is important that the unit be large enough to adequately cool the required space without running at full capacity for extended periods.

When dealing with large air conditioners, especially those used for central cooling and for commercial installations, the term *ton of refrigeration* is most often used instead of a Btu rating. A ton of refrigeration is equivalent to 12,000 Btu per hour. This is derived from the fact that one pound of ice absorbs 144 Btu when it melts. Thus, one ton of ice absorbs 2000 × 144, or 288,000 Btu. When one ton of ice melts in 24 hours, the rate is 288,000 ÷ 24, or 12,000 Btu per hour.

In residences and offices, a one-ton unit will usually take care of 5000 to 7000 cubic feet of space, depending on the number of occupants. A one-ton unit in a theater will take care of approximately 15 seats. Thus, for a theater seating 1500 people, a 100-ton machine would be needed. The volume of such a theater is around 800,000 cubic feet, meaning that one ton of refrigeration is required for each 8000 cubic feet in an average theater.

Ventilation

Practically all rooms and buildings in which man lives have a certain amount of natural ventilation, called *infiltration.* Air seeps through cracks in the floors and walls, and around windows and doors, amounting to a surprising number of cubic feet of air per hour. The most practical method of estimating infiltration is to base it on the cubical content of the space to be cooled. For the average home, office, and small shop where only a limited number of persons gather, natural ventilation is usually sufficient. Table 1 lists the natural changes that can be expected for normal construction.

A fault commonly found with air conditioning is that drafts are easily detected. Where human comfort is concerned, the conditioned air must be evenly and thoroughly distributed without a trace of draft.

Estimating Infiltration

If it is desirable to estimate the sensible heat load of a space, room, or building, use the following formula:

$$H_s = \frac{V(t_o - t)}{n \times 3360}$$

where,

- H_s is the sensible heat in Btu per minute from infiltration,
- V is the volume of the space, room, or building in cubic feet,
- t_o is the outdoor dry-bulb temperature assumption,
- t is the indoor dry-bulb temperature to which the space is to be cooled,
- n is the number of hours required to effect one complete air change.

The total heat load of infiltration can be computed by means of the following formula:

$$H_t = \frac{V(H_o - H_l)}{n \times 810}$$

- H_t is the total heat load in Btu per minute due to infiltration.
- H_o is the total heat per pound of air at the outdoor wet-bulb temperature,

H_I is the total heat per pound of air at the indoor wet-bulb temperature.

If it is desired to estimate the latent heat load, use the following formula:

$$H_l = H_t - H_s$$

where,
H_l is the latent heat load,
H_t is the total heat load,
H_s is the total sensible heat load.

Refrigeration Demand

Table 2 may be used to calculate the amount of refrigeration required to cool the air and reduce the humidity. Note that the table lists the load demand in Btu per cubic foot of air to be cooled a certain number of degrees and removing moisture to a specified humidity level.

Table 1 lists the number of air changes dependent on the exposure. For excellent construction, the minimum infiltration figure can be employed, and for old or poor construction, the maximum factor should be used. This table can be used in estimating the cooling and condensation load by multiplying the number of air changes by the volume of the room, and then multiplying this result by the proper factor from Table 2. This will give the total Btu cooling load required, and provides a very convenient and reasonably accurate method.

K Factor

For greater accuracy in estimating the total cooling load, the use of the *K factor*, or coefficient of heat transfer, may be used. The K factor is the amount of heat (in Btu) which will pass through one square foot of wall, ceiling, floor, window, or door in one hour with a 1°F temperature difference existing between the two surfaces between which the heat is being transferred.

When windows have an eastern, southern, or western exposure, they should be provided with awnings; otherwise the direct rays of the sun will give a very high K effect. In the case of exposed east or west windows without awnings, a K factor of 160 Btu per square foot of glass area per hour per degree temperature differential should be used. For southern exposure, windows should be given a K factor of 75, if not

Exposure	Changes per Hour
One Side	½ to 1
Two Sides	¾ to 1½
Three Sides	1 to 2
Four Sides	1 to 2
Inside Room	½ to ¾

Table 1. Infiltration.

Table 2. Refrigeration Load for Air Cooling.

Relative Humidity	Temperature Reduction Increments, °F										
%	5°	6°	7°	8°	9°	10°	11°	12°	13°	14°	15°
40	0.045	0.085	0.120	0.155	0.195	0.235	0.280	0.320	0.365	0.415	0.460
45	0.123	0.165	0.205	0.245	0.290	0.335	0.383	0.428	0.482	0.530	0.580
50	0.200	0.245	0.290	0.335	0.385	0.435	0.485	0.535	0.590	0.645	0.700
55	0.280	0.352	0.377	0.425	0.480	0.528	0.585	0.642	0.703	0.765	0.825
60	0.360	0.410	0.465	0.515	0.575	0.630	0.685	0.750	0.815	0.885	0.950
65	0.438	0.488	0.543	0.598	0.663	0.720	0.783	0.850	0.918	0.990	1.170
70	0.515	0.565	0.620	0.680	0.745	0.810	0.875	0.950	1.020	1.095	1.175
75	0.587	0.648	0.703	0.763	0.833	0.895	0.968	1.055	1.125	1.202	1.295

Exposure	K
North	1.13
East	160.00
South	75.00
West	160.00

Table 3. K Factors for Exposed Windows and Doors.

provided with awnings. These factors are listed in Table 3. When awnings are used over windows and doors, the K factors listed in Table 4 should be used, but twice the temperature difference should be employed to calculate the heat leakage.

Awnings should be used wherever possible in order to reduce the heat load imposed on the air-conditioning equipment and thus save operation costs and initial installation costs. The use of awnings will usually allow the installation of a smaller unit.

When calculating the heat load due to solar radiation, it should be kept in mind that the sun's rays do not strike the east, south, and west windows at the same time. It is a rather difficult matter to determine just how many hours the sunlight will strike any particular exposure. The safest procedure is to select the side of the building (except north) that has the largest glass area and to use this side as the one being continuously exposed to the direct sun. The remaining glass areas should then be estimated for heat load by the use of the K factors listed in Table 3. The same procedure can be followed for windows with awnings, except Table 4 would be used.

In many cases, awnings cannot be employed because of the building design. Here, light-colored shades or other light-impervious materials can be used to reduce the effect of solar radiation. Where such shades or screens are used, a 50% reduction of the heat load caused by the exposed glass areas can be expected and this reduced figure used in the calculations.

All motors generate heat, and when they are installed in air-conditioned areas, the heat they generate must be included in estimating the total heat load. Table 5 lists the heat generated per hour by motors of various sizes.

Electric lights and appliances also generate heat. Incandescent lights dissipate approximately 3.42 Btu per hour for each rated watt. The same figure can be used for determining the heat load added by appliances. The nameplate usually carries wattage rating.

Heat Transmission Through Walls, Ceilings, and Floors

To accurately determine the size of cooling unit needed for a particular installation, it is important to consider the heat transmission through the walls, ceilings, and floors of the space to be cooled. The coefficient of heat transfer for nearly every type and combination of

Table 4. K Factors for Unexposed or Shaded Windows and Doors.

Glass Thickness	K
1	1.13
2	0.46
3	0.29
4	0.21

Motor HP	BTU per Hour
1/20	425
1/10	680
1/8	750
1/6	817
1/4	1020
1/3	1290
1/2	1870
3/4	2750
1	3410

Table 5. Motor Heat Load.

wall, floor, and ceiling is listed in the tables which follow. Partition walls have been included so that the heat leakage through them can also be estimated where necessary.

The values in the tables may not always agree exactly with data included in various handbooks, for the factors given apply mainly to smaller installations instead of for large plants installed in theaters, auditoriums, factories, etc. The tables list heat leakage in Btu per hour per degree (°F) temperature difference per square foot of exposed surface.

In making use of the data, it must be remembered that the outside area is used as a basis for estimating. Ceiling and floor construction must be determined, and in many instances, the various walls may be different construction, especially if one or more is a partition wall. The following formula can be used to determine the heat leakage:

$$H = KA\,(t_1 - t_2)$$

where

K is the K factor

A is the area of the wall, floor, or ceiling

(t_1-t_2) is the temperature difference between the inside and outside.

MOISTURE

Moist or properly humidified air serves to maintain paper, wood, plaster, rugs, and cloth at the correct moisture balance, so that such materials are kept in the proper condition. Dehydration increases the fragility of most materials and, in many cases, causes a change in texture and color. Under ordinary conditions, when the outdoor

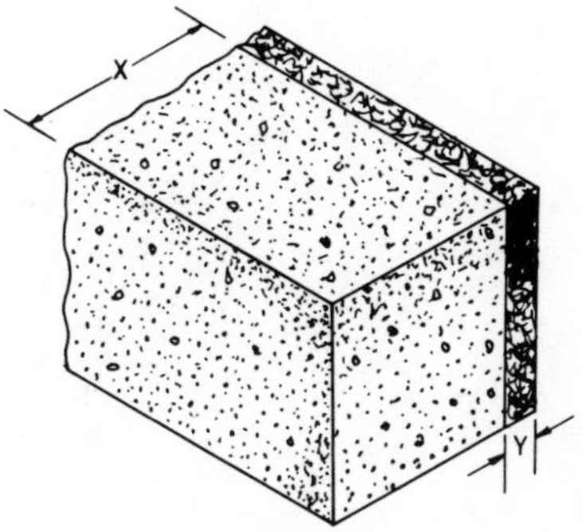
X
Y

K=X+Y
X-WALL
Y-INTERIOR CONSTRUCTION

Concrete Wall
(No Exterior Finish)
Values of K in Btu/Hr/1°F/Sq. Ft.

Wall Construction (Y)	Thickness (X)				
	6″	8″	10″	12″	16″
Plain wall—no interior finish	.58	.51	.46	.41	.34
1/2″ plaster—direct on concrete	.52	.46	.42	.38	.32
1/2″ plaster on wood lath, furred	.31	.29	.26	.24	.22
3/4″ plaster on metal lath, furred	.34	.32	.29	.27	.24
1/2″ plaster on 3/8″ plasterboard, furred	.32	.30	.27	.25	.22
1/2″ plaster on 1/2″ board insulation, furred	.21	.20	.19	.18	.17
1/2″ plaster on 1″ corkboard, set in 1/2″ cement	.16	.15	.14	.14	.13
1/2″ plaster on 1 1/2″ corkboard, set in 1/2″ cement	.15	.14	.14	.13	.12
1/2″ plaster on 2″ corkboard, set in 1/2″ cement	.13	.12	.11	.10	.09
1/2″ plaster on wood lath on 2″ fur., 1 5/8″ gypsum fill	.20	.19	.18	.17	.16

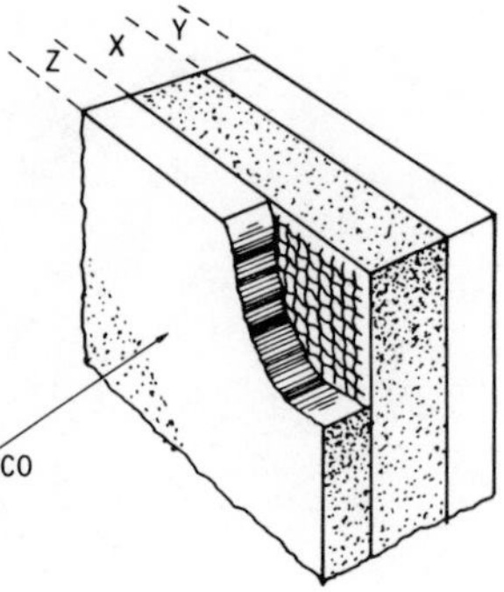
Z
X
Y
EXTERIOR 1" STUCCO
ON WIRE MESH
X= CONCRETE
Y=INSULATION
Z=EXTERIOR STUCCO 1"
K=X + Z + Y

Concrete Wall
(Exterior Stucco Finish)
Values of K in Btu/Hr/1°F/Sq. Ft.

Wall Construction (Y)	Thickness of Concrete (X)					
	6"	8"	10"	12"	16"	18"
Plain walls—no interior finish	.54	.48	.43	.39	.33	.28
½" plaster direct on concrete	.49	.44	.40	.36	.31	.27
½" plaster on wood lath, furred	.31	.29	.27	.25	.23	.22
¾" plaster on metal lath, furred	.32	.30	.28	.26	.24	.23
½" plaster on ⅜" plasterboard, furred	.31	.29	.27	.25	.23	.22
½" plaster on wood lath on 2" furring strips with 1⅝" cellular gypsum fill	.20	.19	.18	.17	.16	.15
½" plaster on ½" board insulation, furred	.22	.21	.20	.19	.18	.17
1"	.17	.16	.15	.145	.140	.13
½" plaster on 1½" sheet cork set in cement	.145	.140	.135	.130	.12	.11
2"	.12	.118	.115	.110	.105	.100

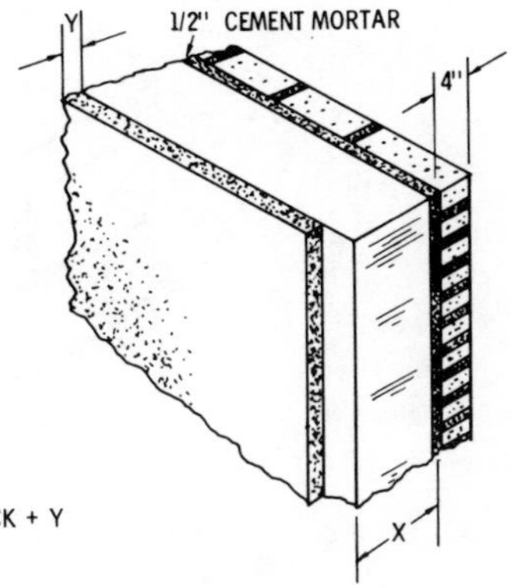

Y
1/2" CEMENT MORTAR
4"
X
X - CONCRETE
Y - INTERIOR FINISH
K= X + 1/2" CEMENT + 4" BRICK + Y

Concrete Wall
(Brick Veneer)
Values of K in Btu/Hr/1°F/Sq. Ft.

Wall Construction (Y)	Concrete (X)			
	6"	8"	10"	12"
Plain wall—no interior finish	.39	.36	.33	.30
½" plaster direct on concrete	.37	.34	.30	.29
½" plaster on wood lath, furred	.24	.23	.21	.20
¾" plaster on metal lath, furred	.27	.25	.23	.22
½" plaster on wood lath, on 2" furring strips, with 1⅝" cellular gypsum fill	.18	.17	.16	.15
½" plaster on ⅜" plasterboard, furred	.25	.24	.22	.21
½" plaster on ½" board insulation, furred	.17	.16	.14	.12
½" plaster on 1" board insulation, furred	.15	.14	.12	.11
½" plaster on 1½" board insulation, furred	.13	.12	.11	.10
½" plaster on 2" board insulation, furred	.11	.10	.09	.08

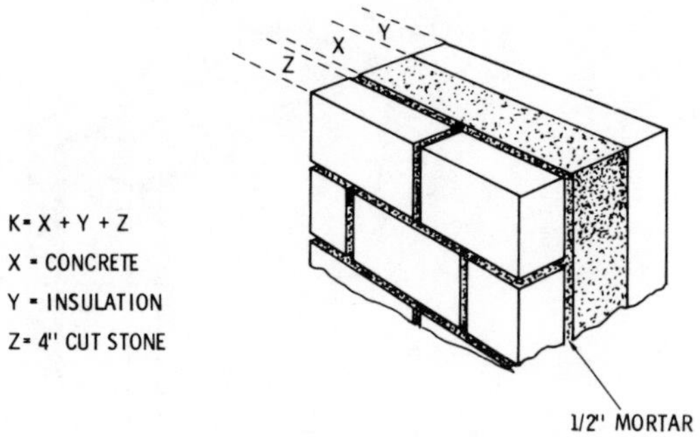
Y
X
Z
K= X + Y + Z
X = CONCRETE
Y = INSULATION
Z= 4" CUT STONE
1/2" MORTAR

Concrete Wall
(4 in. Cut Stone)

Values of K in Btu/Hr/1°F/Sq. Ft.

Wall Construction (Y)	Thickness of Concrete (X)				
	6″	8″	10″	12″	16″
Plain walls—no interior finish	.46	.42	.38	.34	.30
½″ plaster direct on concrete	.42	.38	.34	.32	.28
¾″ plaster on metal lath, furred	.29	.27	.26	.24	.22
½″ plaster on wood lath, furred	.28	.26	.25	.23	.21
½″ plaster on ⅜″ plasterboard, furred	.28	.26	.25	.23	.21
½″ plaster on wood lath on 2″ furring strips, filled with 1⅝″ gypsum	.18	.175	.17	.16	.15
½″ plaster on ½″ board insulation, furred	.21	.20	.19	.18	.17
½″ plaster on 1″ board insulation, furred	.16	.155	.15	.14	.13
½″ plaster on 1″ sheet cork, set in ½″ cement or mortar	.15	.145	.140	.135	.130
1½″	.138	.135	.130	.125	.12
2″	.115	.110	.108	.105	.100

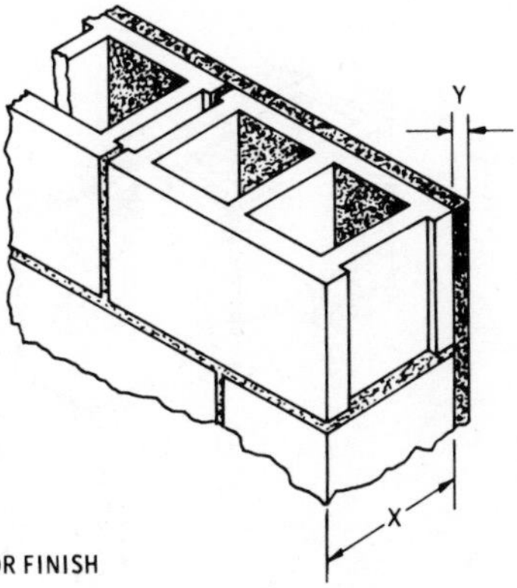
Y
X
K = X + Y
X-WALL
Y-INTERIOR FINISH

Cinder and Concrete Block Wall

Values of K in Btu/Hr/1°F/Sq. Ft.

Wall Construction (Y)	Thickness (X) and kind of blocks			
	Concrete		Cinder	
	8″	12″	8″	12″
Plain wall—no interior finish	.46	.34	.31	.23
½″ plaster—direct on blocks	.42	.32	.29	.21
½″ plaster on wood lath, furred	.28	.23	.22	.17
¾″ plaster on metal lath, furred	.29	.24	.23	.17
½″ plaster on ⅜″ plasterboard, furred	.28	.23	.22	.17
½″ plaster on ½″ board insulation, furred	.21	.18	.17	.14
½″ plaster on 1″ board insulation, furred	.16	.14	.14	.11
½″ plaster on 1½″ sheet cork set in ½″ cement	.13	.12	.12	.10
½″ plaster on 2″ sheet cork set in ½″ cement	.11	.10	.10	.09
½″ plaster on wood lath, on 2″ fur. strips, 1⅝″ Gypsum fill	.18	.16	.16	.13

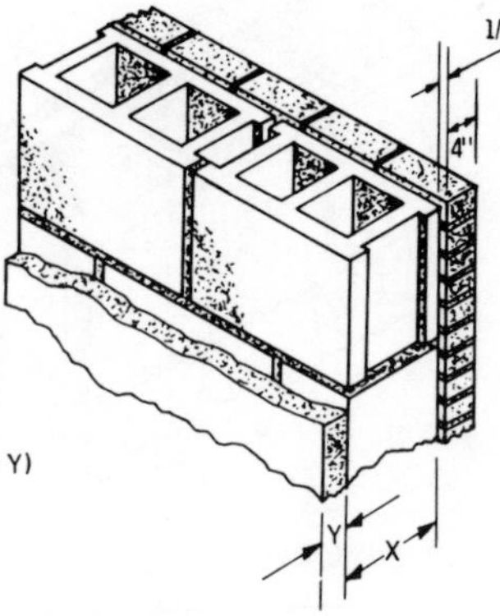
1/2" MORTAR
4"
X-WALL & EXTERIOR
Y-INTERIOR FINISH
K FACTORS GIVEN FOR
TOTAL WALL THICKNESS
(X+1/2" MORTAR + 4" BRICK + Y)
Y
X

Cinder and Concrete Block Wall (Brick Veneer)

Values of K in Btu/Hr/1°F/Sq. Ft.

Wall Construction (Y)	Thickness (X) and kind of blocks			
	Concrete		Cinder	
	8″	12″	8″	12″
Plain wall—no interior finish	.33	.26	.25	.18
½″ plaster—direct on blocks	.31	.25	.23	.18
½″ plaster on wood lath, furred	.23	.19	.18	.15
¾″ plaster on metal lath, furred	.24	.20	.19	.15
½″ plaster on wood lath, on 2″ fur. strips, 1⅝″ gypsum fill	.16	.14	.14	.12
½″ plaster on ½″ board insulation, furred	.18	.16	.15	.13
½″ plaster on 1″ board insulation, furred	.14	.13	.12	.10
½″ plaster on ⅜″ plasterboard, furred	.23	.19	.18	.15
½″ plaster on 1½″ sheet cork, set in ½″ cement mortar	.12	.11	.11	.09
½″ plaster on 2″ sheet cork, set in ½″ cement mortar	.10	.09	.09	.08

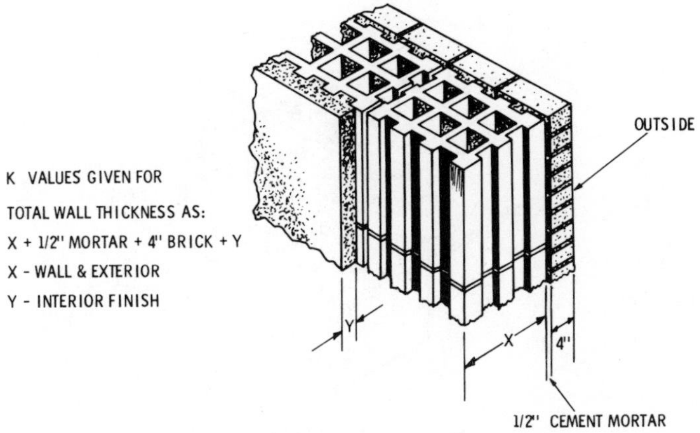
K VALUES GIVEN FOR
TOTAL WALL THICKNESS AS:
X + 1/2" MORTAR + 4" BRICK + Y
X - WALL & EXTERIOR
Y - INTERIOR FINISH
OUTSIDE
Y
X
4"
1/2" CEMENT MORTAR

Brick Veneer on Hollow Tile Wall (Brick Veneer)

Values of K in Btu/Hr/1°F/Sq. Ft.

Wall Construction (Y)	Tile Thickness (X)				
	4″	6″	8″	10″	12″
Plain wall—no interior finish	.29	.27	.25	.22	.18
½″ plaster on hollow tile	.28	.25	.24	.21	.17
½″ plaster on wood lath, furred	.20	.19	.18	.17	.15
¾″ plaster on metal lath, furred	.21	.20	.19	.18	.16
½″ plaster on ⅜″ plasterboard, furred	.20	.19	.18	.17	.15
½″ plaster on wood lath, on 2″ furring strips, with 1⅝″ cellular gypsum fill	.14	.13	.12	.11	.10
½″ plaster on ½″ board insulation, furred	.16	.15	.14	.13	.12
½″ plaster on 1″ board insulation, furred	.14	.13	.12	.11	.10
½″ plaster on 1½″ board insulation, furred	.13	.12	.11	.10	.09
½″ plaster on 2″ board insulation, furred	.12	.11	.10	.09	.08

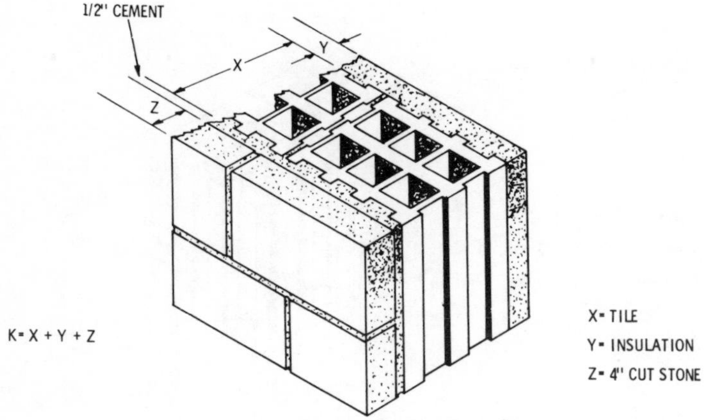
1/2" CEMENT
X
Y
Z
K = X + Y + Z
X = TILE
Y = INSULATION
Z = 4" CUT STONE

Hollow Tile Wall
(4 in. Cut Stone Veneer)

Values of K in Btu/Hr/1°F/Sq. Ft.

Wall Construction (Y)	Thickness of Tile (X)			
	6″	8″	10″	12″
Plain walls—no interior finish	.30	.29	.28	.22
½″ plaster—direct on hollow tile	.28	.27	.26	.21
¾″ plaster on metal lath, furred	.22	.21	.20	.18
½″ plaster on wood lath, furred	.21	.20	.19	.17
½″ plaster on ⅜″ plasterboard, furred	.21	.20	.19	.17
½″ plaster on ½″ cork board set in ½″ cement	.17	.16	.15	.14
½″ plaster on 1″ corkboard set in ½″ cement	.14	.13	.13	.12
½″ plaster on 1½″ corkboard set in ½″ cement	.13	.12	.11	.10
½″ plaster on 2″ corkboard set in ½″ cement	.11	.10	.09	.08
½″ plaster on wood lath, 2″ furring & 1⅝″ gypsum fill	.15	.14	.13	.12
½″ plaster on ½″ board insulation	.18	.17	.16	.15

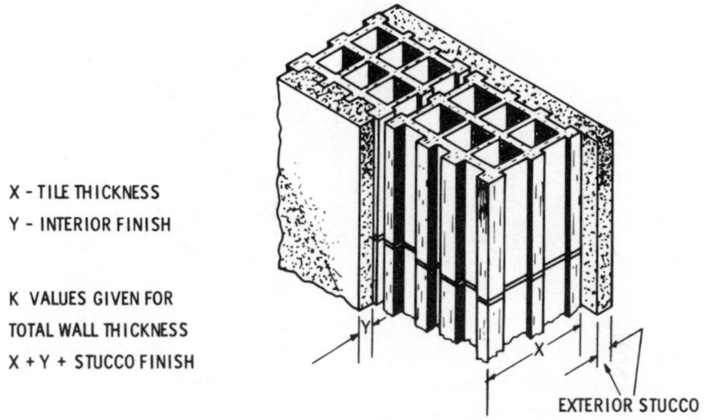
X - TILE THICKNESS
Y - INTERIOR FINISH
K VALUES GIVEN FOR
TOTAL WALL THICKNESS
X + Y + STUCCO FINISH
Y
X
EXTERIOR STUCCO

Hollow Tile Wall
(Stucco Exterior)
Values of K in Btu/Hr/1°F/Sq. Ft.

Wall Construction (Y)	Tile Thickness (X)				
	6″	8″	10″	12″	16″
Plain wall with stucco—no interior finish	.32	.30	.28	.22	.18
½″ plaster—direct on hollow tile	.31	.29	.27	.21	.17
½″ plaster on wood lath, furred	.22	.20	.18	.16	.12
¾″ plaster on metal lath, furred	.23	.21	.19	.17	.13
½″ plaster on ⅜″ plasterboard, furred	.21	.20	.18	.16	.12
½″ plaster on wood lath, on 2″ furring strips, with 1⅝″ cellular gypsum fill	.16	.15	.14	.13	.12
½″ plaster on ½″ board insulation, furred	.15	.14	.13	.12	.11
½″ plaster on 1″ board insulation, furred	.13	.12	.11	.10	.09
½″ plaster on 1½″ board insulation, furred	.12	.11	.10	.09	.08
½″ plaster on 2″ board insulation, furred	.11	.10	.09	.08	.07

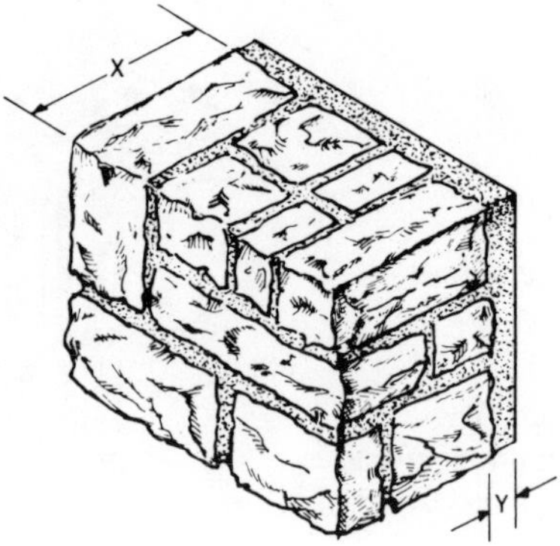
X
Y

K = X + Y

X - WALL
Y - INTERIOR CONSTRUCTION

Limestone or Sandstone Wall

Values of K in Btu/Hr/1°F/Sq. Ft.

Wall Construction (Y)	Thickness (X)			
	8″	10″	12″	16″
Plain wall—no interior finish	.56	.50	.48	.39
½″ plaster—direct on stone	.50	.45	.42	.36
¾″ plaster on metal lath, furred	.33	.31	.29	.26
½″ plaster on wood lath, furred	.31	.29	.28	.25
½″ plaster on ⅜″ plasterboard, furred	.31	.30	.28	.25
½″ plaster on wood lath on 2″ fur. strips, 1⅝″ gypsum fill	.20	.19	.18	.17
½″ plaster on ½″ board insulation, furred	.23	.22	.21	.19
½″ plaster on 1″ board insulation, furred	.17	.16	.15	.14
½″ plaster on 1½″ corkboard, set in ½″ cement	.14	.14	.13	.12
½″ plaster on 2″ corkboard, set in ½″ cement	.11	.11	.10	.10
½″ plaster on wood lath, on 2″ fur. strips, 2″ gypsum fill	.19	.18	.17	.16

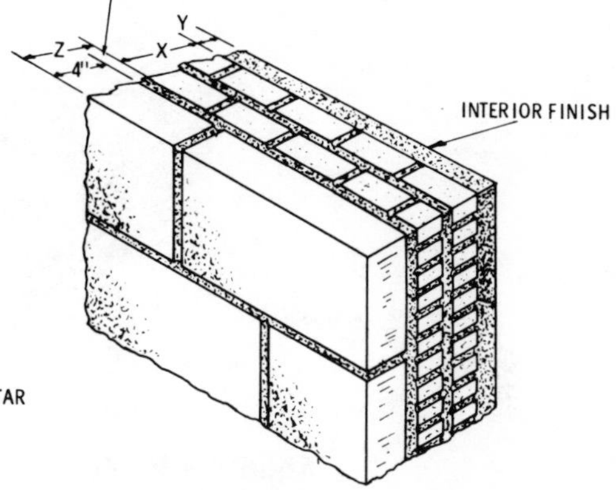
1/2" MORTAR
Y
Z
X
4"
INTERIOR FINISH
K = X + Y
X = WALL
Y = INTERIOR
Z = 4" STONE 8-1/2" MORTAR

Brick Wall
(4 in. Cut Stone)

Values of K in Btu/Hr/1°F/Sq. Ft.

Wall Construction (Y)	Thickness (X)						
	8"	9"	12"	13"	16"	18"	24"
Plain wall—no interior finish	.33	.29	.26	.25	.22	.20	.16
½" plaster—direct on brick	.31	.28	.25	.24	.21	.19	.15
¾" plaster on metal lath, furred	.23	.22	.20	.19	.18	.17	.15
½" plaster on wood lath, furred	.22	.21	.19	.18	.17	.15	.12
½" plaster on ⅜" plasterboard, furred	.23	.22	.20	.19	.18	.17	.14
½" plaster, lath, 2" furring strips, 1⅝" gypsum fill	.16	.15	.14	.13	.12	.11	.10
½" plaster on ½" sheet cork	.18	.16	.15	.14	.13	.12	.11
½" plaster on 1" sheet cork	.14	.13	.12	.11	.10	.10	.09
½" plaster on 1½" sheet cork	.12	.12	.11	.11	.10	.10	.09
½" plaster on 2" sheet cork	.11	.10	.10	.09	.09	.08	.08

K= X+Y
X= WALL
Y= INTERIOR CONSTRUCTION
X
Y

Solid Brick Wall
(No Exterior Finish)

Values of K in Btu/Hr/1°F/Sq. Ft.

Wall Construction (Y)	Thickness (X)						
	8″	9″	12″	13″	16″	18″	24″
Plain wall—no interior finish	.39	.37	.30	.29	.24	.23	.18
½″ plaster—direct on brick	.36	.35	.28	.27	.23	.20	.16
¾″ plaster on metal lath, furred	.26	.25	.23	.21	.19	.18	.15
½″ plaster on wood lath, furred	.25	.24	.21	.19	.18	.17	.14
½″ plaster on ⅜″ plasterboard, furred	.24	.23	.20	.18	.17	.16	.14
½″ plaster on ½″ board insulation, fur.	.20	.19	.18	.17	.16	.15	.13
½″ plaster on ½″ sheet cork, set in ½″ cement	.18	.17	.16	.15	.14	.13	.12
½″ plaster on 1″ sheet cork, set in ½″ cement	.15	.14	.13	.12	.11	.10	.09
½″ plaster on 1½″ sheet cork, set in ½″ cement	.14	.13	.12	.11	.10	.09	.08
½″ plaster on 2″ sheet cork, set in ½″ cement	.13	.12	.11	.10	.09	.08	.07
½″ plaster on 1½″ split furring, tiled	.28	.26	.23	.22	.20	.19	.17

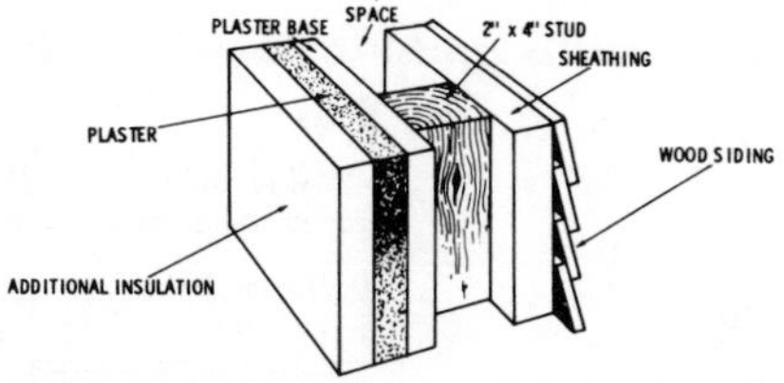
PLASTER BASE
SPACE
2' x 4' STUD
SHEATHING
PLASTER
WOOD SIDING
ADDITIONAL INSULATION

Wood Siding Clapboard Frame or Shingle Walls

Values of K in Btu/Hr/1°F/Sq. Ft.

Type of Sheathing	Insulation between studding	Plaster Base						
		Wood lath	Metal lath	3/8″ plaster board	1/2″ rigid insulation	1″ rigid insulation	1 1/2″ sheet cork	2″ sheet cork
1/2″ plasterboard	None	.30	.31	.30	.22	.16	.12	.10
	Cellular gypsum, 18#	.12	.125	.12	.10	.09	.08	.07
	Flaked gypsum, 24#	.10	.108	.10	.09	.08	.07	.06
	1/2″ flexible insulation	.16	.17	.16	.14	.11	.09	.08
1″ wood	None	.26	.28	.26	.20	.15	.12	.10
	Cellular gypsum, 18#	.11	.118	.11	.10	.09	.08	.07
	Flaked gypsum, 24#	.10	.108	.10	.09	.08	.07	.06
	1/2″ flexible insulation	.15	.16	.15	.13	.11	.09	.08
1/2″ rigid insulation (board form)	None	.22	.23	.22	.17	.14	.11	.09
	Cellular gypsum, 18#	.10	.11	.10	.09	.08	.07	.06
	Flaked gypsum, 24#	.09	.10	.09	.08	.07	.06	.05
	1/2″ flexible insulation	.14	.148	.14	.12	.10	.08	.07

Interior Walls and Plastered Partitions (No Fill)

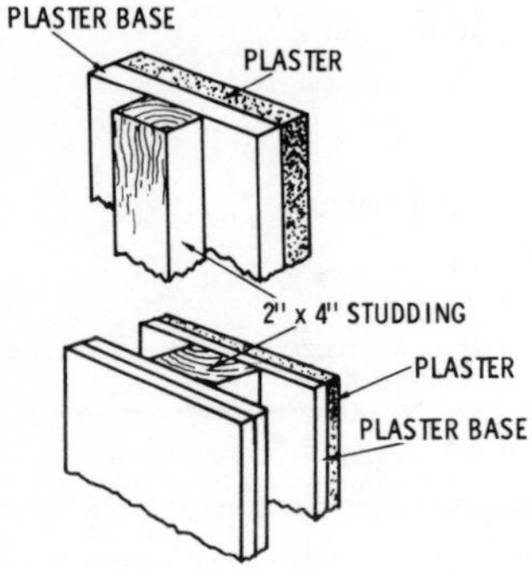
PLASTER BASE
PLASTER
2" x 4" STUDDING
PLASTER
PLASTER BASE

Values of K in Btu/Hr/1°F/Sq. Ft.

Plaster base	Single partition 1 side plastered	Double partition 2 sides plastered
Wood lath	.60	.33
Metal lath	.67	.37
⅜" plasterboard	.56	.32
½" board (rigid) insulation	.33	.18
1" board (rigid) insulation	.22	.12
1½" corkboard	.16	.08
2" corkboard	.12	.06

Frame Floors and Ceilings (No Fill)

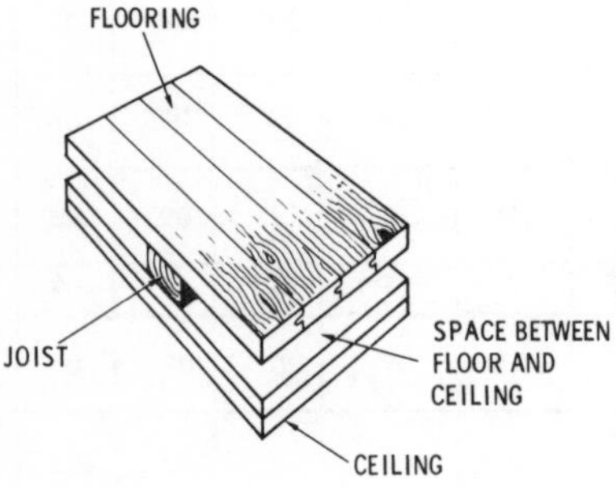
FLOORING
JOIST
SPACE BETWEEN FLOOR AND CEILING
CEILING

Values of K in Btu/Hr/1°F/Sq. Ft.

Type of ceiling with ½" plaster	No flooring	1" yellow pine flooring	1" yellow pine flooring and ½" fiberboard	1" sub oak or maple on 1" yellow pine flooring
No ceiling		.45	.27	.32
Metal lath	.66	.24	.20	.24
Wood lath	.60	.27	.19	.23
⅜" plasterboard	.59	.28	.20	.24
½" rigid insulation	.34	.20	.15	.17

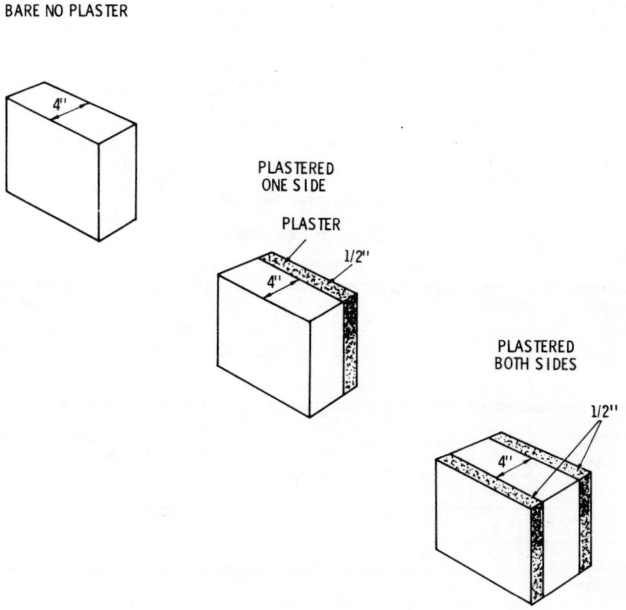
BARE NO PLASTER
4''
PLASTERED
ONE SIDE
PLASTER
1/2''
4''
PLASTERED
BOTH SIDES
1/2''
4''

Masonry Partitions

Values of K in Btu/Hr/1°F/Sq. Ft.

Wall	Bare no plaster	One side plastered	Two Sides plastered
4″ brick	.48	.46	.41
4″ hollow gypsum tile	.29	.27	.26
4″ hollow clay tile	.44	.41	.39

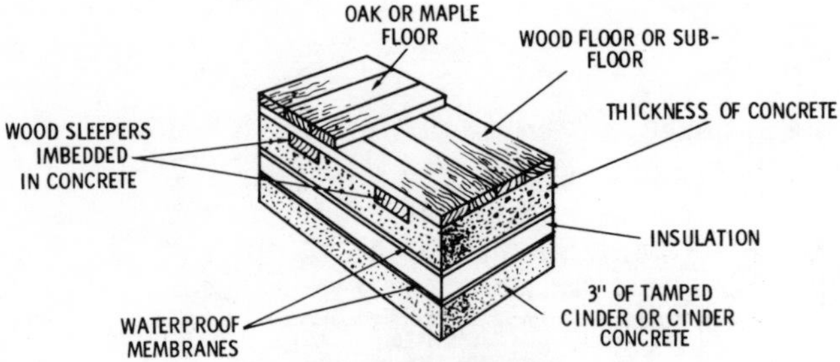
OAK OR MAPLE FLOOR
WOOD FLOOR OR SUB-FLOOR
THICKNESS OF CONCRETE
WOOD SLEEPERS IMBEDDED IN CONCRETE
INSULATION
3" OF TAMPED CINDER OR CINDER CONCRETE
WATERPROOF MEMBRANES

Concrete Floors (On Ground)

Values of K in Btu/Hr/1°F/Sq. Ft.

Type of insulation	Thickness of Y	Concrete thickness X	No flooring	Tile or terrazzo floor	1″ yellow pine on wood sleepers	Oak or Maple on yellow pine sub floor on sleepers
None and no cinder base (Z)	0	4	1.02	.93	.51	.37
	0	5	.94	.83	.48	.36
	0	6	.86	.79	.46	.34
	0	8	.76	.70	.43	.32
None with 3″ cinder base (Z)	0	4	.58	.55	.51	.31
	0	5	.55	.52	.47	.30
	0	6	.52	.49	.44	.29
	0	8	.50	.47	.40	.28
1″ rigid insulation with 3″ cinder base (Z)	1″	4	.22	.21	.18	.165
	1″	5	.21	.20	.175	.16
	1″	6	.20	.19	.17	.155
	1″	8	.19	.18	.16	.15
2″ corkboard insulation with 3″ cinder base (Z)	2″	4	.125	.12	.115	.105
	2″	5	.12	.115	.11	.10
	2″	6	.115	.11	.105	.095
	2″	8	.11	.105	.10	.09

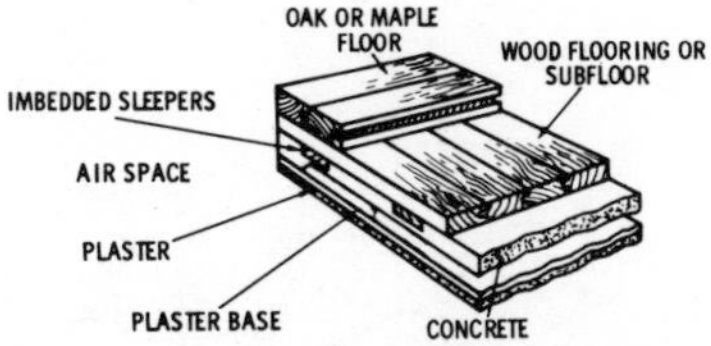
OAK OR MAPLE FLOOR
WOOD FLOORING OR SUBFLOOR
IMBEDDED SLEEPERS
AIR SPACE
PLASTER
PLASTER BASE
CONCRETE

Concrete Floors and Ceilings

Values of K in Btu/Hr/1°F/Sq. Ft.

Type of ceiling and base	Thick-ness	Floor Type			
		No flooring	Tile or Terrazzo floor	1″ yellow pine on wood sleepers	Oak or maple on yellow pine, sub-floor on sleeper
No ceiling	4	.60	.57	.38	.29
	6	.54	.52	.35	.28
	8	.48	.46	.33	.26
	10	.46	.44	.31	.25
½″ plaster direct on concrete	4	.56	.53	.38	.30
	6	.51	.49	.35	.29
	8	.47	.44	.33	.27
	10	.42	.41	.31	.26
Suspended or furred ceiling on ⅜″ plasterboard ½″ plaster	4	.33	.32	.24	.20
	6	.31	.30	.23	.19
	8	.29	.28	.22	.18
	10	.28	.27	.21	.17
Suspended or furred metal lath ¾″ plaster	4	.36	.34	.26	.21
	6	.34	.32	.24	.20
	8	.32	.30	.23	.19
	10	.31	.29	.22	.18
Suspended or furred on ½″ rigid insulation ½″ plaster	4	.22	.21	.20	.18
	6	.21	.20	.19	.17
	8	.20	.19	.18	.16
	10	.19	.18	.17	.15
1½″ cork-board in ½″ mortar ½″ plaster	4	.16	.15	.14	.13
	6	.15	.14	.13	.12
	8	.14	.13	.12	.11
	10	.13	.12	.11	.10

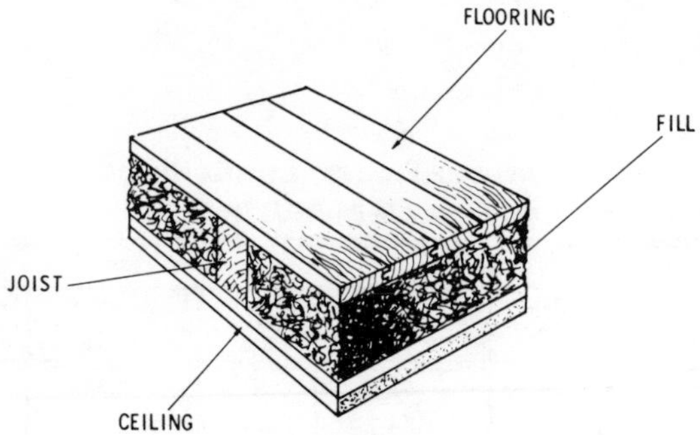
FLOORING
FILL
JOIST
CEILING

Frame Floors and Ceilings (With Fill)

Values of K in Btu/Hr/1°F/Sq. Ft.

Type of ceiling with ½″ plaster	Insulation or fill between joists	No flooring	1″ yellow pine flooring	1″ yellow pine flooring and ½″ fiberboard	1″ sub-oak, maple on 1″ yellow pine flooring
Wood lath	½″ flexible insulation	.22	.15	.12	.14
Wood lath	½″ rigid insulation	.24	.16	.13	.15
Wood lath	Flaked 2″ gypsum	.15	.12	.105	.12
Wood lath	Rock wool, 2″	.12	.095	.08	.09
Corkboard ½″	None	.155	.115	.10	.105
Corkboard ½″	None	.12	.10	.085	.095

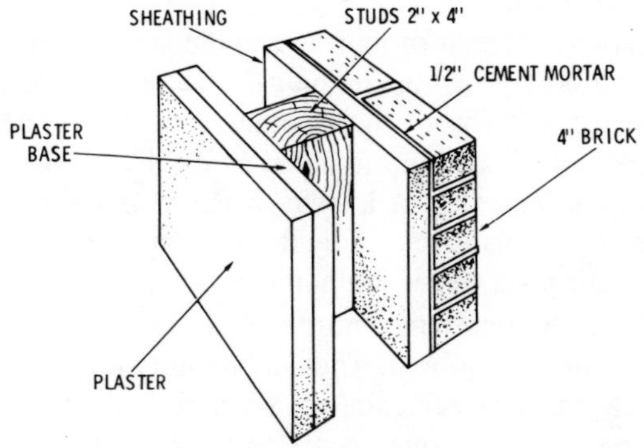
SHEATHING
STUDS 2" x 4"
1/2" CEMENT MORTAR
PLASTER BASE
4" BRICK
PLASTER

Brick Veneer
(Frame Walls)
Values of K in Btu/Hr/1°F/Sq. Ft.

Type of sheathing	Insulation between studding	Plaster Base							
		Wood lath	Metal lath	½" fiber board	⅜" plaster board	½" cork board	1" cork board	1½" cork board	2" cork board
1" wood sheathing	None	.26	.27	.20	.27	.18	.15	.12	.095
	½" fiberboard	.17	.18	.14	.16	.14	.11	.09	.08
	Flaked gypsum 24#	.095	.10	.09	.10		.08	.07	.06
	Cellular gypsum 18#	.11	.15	.10	.11		.09	.08	.07
½" plaster-board sheathing	None	.32	.34	.24	.30		.16	.13	.11
	½" fiberboard	.18	.19	.14	.17		.11	.09	.08
	Flaked gypsum 24#	.10	.11	.09	.10		.08	.07	.06
	Cellular gypsum 18#	.12	.12	.10	.12		.09	.08	.07
½" fiber-board (rigid) insulation sheathing	None	.25	.26	.19	.26		.15	.12	.10
	½" fiberboard	.16	.17	.13	.15		.10	.09	.08
	Flaked gypsum 24#	.09	.09	.08	.09		.07	.07	.06
	Cellular gypsum 18#	.10	.105	.09	.10		.08	.07	.06

temperature makes it necessary to use the heating system, materials in the heated space will give up their moisture to the air. This fact can be shown by the condensation on windows during frosty days of the heating season. During this time, rugs, plaster, wood, etc., give up a certain portion of their moisture content.

Humidification (addition of moisture to the air) is usually necessary during such periods. When the humidity is increased, the materials in the heated space will absorb the amount of moisture lost during the dehydrating period. The absorption, and especially the reabsorption, of moisture by these materials is known as the *moisture regain.*

Human comfort is also affected by the amount of moisture in the air. During the heating season, if the humidity is allowed to decrease too much, the dry air becomes uncomfortable and feels chilly, even though its temperature may be normal. Thus, a humidity level of 30% to 50% must be maintained to ensure comfort. When the humidity is kept at this level, condensation or frosting occurs on the windows and door glass. This is objectionable for many reasons. If the condensation is appreciable and remains for extended periods, it will damage or destroy the finish on the window frames and sills and will create rust and rot. Putty is also impaired and may have to be replaced periodically.

Condensation can be prevented by maintaining a lower relative humidity, but this would result in discomfort as well as damage to other material in the heated space. Instead, condensation can be reduced or eliminated by the use of double or triple window glass.

Double glass, with dead-air space between, will result in the inside glass having a higher surface temperature than where a single glass is employed, so that little or no condensation will occur under normal conditions. Table 6 lists the temperature of the glass at which condensation takes place for various relative humidities.

COOLING EQUIPMENT

It is possible to select an air-conditioning unit or units to suit practically any type or size of building. The cooling apparatus may be entirely separate from the heating equipment, may be added to the equipment, or may be included as part of the heating plant. In the last case, a true air-conditioning system is the result, giving the occupants heat in winter and cooling in the summer without the necessity of switching from one to the other. It is all taken care of by automatic controls that require only the setting of a thermostat to the temperature desired.

Table 6. Temperature and Relative Humidity at Which Condensation Occurs on Window Glass.

Outdoor Temperature °F	Temperature of Inner Surface		Percent of Relative Humidity at Which Condensation Forms	
	Single glass	Double glass	Single glass	Double glass
—10	13	46	12	45
0	20	49	18	49
10	28	52	23	54
20	34	54	30	60
30	42	57	38	68
40	48	60	48	75
50	55	63	64	83
60	62	66	80	91
70	69	69	99	99

Fig. 6 shows an air-conditioning cooling system for use in conjunction with a warm-air furnace. The unit shown contains the condenser, compressor, and fan to cool the condenser. This unit is normally installed in some remote location where the heat it must dissipate will not interfere with the space to be cooled. A unit of this type is often installed outside the building. The cooling coil is a separate unit and is installed in the furnace plenum. Flexible copper refrigerant lines connect the cooling coil and the condensing unit, being brazed in place. These lines are available in various lengths. Special couplings permit both the cooling coils and the condensing unit to be fully charged and hermetically sealed at the factory. The connections to the condenser unit are shown at the lower left corner of the cabinet in Fig. 6.

Many modern refrigerating units incorporate a sealed compressor similar to the one shown in Fig. 7. This type of compressor is quite popular for the smaller air-conditioning installations such as those for homes and small commercial buildings. The built-in motor contributes to the quiet operation of a compressor of this type, since a separate motor and drive belt are unnecessary. Spring suspension absorbs nearly all vibration. This particular model features a three-stage muffler and three expansion chambers to reduce noise level and discharge-gas pulsations.

Where it is impractical or undesirable to install cooling coils in the plenum of a warm-air furnace, individual cooling cabinets can be used

JANITROL

Courtesy *Janitrol Division, Midland-Ross Corp.*

Fig. 6. Condenser-compressor unit for use with an existing warm-air furnace.

in various locations in the space to be cooled. Fig. 8 shows such a unit. These cabinet-type units provide custom-tailored conditioned air to meet nearly any requirement for individual rooms. Many sizes and styles are available for floor, wall, and ceiling installation. These units are also available with various types and sizes of heating coils to provide year-round air conditioning. Each unit can be individually controlled by selecting the proper control from the many that can be supplied. The unit in Fig. 8 has the back panel removed to show the cooling-coil assembly, the blower motor and fans, and the removable air filter.

For air-conditioning commercial and industrial buildings, larger units are necessary. Roof-mounted systems are becoming quite popular, especially where inside space is at a premium. Combining both heating and cooling, this type of installation eliminates the need for a heating and cooling room. Fig. 9 shows the interior of a typical roof-mounted

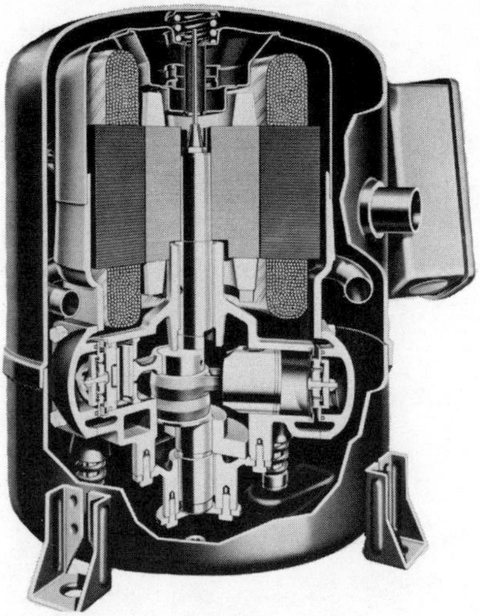

Courtesy *Trane Company*

Fig. 7. A sealed compressor is used in many small air-conditioning systems.

Courtesy *Dunham-Bush, Inc.*

Fig. 8. A cabinet-type cooling unit with back cover removed.

unit in which a gas burner is incorporated for heating purposes. Models are available without the heating section where summer cooling only is required. Electric heat is also available instead of gas. The unit is completely self-contained in a weatherproof and insulated cabinet.

Courtesy *Dunham-Bush, Inc.*

Fig. 9. Interior view of a roof-mounted air conditioner. This model features a gas-fired heating unit as well as a refrigeration unit.

Systems from 15 to 35 tons of refrigeration are manufactured to fill a wide range of air-conditioning needs.

The units described previously have all utilized air as the cooling medium for the condenser. Water-cooled condensers are required or are desirable for some air-conditioning installations. The piping diagram of a typical roof-mounted water-cooled packaged air-conditioning unit is illustrated in Fig. 10.

A water-cooled condenser and compressor unit is shown in Fig. 11. This unit is for use when it is desirable to locate the cooling coils remotely from the compressor assembly.

For cooling very large buildings, a separate compressor room is often necessary. An example of such an installation is shown in Fig. 12. This system is used to cool a large shopping center in eastern New York State.

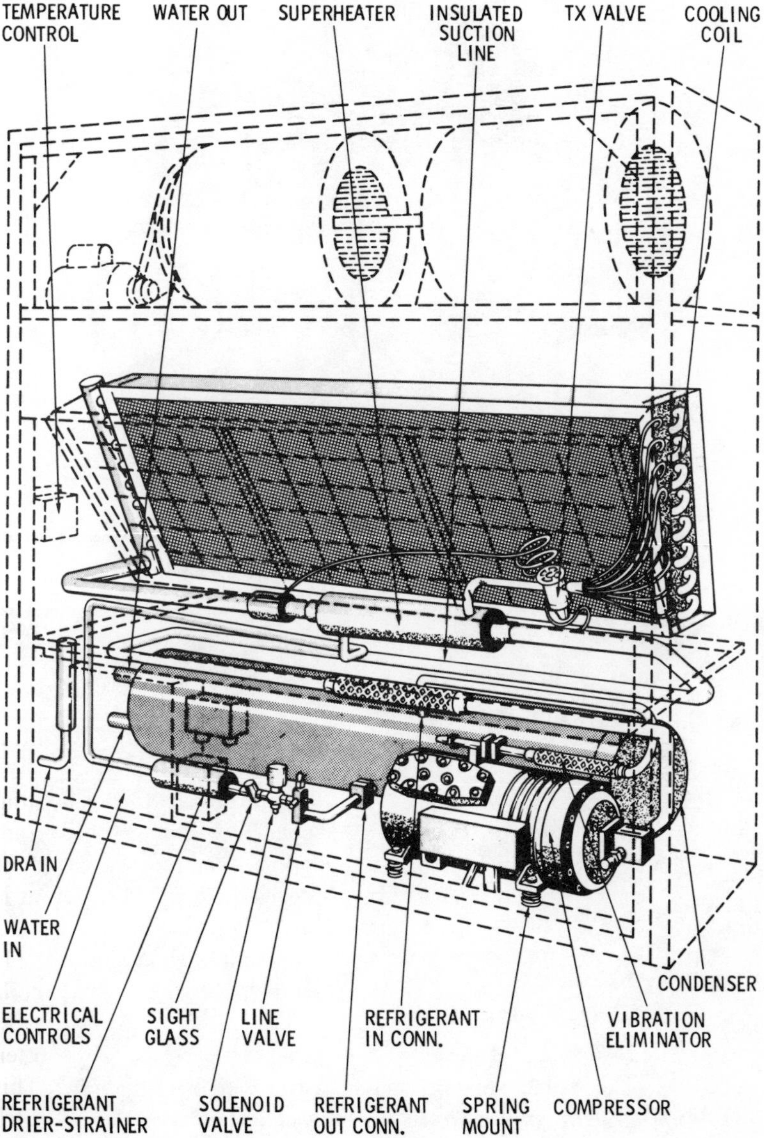
TEMPERATURE CONTROL
WATER OUT
SUPERHEATER
INSULATED SUCTION LINE
TX VALVE
COOLING COIL
DRAIN
WATER IN
CONDENSER
ELECTRICAL CONTROLS
SIGHT GLASS
LINE VALVE
REFRIGERANT IN CONN.
VIBRATION ELIMINATOR
REFRIGERANT DRIER-STRAINER
SOLENOID VALVE
REFRIGERANT OUT CONN.
SPRING MOUNT
COMPRESSOR

Courtesy *Dunham-Bush, Inc.*

Fig. 10. Piping diagram of a typical packaged air-conditioning unit.

USE
R12
ONLY

Courtesy *Dunham-Bush, Inc.*

Fig. 11. A compact industrial-type cooling unit.

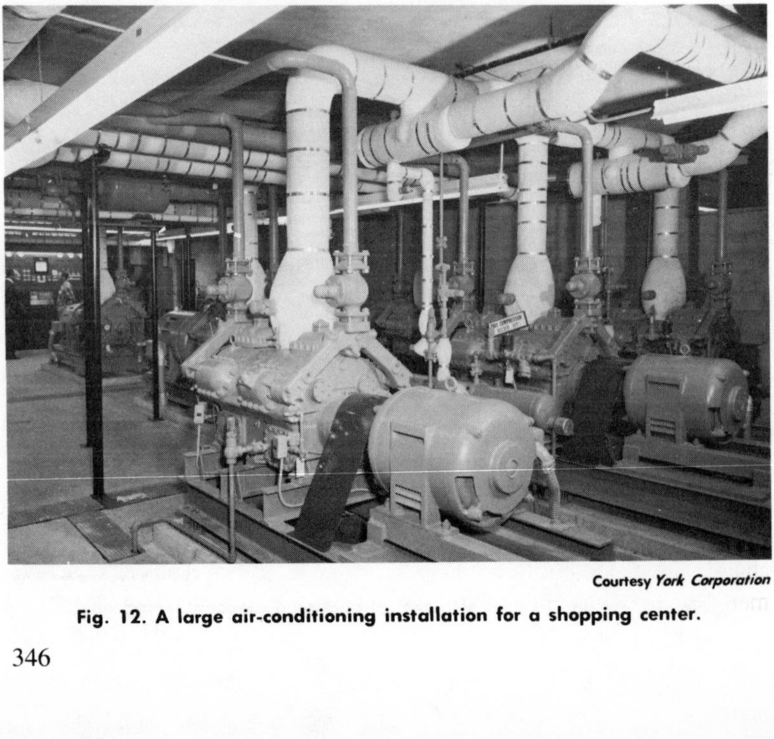

Courtesy *York Corporation*

Fig. 12. A large air-conditioning installation for a shopping center.

CONTROL DEVICES

There are three types of controls used with air-conditioning installations. These are:

1. Manual.
2. Automatic.
3. Semiautomatic.

The manual type of control requires personal attention and regulation at frequent intervals, especially when the load varies continuously. Even if the system is in operation only during certain times of the day, inside and outside conditions usually vary so much that constant changes in adjustment are necessary.

The manual control may consist of an ordinary on-off switch which must be operated by hand to start and stop the actuating mechanism. This type of control is seldom used.

Automatic controls will meet the demands of a varying load almost instantly. An automatic system usually involves the use of several instruments, such as time clocks, thermostats, humidity controllers, and automatic dampers operated either by electricity or by compressed air. Automatic controls provide all the features that are considered necessary or desirable for temperature and humidity regulation as well as economy of operation.

A system using only a thermostat is usually a semiautomatic system and, in most cases, will provide satisfactory operation. Time clocks may be desirable on installations where definite shutdown periods are required. The fresh-air intake from outdoors should always be equipped with adjustable louvers so that the air intake can be regulated for both summer cooling and winter heating. It is desirable to keep the control system on any installation as simple and free from service liabilities as possible.

Thermostats

The device which operates by changes in room temperature is called a *thermostat.* Some thermostats depend on a bimetallic strip to make and break a set of contacts. In other designs, the bimetallic strip is coiled into a loose helix, with the end of the outer coil rigidly fixed to the case. The inner end of the coil is free to move in accordance with temperature changes, and to this free end is attached a glass capsule containing mercury. The ends of two wires project into and are sealed in one end of the capsule so that, when it is tilted, the mercury flows to the low end. If the low end contains the projecting wire ends, the circuit is completed; if tilted the other way, the circuit is broken. The mercury bridges the gap between the two wires and acts as a metallic conductor.

The capsule is usually charged with some inert gas so that no arcing will occur when the electrical circuit makes and breaks. Such devices are made to handle currents up to and including that taken by a 1-HP motor. For larger motors, a relay used in conjunction with a thermostat is employed.

Another variety of thermostat depends on the pressure-temperature relationship of a gas for its operation. In this type, a bulb is charged with refrigerant and sealed. A small tube connects the charged bulb to an expandable diaphragm or drum. With an increase in temperature, there will be a corresponding increase in pressure. By means of levers, springs, and adjustments, the device can be set to cut in or out at any desired point.

A thermostat may be used to stop and start an air-conditioning system equipped with either an automatic constant-pressure expansion valve or a thermostatic expansion valve. If cooling only is desired, the simplest arrangement is to connect a single-pole, single-throw thermostat in series with the solenoid coil of a magnetic liquid-line stop valve in the liquid line of the evaporator expansion valve. Such a thermostat should be connected in series with a snap switch to make sure the thermostat does not operate the equipment during shutdown periods.

For circulating cold water or brine, a thermostat may be connected in series with the solenoid coil of the circulating-pump starting switch. With this arrangement, one thermostat can control the circulation of the cooling medium to a number of cooling cabinets.

Humidity Controls

The instrument used to automatically control humidity is known as a *hygrostat* and is available under a variety of trade names. The element in the control is some hygroscopic material such as wood, human hair, paper, etc. When the material absorbs water from the air, it expands; when the material gives up moisture, it contracts. It is this action that is used to make and break an electrical circuit controlling the humidifying or dehumidifying equipment.

Time Clocks

Time clocks are merely clock-operated switches placed in series with the thermostat to automatically prevent operation except during certain desired periods of time.

Damper Motors

Motors for operating dampers are provided when automatic regulation of air flow is required in heating, ventilating, and air-conditioning systems and supply ducts The motors are usually powered by electricity (compressed air is sometimes used) and controlled by thermostats. The dampers can be positioned by the motors as required and in accordance with the indication from the governing device.

Dampers can also be operated by thermostatic elements in which the expansion and contraction of a liquid or gas supplies the power.

Contactors, Starters, and Relays

Most condensing units having motors above 1-1/2 HP require a starter or contactor. This device will usually be furnished as part of the unit. Relays are necessary in many control and pilot circuits. These electrically operated switches are similar in design to contactors, but generally lighter in construction and carry a smaller current.

Limit Switches

Limit switches are often used to prevent the fan motor in a heating installation from circulating cold air when the heat is off, to prevent a humidifier from operating when the heat is off, or to hold refrigeration off when the heat is on, etc

SERVICING

A malfunctioning system may be caused by one part of the system or a combination of several parts. For this reason, it is necessary and advisable to check the more obvious causes first. Each part of the system has a definite function to perform, and if this part does not operate properly, the performance of the entire air-conditioning system will be affected.

To simplify servicing, especially a system that is unfamiliar, the serviceman should remember that the refrigerant, under proper operating conditions, travels through the system in one specified direction. Remembering this, the path of the refrigerant can be traced through any refrigerating system. For instance, beginning at the receiver, the refrigerant will pass through the liquid shutoff valve. If this valve is partially closed, or is completely closed or clogged, no refrigeration is possible, even though the unit itself can and will operate.

Strainers and Filters

Most refrigerating units have a strainer or filter in the liquid line. If this device becomes clogged or filled with dirt, the liquid refrigerant will be unable to pass. Filters are usually designed to hold a certain amount of dirt, scale, metallic particles, etc., and still function, but sometimes through carelessness, an excessive amount of extraneous matter may be allowed to enter the system during assembly or servicing. The remedy is obvious—the dirt and foreign matter must be removed to permit the refrigerant to pass through the filter.

Copper Tubing

Many smaller air-conditioning units make use of copper tubing for the liquid and suction lines. Since copper is not structurally as strong as iron or steel, it is relatively easy to collapse the copper line by an accidental blow, thus preventing the circulation of the refrigerant. When this happens, a new line or section of the line must be installed.

Expansion Valves

Expansion valves must be properly adjusted and in perfect working order to maintain the correct pressure in the low side of the line. In many cases, oil from the compressor crankcase enters and remains in the evaporator, occupying the space intended for the refrigerant. Naturally, this reduces the refrigerating effect and service problems develop. The oil must be drained from the evaporator coil, and the compressor crankcase inspected to make sure the oil level is correct.

Compressors

Compressors will start developing trouble after long use as the pistons and rings, as well as the other components, become worn. This wear allows some of the high-pressure gas to leak by the pistons and rings, so that proper compression cannot be obtained, and the compressor becomes inefficient and unable to take care of the load. New rings, pistons, and connecting rods usually are all that are required to bring the compressor back to its proper performance. Suction and discharge valves may stick, crack, or fail entirely. Sometimes, a slight warpage or piece of dirt under them results in improper operation. Seals around the crankshaft may leak and require repacking or replacement.

Condensers

The condensers may become dirty and inefficient, thereby resulting in high pressure, loss of efficiency, and an increase in the power requirement.

Ducts

Some air-conditioning ducts are provided with insulation on the inside which may tear loose and flap, resulting in noise.

Water Jets

Water jets or spray heads (if used) may become clogged or worn. Automatic water valves may get out of adjustment, and the flow of water may become restricted because of clogging of the water strainer.

SERVICING CHECKS

The proper diagnosis of air-conditioning system service problems can only be determined by intelligent thought, patience, and diligence. Each and every contributory factor must be considered and eliminated before going to some other cause. The various important steps in checking a system are listed as follows:

1. Determine the refrigerant used in the system.This is an important factor, since each refrigerant has its own operating characteristics, such as pressure-temperature differences.
2. Install a gauge test set. If the unit is a large one, it will have a set of gauges as part of the equipment. The gauge test will indicate the condition of the frigerant by checking the pressure-temperature relations. Put a thermometer on the evaporator coil, near the expansion valve, and obtain a reading. This temperature reading, along with a back-pressure reading, will indicate the refrigerant condition. Then check for improper expansion-coil pressure.
3. A low-pressure reading may be caused by a shortage of refrigerant, the presence of ice or water in the adjustment side of the expansion valve, moisture in the refrigerant system, plugged screens, partly closed liquid shutoff or service valves, excessive charge, or air in the system. Improper valve settings, blocked air circulation over the condenser, cooling air passing over the condenser at too high a temperature, reversed rotation of the

motor, bent fan blades, or a clogged condenser are indicated by high head pressures.

4. Frost or a sweat line on the coil should be noted. A coil is only active up to its frost or sweat line. All lengths of tubing beyond the frost or sweat line are inactive, since they do not receive liquid refrigerant. For greatest efficiency, the entire coil must have frost or sweat, and if the liquid line is obstructed, the screen plugged, or the expansion improperly set, the proper amount of liquid refrigerant will not enter the evaporator, and only a limited portion of the coil will frost or sweat, depending on the type of liquid refrigerant used.
5. An increase or overabundance of refrigerant will cause the entire coil, and possibly the suction line, to frost or sweat. This may also be caused by a leaky needle, improper adjustment, ice in the adjustment side of the expansion valve, or fused thermostat contacts. Most complaints with regard to excessive or insufficient frosting are due to weather conditions.
6. Improper operation will occur if refrigerant has escaped from the system. The first thing to do when leakage has occurred is to detect and repair the leak. Indications of refrigerant shortage include hissing at the expansion valve, a warm or hot liquid line, little or no frost on the expansion valve or coil, continuous operation, low head pressure, and bubbles in the sight glass if it is inserted into the liquid line and used to test for refrigerant shortage.
7. Bubbles may form in the sight glass if the head pressure is under 120 pounds with F-12, and under 100 pounds with methyl chloride. An excellent indication of leakage is the presence of oil on a joint or fitting; methyl chloride and F-12 dissolve oils, and when a leak occurs, the escaped refrigerant evaporates in the atmosphere, leaving the oil behind. Refrigerant must be added to system having a refrigerant shortage until the bubbles in the sight glass cease, or until the hissing sound at the expansion valve is eliminated.
8. Check for improper installation. Compressor units and low sides must be level. Tubes and fittings forming the liquid and suction lines must be the proper size. Baffles, ducts, and eliminators must be properly located and not obstructed. Ducts must be insulated properly, or short circuiting of air currents may take place,

causing the ducts to sweat. Liquid and suction lines must be checked for pinches, sharp or flattened bends, and obstructions. Lines should not be run along the ceiling in a hot room, and the lines should not run adjacent to any active hot-water or steam pipes.

9. The location and installation of the thermostat must be checked. The thermostatic-switch bulb should be installed in a location where average temperature conditions exist. The thermostatic bulb, or switch itself (if self-contained), should not be placed where an inrush of warm air, such as that caused by the opening of doors or windows, will cause the mechanism to cut in prematurely. Use thermostatic control switches with a minimum of tubing—gas will condense in long runs of tubing and the condensate will produce erratic operation. If a thermostatic bulb is used to control a liquid bath or brine, the bulb should be housed in a dry well or cavity, which should be located where average temperature conditions will be stable.

10. The condition of the thermostatic bulb should be checked. Apply heat gently by means of a cloth saturated with hot water, or else hold the bulb in the hand. If the contacts do not close after applying heat, the thermostatic element is discharged. Dirty or oxidized contacts can also cause defective operation.

11. Starter fuses should be checked and, if blown, replaced with fuses of the proper size. Determine the cause of the trouble—shortage of oil, air in the system, overcharge, misalignment, high back pressure, lack of air or water over the condenser or motor—any of which may cause the fuses to blow. If the trouble is not corrected, serious damage can result in a relatively short time.

12. Filters and/or screens must be checked. A clogged screen or filter, or a pinched or clogged liquid line, will produce the same trouble as a leaky or stuck expansion valve, depending on the degree of the obstruction.

13. A leaky or stuck expansion-valve needle will be indicated by a low pressure in the evaporator side and continuous operation of the unit. Sometimes a low-pressure control is wired in series with the thermostatic control and motor. When this is done, the low-pressure control will cut out, and the unit will remain inoperative.

COMMON TROUBLES IN AUTOMATIC EXPANSION VALVE SYSTEMS WITH THERMOSTATIC CONTROL

Refrigerant Shortage

1. Warm or hot liquid line.
2. Hissing sound at expansion valve.
3. Low head or condensing pressure.
4. Evaporator not entirely chilled.
5. Poor refrigeration.
6. Low-side pressure may be high if only gas is entering the evaporator.

Poor Refrigeration

1. Heavy coat of frost or ice on evaporator.
2. Refrigerant shortage.
3. Thermostat not adjusted properly.
4. Thermostat defective.
5. Thermostat shifted and not level.
6. Thermostat shielded by covering.
7. Thermostat in draft.
8. Stuck expansion valve.
9. Expansion valve set too low; allows only a portion of the evaporator to be effective.
10. Liquid line pinched or restricted.
11. Suction line pinched or restricted.
12. Compressor valves defective, broken, or sticking.
13. Strainer on liquid or suction line clogged.
14. Partially closed liquid- or suction-line valves.
15. Ice or moisture in adjustment side of expansion valve.
16. Ice freezing in seat of expansion valve.
17. High head pressure.
18. Compressor losing efficiency through wear.
19. Compressor may be too small.

Compressor Discharge Valve Defective

1. Low head pressure.
2. Poor refrigeration.
3. Compressor excessively warm.
4. If compressor is stopped, pressure will equalize.

Expansion-Valve Needle Stuck Open

1. Poor refrigeration.
2. High head pressure if stuck partially open.
3. Low head pressure and high back pressure if stuck fully open.
4. Frosted or sweating suction line.
5. Hissing sound at expansion valve.
6. Impossible to adjust for higher or lower back pressures.
7. On methyl chloride and F-12 units, this may be caused by moisture freezing at seat of expansion valve.
8. Improper oil, freezing at seat.
9. Improper oil and high compressor temperatures may result in carbonization which may build up at the expansion valve, especially if filters are defective.

Expansion Valve Needle Stuck Closed

1. No refrigeration if shut tight.
2. Little refrigeration if stuck partially shut.
3. Evaporator will be pumped down so that low side will show an unduly low pressure.
4. A pinched liquid line, plugged filter, or closed hand valve will give the same symptoms.
5. On methyl chloride and F-12 units, this may be caused by moisture freezing in the expansion valves.

High Head Pressure

1. Air in system.
2. Excessive refrigerant charge.
3. Air or water going through condenser at too high a temperature.
4. Air circulation over condenser blocked, if the unit is an air-cooled type.
5. If the unit is of the water-cooled type, the water may be turned off or restricted.
6. Rotation of motor reversed.
7. If a higher setting is used on the expansion valve, the head pressure will be higher, and vice versa.

Unable to Adjust Valve

1. Refrigeration shortage.
2. Compressor valve broken or defective.

3. Partially plugged screens or filters.
4. Liquid line pinched almost closed.
5. Stoppage in fitting or restriction in liquid line.
6. Needle or seat eroded and leaky.
7. Oil-logged coil.
8. Ice in adjustment side of expansion valve.

Low Head Pressure

1. Refrigerant shortage.
2. Worn pistons.
3. Head or clearance gasket too thick.
4. Suction valve worn, split, or stuck.
5. Low setting on expansion valve.
6. Gasket between cylinders blown.

Suction Line and Drier Coil Frosted or Sweating

1. Expansion valve stuck open or leaky.
2. Needle seat eroded or corroded.
3. Valve set at too high a back pressure.
4. Ice or moisture in adjustment side of expansion valve.
5. Thermostat out of order or poorly located.
6. Fan not operating, so that air is not blown over coils.
7. No water, or water pump not operating to pass water over evaporator.

COMMON TROUBLES IN THERMOSTATIC EXPANSION-VALVE SYSTEMS WITH LOW-SIDE OR THERMOSTATIC CONTROL

Refrigerant Shortage

1. Continuous operation.
2. Low head pressure.
3. Poor refrigeration.
4. Warm or hot liquid line.
5. Evaporator coils not chilled throughout entire length.
6. Hissing at expansion line.

Poor Refrigeration

1. Heavy coating of ice or frost on evaporator coils.
2. Expansion valve set too high.
3. Refrigerant shortage.
4. Thermostat bulb placed where there is little change in coil temperature.
5. Thermostat bulb placed where it is in a cold pocket and not affected by average conditions.
6. Expansion valve set too low.
7. Pigtail of expansion valve improperly placed, so that maximum coil surface is not used.
8. Compressor valves defective, broken, or sticking.
9. Liquid line pinched.
10. Suction line pinched.
11. Strainer clogged.
12. Suction line too small for job.
13. Partially closed liquid-or suction-line hand valves.
14. Compressor too small for job.
15. Moisture in methyl chloride or F-12 refrigerant.

Compressor Discharge Valve Defective

1. Low head pressure.
2. Poor refrigeration.
3. When compressor is stopped, pressures equalize.

Expansion Valve Needle Stuck Open

1. Continuous operation.
2. Poor refrigeration.
3. High head pressure.
4. Hissing sound at expansion valve.
5. Moisture in methyl chloride or F-12 refrigerant.

Low Head Pressure

1. Shortage of refrigerant.
2. Worn pistons in compressor.
3. Warped, split, or stuck discharge valve.
4. Suction valve warped, split, or stuck.
5. Expansion-valve needle stuck wide open.
6. Gasket between cylinder blown.
7. Thermostatic bulb discharged.

Suction Line and Drier Coil Sweating or Frosted

1. Expansion-valve needle stuck open.
2. Expansion-valve needle or seat eroded and leaky.
3. Expansion valve set too high above cutout point.
4. Control-switch points fused together.
5. Low-side control switch locked in operating position.

Expansion Valve Needle Stuck Partially Closed

1. Little or no refrigeration.
2. The high-pressure safety cutout may trip.
3. Evaporator will be pumped down and show a low pressure at below cut-in temperature.
4. If the liquid line is plugged, the thermostatic bulb discharged, or the capillary tube pinched, the result will be the same as a needle stuck shut.

High Head Pressure

1. The high-pressure safety cutout may cause system to be stopped.
2. Air in the system.
3. Excessive refrigerant charge.
4. Air or water passing over condenser at too high a temperature.
5. If unit is water-cooled, flow may be restricted or turned off.
6. If a high setting is used on the expansion valve (resulting in high back pressure), the head pressure will be higher than if he suction pressure were low.
7. Rotation of fan for cooling condenser reversed.
8. Fan blades bent or air passing over condenser restricted.

Expansion Valve Cannot Be Adjusted

1. Oil-logged evaporator.
2. Shortage of refrigerant.
3. Compressor valve broken or stuck.
4. Partially plugged screen in filter.
5. Liquid line pinched.
6. Stoppage in fitting or restriction in liquid line.
7. Stop-valve seat dropped and sealing open.
8. Charge lost in valve bulb.
9. Valve bulb loosened in its cradle by frost action; not making proper contact.
10. Valve covered; not open to atmospheric conditions.
11. Valve in too cold a location.

Table 7. Pressure in Pounds Per Square In. (Gauge) or Inches of Vacuum Corresponding to Temperature in Degrees F. for Various Common Refrigerants.

Tem., °F.	Ammonia NH_3	Sulfur dioxide SO_2	Methyl chloride CH_3Cl	Ethane C_2H_6	Propane C_3H_8	Ethyl chloride C_2H_5Cl	Carbon dioxide CO_2	Butane C_4H_{10}	Iso-butane C_4H_{10}	Freon F-12 CCl_2F_2	Carrene CH_2CL_2	Methyl formate $C_2H_4O_2$
− 40	8.7″	23.5″	15.7″	99.8#	1.5#		131.1#			11.0″		
− 35	5.4″	22.4″	14.4″	109.8#	3.4#		156.3#			8.4″		
− 30	1.6″	21.1″	11.6″	120.3#	5.6#		163.1#			5.5″		
− 25	1.3#	19.6″	9.2″	132.0#	8.0#		176.3#			2.3″		
− 20	3.6#	17.9″	6.1″	144.8#	10.7#	25.3″	205.8#		14.6″	0.5#		
− 15	6.2#	16.1″	2.3″	157 #	13.6#	24.5″	225.8#		13.0″	2.4#		
− 10	9.0#	13.9″	0.2#	172 #	16.7#	23.6″	247.0#		11.0″	4.5#	28.1″	
− 5	12.2#	11.5″	2.0#	187 #	20.0#	22.6″	269.7″		8.8″	6.8#	27.8″	
0	15.7#	8.8″	3.8#	204 #	23.5#	21.5″	293.9#	15.0″	6.3″	9.2#	27.5″	26.5″
+ 5	19.6#	5.8″	6.2#	221 #	27.4#	20.3″	319.7#	12.2″	3.3″	11.9#	27.1″	25.9″
+ 10	23.8#	2.6″	8.6#	239 #	31.4#	18.9″	347.1#	11.1″	0.2″	14.7#	26.7″	25.4″
+ 15	28.4#	0.5#	11.2#	257 #	35.9#	17.4″	376.3#	8.8″	1.6#	17.7#	26.2″	24.7″
+ 20	33.5#	2.4#	13.6#	227 #	40.8#	15.8″	407.3#	6.3″	3.5#	21.1#	25.6″	24.0″
+ 25	39.0#	4.6#	17.2#	292 #	46.2#	14.0″	440.1#	3.6″	5.5#	24.6#	24.9″	23.1″
+ 30	45.0#	7.0#	20.3#	320 #	51.6#	12.2″	474.9#	0.6″	7.6#	28.5#	24.3″	22.3″
+ 35	51.6#	9.6#	24.0#	343 #	57.3#	10.1″	511.7#	1.3#	9.9#	32.6#	23.5″	21.1″

Table 7. (Cont'd.) Pressure in Pounds Per Square In. (Gauge) or Inches of Vacuum Corresponding to Temperature in Degrees F. for Various Common Refrigerants.

+ 40	58.6#	12.4#	28.1#	368 #	63.3#	8.0"	550.7#	3.0#	12.2#	37.0#	22.6"	20.0"
+ 45	66.3#	15.5#	32.2#	390 #	69.9#	5.4"	591.8#	4.9#	14.8#	41.7#	21.7"	18.7"
+ 50	74.5#	18.8#	36.3#	413 #	77.1#	2.3"	635.3#	6.9#	17.8#	46.7#	20.7"	17.3"
+ 55	83.4#	22.4#	41.7#	438 #	84.6#	0.3#	681.2#	9.1#	20.8#	52.0#	19.5"	15.7"
+ 60	92.9#	26.2#	46.3#	466 #	92.4#	1.9#	729.5#	11.6#	24.0#	57.7#	18.2"	14.0"
+ 65	103.1#	30.4#	53.6#	496 #	100.7#	3.3#	780.4#	14.2#	27.5#	63.7#	16.7"	11.9"
+ 70	114.1#	34.9#	57.8#	528 #	109.3#	6.2#	834.0#	16.9#	31.1#	70.1#	15.1"	9.8"
+ 75	125.8#	39.8#	64.4#	569 #	118.5#	8.3#	890.4#	19.8#	35.0#	76.9#	13.4"	7.3"
+ 80	138.3#	45.0#	72.3#	610 #	128.1#	10.5#	949.6#	22.9#	39.2#	84.1#	11.5"	4.9"
+ 85	151.7#	50.9#	79.4#	657 #	138.4#	12.9#	1011.3#	26.2#	43.9#	91.7#	8.4"	2.4"
+ 90	165.9#	56.5#	87.3#	693 #	149 #	15.4#		29.8#	48.6#	99.6#	7.3"	0.5#
+ 95	181.1#	62.9#	95.6#		160 #	18.2#		33.2#	53.7#	108.1#	5.0"	2.1#
+100	197.2#	69.8#	102.3#		172 #	21.0#		37.5#	59.0#	116.9#	2.4"	3.8#
+105	214.2#	77.1#	113.4#		185 #	24.3#		41.7#	64.6#	126.2#	0.19#	5.8#
+110	232.2#	85.1#	118.3#		197 #	27.3#		46.1#	70.4#	136.0#	1.6#	7.7#
+115	251.5#	93.5#	128.6#		207.6#	31.6#			76.7#	146.5#	3.1#	10.4#
+120	271.7#	106.4#	139.3#		218.3#	35.5#			84.3#	157.1#	4.7#	13.1#
+125	293.1#	111.9#	150.3#		232.3#	39.5#			90.1#	168.6#	6.6#	15.7#
+130	315.6#	121.9#	161.3#		246.3#	44.0#			97.3#	180.2#	8.4#	18.2#

Note " inches of mercury
Note # PSIG

REFRIGERANT PRESSURES

Table 7 lists twelve of the most common refrigerants which can be encountered in almost any unit. The column at the left in the table is temperature in degrees Fahrenheit. Each refrigerant is listed in separate columns, and the pressures given are listed in *pounds of pressure per square inch (gauge),* or in *inches of mercury* if the pressures are below zero pounds gauge pressure. The temperatures are given in five-degree increments within the range of normal practice. The very low temperatures are seldom employed, but have been included for convenience. If the desired temperature lies between any two of the given temperatures, a little arithmetic can be used to determine the corresponding pressure with sufficient accuracy to set the valves or controls. To determine the vaporization pressure in the evaporator, the temperature of the liquid must be decided on.

For convenience, let us select ammonia as the refrigerant and decide to have the ammonia boil at 10° above zero F. With an ammonia system, the refrigerating plant will probably be located in the basement or on the roof and will be used to cool brine. The cooled brine will be pumped to the cooling coils at about 25°F. First, run down the temperature column until the 10°line is found, then follow the line horizontally to the right, and under the heading of "Ammonia," read the pressure of 23.8 pounds. The head of condensing pressures can then be checked. This is done by taking the temperature of the outlet water, if the refrigerating unit is water-cooled. When we assume that the water is leaving the condenser at a temperature of 90° F, a reference to the table indicates that the head pressure should be in the vicinity of 166 pounds. Pressures will usually be a few pounds higher in actual practice.

If air-cooled apparatus is encountered, the temperature of the air passing over the condenser must first be determined. Let us assume that the air passing over the condenser is at a temperature of 70°F. In Table 8, it will be observed that a temperature column is also given which is the temperature of the liquid in the evaporator. In our particular case, the ammonia is evaporating at a temperature of 10°F. Therefore, find this specific temperature in the left-hand column and then continue horizontally from that point to the column headed 70°F. At this intersection, the figure "35" is given. This figure (35) is added to the temperature of the air (70), which gives a total of 105°F. Turn to Table 7 and find the 105° line, which is the pressure corresponding to the temperature that exists in the condenser. Table 7 gives the temperature at which boiling or vaporization of each particular refrigerant takes place; the pressure at which this occurs is listed under the refrigerant type.

If the refrigerant is ammonia at 15.7 *psig*, the liquid will boil at 0°F. If it is desired to have a somewhat colder liquid, alter the pressure. This is accomplished in the refrigerating apparatus by adjusting the temperature control, if a thermostat is the control feature (such as with a low-side float), or by adjusting the thermostat or pressurestat in either a high-side or low-side float system. The continued operation will lower the evaporator temperature and the temperature of the room in which the evaporator is located. With an expansion-valve system, if a thermostat is employed to stop and start the system, the thermostat can be adjusted for a higher or lower temperature. Before and after the adjustment, it is important that the frost line on the evaporator be checked.

If the expansion valve is opened to admit more refrigerant so that the pressure is raised and its boiling temperature increased, the suction line may frost back. Also, if the expansion valve is closed so that less refrigerant enters the evaporator, the compressor will be unable to hold the evaporator at a lower pressure and a lower temperature. When this is done, only a portion of the coil maybe effective. A coil must contain liquid refrigerant to be able to refrigerate. Therefore, if only part of the coil is frosted, the other part might just as well be taken out, since it is doing no work. The best plan is to maintain the highest pressure possible so that the compressor will work on the densest vapor without having the suction line frost back.

It may be desirable, however, to make a change in the suction pressure so that the temperature of the coils can be altered. Thus, with ammonia, if it is desired that the coil temperature is to be 20°F, the expansion valve must be adjusted to maintain a constant pressure of 33.5 *psig* in the evaporator. If a 40° temperature is desired, a 58.6 *psig* reading on the low side would have to be maintained.If a 40° temperature were required in the evaporator of a methyl-chloride system, the expansion valve would have to be adjusted to hold a constant pressure of 28.1 *psig*.

Table 7 lists the temperature of the liquid refrigerant boiling at the constant pressure given. One factor cannot be varied without varying the other. The only way to obtain another boiling temperature is to adjust the expansion valve to maintain the pressure that corresponds to the temperature desired. If the liquid is contained in a vessel or cylinder, and the vessel is closed (such as in a supply drum), it will be observed that the pressure in such a vessel will be from one to five pounds greater than the temperature of the drum. This is due to the fact that pressure has built up and prohibits further evaporation. If a system contains an unknown refrigerant, such as one without an odor, the kind may be determined by taking a pressure reading of the refrigerant at rest; that is, while the unit is not operating. That pressure which corresponds nearest to the reading will be that particular refrigerant tested.

Table 7 is based on determining, from a pressure reading of the low side or evaporator, just what the temperature of the boiling liquid may be. With a methyl-chloride system, if the low-pressure gauge reads 2 *psig*, the temperature corresponding to this pressure is given as -5°F in the table.

With the temperature of the evaporator known (-5°F), refer to Table 8. The first column in the table on the left-hand side is the evaporator

Table 8. Factors to be Added to the Initial Temperatures of the Coolant In Order That Condenser Temperatures May Be Determined.

Evap.	Initial Coolant Temp. °F.				
°F.	60°	70°	80°	90°	100°
—30	15	15	15	10	10
—25	15	15	15	15	10
—20	20	20	15	15	15
—15	20	20	20	15	15
—10	20	20	20	15	15
— 5	20	20	20	15	15
0	25	25	20	20	15
+ 5	30	30	30	25	20
+10	35	35	30	25	20
15	40	35	30	25	20
20	40	35	30	25	25
25	45	40	35	30	30
30	50	45	40	35	35
35	50	50	45	45	40

temperature. The other columns refer to the temperature of the water or air passing over the condenser. Assume, for instance, that the room temperature is 70°F, and that the air is passing over the condenser at this temperature. Since the evaporator temperature is known, find this temperature in the first column and then find the proper column for the coolant temperature (70°F). At the intersection under these conditions, the figure 20 is given. This is added to the initial temperature, thus: 20° + 70° = 90°. Take this figure (90°) and refer to Table 7. Run down the temperature column to 90° and then across to the methyl-chloride column. It will be found that a head or condensing pressure of about 87.3 pounds can be expected. If the actual pressure reading varies from this, check the condition of the apparatus.

It is essential to obtain a pressure reading on the low or evaporator side so that the temperature of the refrigerant can be determined. If the reading indicates too high a temperature, the expansion valve will have to be adjusted to maintain the proper pressure and, of course, the proper evaporating temperature. Head pressures are important; excessive pressures indicate something out of the ordinary.

AIR-CONDITIONING SYSTEM TROUBLE CHART

Compressor Will Not Start

Possible Cause	Possible Remedy
Thermostat setting too high.	Reset thermostat below room temperature.
High head pressure.	Reset starter overload and determine cause of high head pressure.
Defective pressure switch.	Repair or replace pressure switch.
Loss of refrigerant charge.	Check system for leaks.
Compressor frozen.	Replace compressor.

Compressor Short Cycles

Possible Cause	Possible Remedy
Defective thermostat.	Replace thermostat.
Incorrect setting on low side of pressure switch.	Reset low-pressure switch differential.
Low refrigerant charge.	Check system for leaks; repair and add refrigerant.
Defective overload.	Replace overload.
Dirty or iced evaporator.	Clean or defrost evaporator.
Evaporator blower and motor belts slipping.	Tighten or replace belts.
Dirty or plugged filters.	Clean or replace air filters.

Compressor Runs Continuously

Possible Cause	Possible Remedy
Excessive load.	Check for excessive outside air infiltration and excessive source of moisture.
Air or noncondensable gases in the system.	Purge System.
Dirty condenser.	Clean condenser.
Condenser blower and motor belts slipping.	Tighten or replace belts.
Thermostat setting too low.	Reset thermostat.
Low refrigerant charge.	Check system for leaks; repair and add refrigerant.
Overcharge of refrigerant.	Purge and remove excess refrigerant.
Compressor valves leaking.	Replace compressor.
Expansion valve or strainer plugged.	Clean expansion valve or strainer.

System Short of Capacity

Possible Cause	Possible Remedy
Low refrigerant charge.	Check system for leaks; repair and add refrigerant.
Incorrct superheat setting of expansion valve.	Adjust superheat to 10° F.
Defective expansion valve.	Repair or replace valve.
Air or noncondensable gases in the system.	Purge system.
Condenser blower and motor belts slipping.	Tighten or replace belts.
Overcharge of refrigerant.	Purge excess refrigerant.
Compressor valves leaking.	Replace compressor valves.
Expansion valve or strainer plugged.	Clean valve or strainer.
Condenser air short-circuiting.	Remove obstructions or causes of short.

Head Pressure Too High

Possible Cause	Possible Remedy
Overcharge of refrigerant.	Purge excess refrigerant.
Air or noncondensable gases in system.	Purge system.
Dirty condenser.	Clean condenser.
Condenser blower and motor belts slipping.	Tighten or replace belts.
Condenser air short-circuiting.	Remove obstructions or causes of short-circuiting air.

Head Pressure Too Low

Possible Cause	Possible Remedy
Low refrigerant charge.	Check system for leaks; repair and add refrigerant.
Compressor valves leaking.	Replace compressor valves.

Suction Pressure Too High

Possible Cause	Possible Remedy
Excessive load on system.	Remove conditions causing excessive load.
Expansion valve is stuck in "Open" position.	Repair or replace expansion valve.
Incorrect superheat setting of expansion valve.	Adjust superheat setting to 10° F.

Suction Pressure Too Low

Possible Cause	Possible Remedy
Low refrigerant charge.	Check system for leaks; repair and add refrigerant.
Expansion valve or strainer plugged.	Clean expansion valve or strainer.
Incorrect superheat setting of expansion valve.	Adjust superheat setting to 10′ .
Evaporator air volume low.	Increase air over evaporator.
Stratification of cool air in conditioned area.	Increases air velocity through grilles.

Compressor Is Noisy

Possible Cause	Possible Remedy
Worn or scored compressor bearings.	Replace compressor.
Expansion valve is stuck in "Open" position or defective.	Repair or replace expansion valve.
Overcharge of refrigerant or air in system.	Purge system.

Possible Cause	Possible Remedy
Liquid refrigerant flooding back to compressor.	Repair or replace expansion valve.
Shipping or hold-down bolts not loosened.	Loosen compressor hold-down bolts so compressor is freely floating in mountings.

Compressor and Condenser Fan Motor Will Not Start

Possible Cause	Possible Remedy
Power failure.	Check electrical wiring back to fuse box.
Fuse blown.	Replace blown or defective fuse.
Thermostat setting too high.	Reduce temperature setting of room thermostat.
Defective thermostat.	Replace or repair thermostat.
Faulty wiring.	Check wiring and make necessary repairs.
Defective controls.	Check and replace defective controls.
Low voltage.	Reset and check for cause of tripping.
Defective dual-pressure control.	Replace the control.

Compressor Will Not Start, But the Condenser Fan Motor Runs

Possible Cause	Possible Remedy
Faulty wiring to compressor.	Check compressor wiring and repair.
Defective compressor motor.	Replace the compressor.
Defective compressor overload (single phase only).	Replace overload.
Defective starting capacitor (single phase only).	Replace capacitor.

Condenser Fan Motor Will Not Start, But Compressor Runs

Possible Cause	Possible Remedy
Faulty wiring to fan motor.	Check fan-motor wiring and repair.
Defective fan motor.	Replace fan motor.

Condenser Fan Motor Runs, But the Compressor Hums and Will Not Start

Possible Cause	Possible Remedy
Low Voltage.	Check line voltage. Determine the location of the voltage drop.
Faulty wiring.	Check wiring and make necessary repairs.
Defective compressor.	Replace compressor.
High head pressure.	Check head pressure and complete operation of system to remove the cause of the high pressure condition.
Failure of one phase (three-phase units only).	Check fuses and wiring.
Defective start capacitor (single phase only).	Replace capacitor.
Defective potential relay (single phase only).	Replace relay.

Compressor Starts, But Cycles on Overload

Possible Cause	Possible Remedy
Low voltage.	Check line voltage. Determine the location of the voltage drop.
Faulty wiring.	Check wiring and make necessary repairs.
Defective running capacitor (single phase only).	Replace capacitor.
Defective overload.	Replace overload.
Unbalanced line (three-phase only).	Check wiring; call power company.

Evaporator Fan Motor Will Not Start

Possible Cause	Possible Remedy
Power failure.	Check electrical wiring back to fuse box.

Index